T0227472

SHORT-CUTTING THE PHOSPHORUS CYCLE IN URBAN ECOSYSTEMS

Short-cutting the phosphorus cycle in urban ecosystems

DISSERTATION
Submitted in fulfilment of the requirements of
the Board for Doctorates of Delft University of Technology
and of the Academic Board of the UNESCO-IHE Institute for Water Education
for the Degree of DOCTOR
to be defended in public
on Friday, 17 June 2005 at 15:30 hours
in Delft, the Netherlands

by

CRC Press
Taylor & Francis Group
Boca Raton London New York

CRC Press is an imprint of the
Taylor & Francis Group, an **informa** business
A BALKEMA BOOK

Published by:
CRCPress/Balkema
P.O. Box 447, 2300 AK Leiden, The Netherlands
e-mail: Pub.NL@taylorandfrancis.com
www.crcpress.com – www.taylorandfrancis.com

First issued in hardback 2017

© 2005 by Taylor & Francis Group, LLC
CRC Press/Balkema is an imprint of the Taylor & Francis Group, an informa business

No claim to original U.S. Government works

ISBN 13: 978-1-138-42441-8 (hbk)
ISBN 13: 978-0-415-38484-1 (pbk)

Visit the Taylor & Francis Web site at
http://www.taylorandfrancis.com

and the CRC Press Web site at
http://www.crcpress.com

To my wife Francina and son Nigel
To my late father who could not witness this final output

The cover depicts the sanitation paradigm shift. An adult human being in a year excretes enough fertiliser required to grow the food he/she consumes in a year. About 2.4 billion people in the world today have no access to basic sanitation, whilst 90% of the world's sewage is not treated, dumping valuable plant nutrients into freshwater sources. More than 200 million under five children mostly in developing countries are malnourished. Soil fertility in the same countries is being depleted and sources of fertilisers are becoming scarce and expensive. How then, are the ever-increasing demands for food and fibre going to be met? A sanitation revolution is pending. This calls for a move away from our constipated mindset and to start rethinking sanitation and management of bio-waste products in the particularly in urban and peri-urban environments. The missing-link has been found or has it?

Photograph of a great thinker sitting on a toilet pedestal is adapted from Narain, S. (2002). The flush toilet is ecologically mindless: Think about it. Down to Earth, Vol 10, No. 19.

Design of cover by Peter Stroo

Contents

Acknowledgements

As one embarks on a PhD research there is need to get a compass and probably a torch as well. This research has not been an exception, my sincere gratitude go to my promoter Prof. dr. ir. Hubert H.G. Savenije, my supervisor Dr Peter Kelderman who have given me invaluable guidance and illumination particularly during some of the darkest hours. Thank you for the friendship, which I would cherish for many years beyond this accomplishment.

My mentor, and colleague Prof. dr. ir. Bane Marjanovic, you have always been an inspiration since we first met in 1997. Thank you for all the resources and support and making me feel welcome at the University of the Witwatersrand. My gratitude is also extended to Dr Zoran Cukic, Ljubica Korac and Vlada Jovic.

My Dutch connection and mentor Prof. dr. ir. Pieter van der Zaag for your zeal and enthusiasm on almost everything, I thank you.

My PhD colleagues at IHE-Delft, now UNESCO-IHE Institute for Water Education with whom we toiled together in search of knowledge and also shared some wonderful and distressing times, Dr Marieke de Groen (*pace-setter*), Lawrence Nyagwambo, Dr Innocent Nhapi, Seyam Ismail and Joylyn Ndoro. Not forgetting Dr Annelies van der Vleuten-Balkema who graduated at Eindhoven Technical University with whom I exchanged a lot of notes but never got to meet.

As much as I did a large part of the digging they were also a number of MSc and BSc students whom I worked with, gave advice and supervised their projects. I am happy that they all struck gold. These are, MSc students: Eng. Ratidzai Madimutsa and Morris Chidavaenzi, BSc students: (*The guinea pigs*) Fredrick Mukonoweshuro, Tapiwa Gavaza, Ella Mashingaizde, Julius Makoni, and Farai Chieza.

My colleagues at the University of Zimbabwe, Department of Civil Engineering, academic and technical staff, particularly Eng. Zvikomborero Hoko (*muYuti*), Eng. Evans Kaseke, Eng. Dave Katale and Eng. Tinozivaishe Zhou.

Staff at the Institute of Water and Sanitation Development (IWSD), particularly Eng. Ngoni Mudege and the late Dr Jerry Ndamba. Ecosan fundis from Mvuramanzi Trust, Edward Guzha and Dave Proudfoot. Not forgetting Dr Peter Morgan of Aquamor the ecosan and VIP toilet god-father in Zimbabwe, your urine samples contained excess nutrients (*What diet are you on?*)

Ecological Alternatives in Sanitation group, Prof Hakan Jonsson, Prof Thor Axel Sternstrom, Dr Jan Olof Drangert, Dr Caroline Schonning (nee Hoglund), Dr Bjorn Vinneras and the international group of participants whom we exchanged such valuable experiences in 1998 (*I hope you still continue peeing into your flower pots?*)

Support from GTZ in financing my attendance of two international meetings on ecological sanitation. Ms Christine Werner the GTZ ecosan project co-ordinator for facilitating the sharing of valuable information and having faith in my contribution to ecosan.

For all those whom I have not mentioned by name particularly staff at UNESCO-IHE, I thank you all. Forgive me as the brain-drain syndrome has started catching up with me.

Finally, I am extremely grateful for the generous sponsorship by the Dutch Government (DGIS) through the SAIL foundation programme. This not only afforded me a chance to do my PhD research but also allowed me to learn more about the Dutch culture (a chance to ride a bicycle after a ten year break), language and food. I am convinced that this research work is not an end in itself but will be useful and valuable to all those who still endeavour to link sanitation, urban agriculture, soil fertility and poverty reduction especially in the developing world.

Abstract

Conventional urban water and wastewater systems owe their design to 19th Century thinking in combating waterborne disease and maintaining hygienic conditions in cities. These systems are not compatible with closed loop systems, which seek to minimise the net use of resources by encouraging local recycling of water and nutrients.

A new sanitation paradigm has emerged based on three fundamental aspects *viz.* rendering human excreta safe, preventing pollution, and using the safe products of sanitised human excreta for agricultural purposes. This approach can be called 'ecological sanitation' and it is a concept, which is in agreement with the so-called Bellagio principles of environmental sanitation.

The ecological sanitation concept has gained momentum in recent years and it is seen as one of the solutions for providing access to adequate sanitation by 2015 to at least half of the 2.4 billion people worldwide who lack this service (millennium development goal). Ecological sanitation provides a link between sanitation, food, agriculture and soil fertility. At present sanitation experts, biologists, agriculturists, medical professionals, are involved in reshaping this concept. However, a number of research questions still remain *viz.* risks and perceptions on handling human excreta and organic solid waste, city planning regulations and design, public health and hygiene, and technological constraints.

This dissertation describes a process of testing the feasibility of the ecological sanitation concept in urban ecosystems (household or neighbourhood scale) by critically assessing the knowledge and nature of phosphorus (P) fluxes and stocks (source to sink) at a global, regional, national, river basin and micro-catchment scale. The hypotheses is that short-cutting or closing open-ended water and nutrient cycles at the lowest appropriate level would ensure sustainable use of limited resources like P at the same time protecting the same resources from possible contamination.

Calculations indicate that the current global economic P reserves would last at least 100 years at a consumption rate of about 14 Mt P/a. Global (macro-scale analysis) activities such as clearing of forests, extensive cultivation and urban waste disposal and drainage systems have enhanced the transport of P from terrestrial to aquatic environments to an estimated 22 Mt P/a. As a consequence of these activities, concentrations of P in rivers and supply of P to lakes and estuaries have increased almost three-fold from an estimated 8 Mt P/a of pre-industrial and intensive agriculture era. These actions have decoupled patterns of supply, consumption and waste production from the natural P-cycle, and have created an open-ended linear flow of P driven by economics and unbalanced world trade.

Estimates in this dissertation indicate that P-based fertiliser application rate in year 2000 for Zimbabwe of 18 940 Mg/a as P is exceeded by the total P losses from the soil (soil erosion, runoff and leaching losses) amounting to 57 160 Mg/a as P. Soil erosion and land degradation in Zimbabwe causes major environmental problems leading to loss of soil fertility.

The Lake Chivero case located in the urbanised catchment of the Harare metropolis in Zimbabwe provides a graphic example of the open-ended P-flows. The calculated P-fluxes and stocks for year 2000 indicate that the Lake received about 676 Mg/a as P compared with an outflow of 265 Mg/a as P. There is a steady accumulation of P in the lake with a large percentage of the P trapped in the sediments below the Lake. The amount of P discharged into the Lake each year corresponds to about 4% of Zimbabwe's agricultural P usage in year 2000.

An analytical modelling approach has been followed to describe and analyse the management options for short-cutting the P-cycle in urban ecosystems, using a micro-catchment area consisting of a high density suburb in Harare.

Using a system's thinking approach and Material Flow and Stock Accounting (MFSA) two compartments or subsystems are defined to enable accounting and analysis of P-bearing materials namely the "household" (consumption or use and excretion or waste) and "agriculture" (soil-plant interaction). The identified flows and stocks are used to draw up the P-balance for the micro-study catchment. The data and information collected is incorporated into a conceptual and mathematical formulation of monthly P-fluxes and stocks. A P-calculator is developed using STELLA, a systems analysis software developed by High Performance Systems Inc. The calculator can be used as a planning and decision-making tool for closing the P-cycle within urban ecosystems. It also provides a means to simulate and evaluate different options in linking household waste P-fluxes to agricultural P-requirements.

Two options are considered as far as short-cutting the P cycle is concerned *viz.* urine or *yellow water* separation, collection, storage and application on agricultural land and composting of organic solid waste fraction from the household subsystem which ideally should be applied to agricultural land to increase the available P-status of the soil and as manure to improve the soil condition, respectively.

The amount of available agricultural land per capita is the major constraint in realising 100% recycling of P in yellow water. The calculations reveal that an adult person living within the micro-catchment requires a land area of about 190 m^2 for assimilation of P contained in his/her excreted annually through the yellow water fraction when growing maize. At present only 29 m^2/person is available for urban agriculture. Only partial recycling of P is therefore possible and according to the Bellagio principles export of excess P from the household subsystem to higher levels of jurisdiction would be necessary i.e. outside the micro-catchment.

A major constraint to implementation of ecological sanitation in most cities is the unavailability of adequate land area to enable assimilation of the by-products of human metabolism. City and town planning requires some rethinking so as to reduce the ecological footprint they exert on the hinterland and other distant areas.

B. Gumbo, 2005

Preface

This research was inspired by a number of sustainability concepts put forward towards the end of the 20th century. Sustainability has been a popular subject in a plethora of literature in the last two decades. However, sustainability is still an elusive subject particularly the relationship between the use of natural resources and development. Some researchers and scientists, donor representatives and rhetoric politicians claim that in certain projects of resource use, sustainable consumption has been reached. What is most fascinating for a civil engineer like myself is how sustainability can be measured? Are there any obvious indicators, which are convincing and can be used globally as benchmarks?

The initial proposal I had submitted to my promoter Prof. Huub Savenije in December 1997 was on water quality and the determination of receiving water quality objectives for the Lake Chivero river basin in Zimbabwe. However after some discussions, Prof. Savenije suggested that it was important to embark on a PhD research, which was not only innovative but also visionary in view of the enormous challenges presented by new paradigms in environmental management. In June 1998, I then submitted a revised proposal, which as usual was too ambitious and spanned into disciplines which were new to me.

In the later part of 1998 I had the pleasure to participate in a month-long workshop on Ecological Alternatives in Sanitation held in Stockholm, Uppsala and Linkoping Universities in Sweden funded by Sida (Swedish International Development Cooperation Agency). I became inspired by the research work and pilot projects that were being conducted and implemented in Sweden, especially the idea of recycling human excreta for agricultural purposes (with a focus on human urine). The research carried out by various universities and institutions consisted of a number of multi-disciplinary projects on health and hygiene, agriculture, social and cultural aspects, economics, decision support tools, and environmental issues.

Hence, I was also drawn up in this web of professionals who were spending a great deal of time talking about one of the most mundane subjects, toilets, faeces and urine. Through the GTZ (Deutsche Gesellschaft fur Teschnische Zusammernarbeit) ecological sanitation project I was afforded a chance to stand in front of two international audiences (Nanning China, 2001 and Lubeck, Germany 2003, 1st and 2nd International Symposia on Ecological Sanitation, respectively) to present my views on ecological sanitation, particularly the holistic approach to nutrient management.

Being a civil engineer for almost 14 years the subject matter was quite fascinating and challenging for me, and naturally, I was encouraged to delve into uncharted waters on subject matters such as soil science, agronomy, human anatomy, health, food and nutrition.

Overall I learnt a great deal, even though I still remain unsatisfied because of the vast intricacies and interactions of nature, which man has no clue on. My consolation is that at least I have a son (and still capable of having a daughter) and I believe some things are better left for the next generation to grapple with.

List of Symbols

Symbol	Description	Dimension	Unit
Δ	difference	depends on specification	
σ	phosphorus content in a material expressed as a ratio	M/M	kg/kg
λ	quantity of dry organic and biodegradable waste generated per person per month	M/T	kg/p.month
β	mean rainfall on a rain-day and scale parameter of exponential distribution	L/T	mm/day
γ	time scale for transpiration S_b/T_{pot}	T	days
γ°	dimensionless time scale for transpiration: ratio between γ and days in month	-	-
Δt	time step	T	day or month
A	area of catchment or system boundary	L^2	m^2
a	water use fraction which determines the volume of water used for on-plot garden irrigation	-	-
A'	intercept of relation between monthly effective rainfall and monthly transpiration	L/T	mm/month
A_c	total area under cultivation	L^2	m^2
B	slope or relation between monthly effective rainfall and monthly transpiration	-	-
b	water use fraction which determines the grey water volume generated	-	-
b_1	water use fraction which determines the grey water volume generated from the kitchen	-	-
b_2	water use fraction which determines the grey water volume generated from the bathroom	-	-
b_3	water use fraction which determines the grey water volume generated from laundry	-	-
C	surface runoff coefficient dependent on land use and amount of effective rainfall	-	-
c	water use fraction which determines the volume of black water arising from toilet flushing	-	-
D	daily interception threshold	L/T	mm/day
d	water use fraction which determines the volume of water consumed by an adult person per day or month either directly or indirectly	-	-
E	evaporation	L/T	mm/month
e	water use fraction which determines the volume of grey water applied on land for garden irrigation purposes	-	-
E_c	water excreted by the human population mainly through respiration and perspiration	L/T	mm/month
E_g	evaporation from drying-out of laundry material and kitchen utensils	L/T	mm/month
E_o	open water evaporation	L/T	mm/month
E_{pot}	potential evaporation	L/T	mm/day
f	water use fraction which determines the volume of grey water which evaporates from laundry fabrics, kitchen utensils and human body after washing and bathing	-	-
G	Giga is a prefix to a unit of measure (10^9)	-	-
h	fraction of storm-water entering the foul sewer system	-	-
H	estimated population of the study area in year 2000	-	-
I	interception	L/T	mm/month

Symbol	Description	Dimension	Unit
M_{fb}	quantity of food and beverage per food group consumed per person per month	M/T	kg/p.month
M_{im}	total mass of imported manure per annum	M/T	kg/a
M_{mf}	recommended mass of mineral fertiliser applied per month as '*maizefert*' or Compound D	M/T	kg/month
M_{sd}	quantity of soap and detergent used per person per month	M/T	kg/p. month
Mt	Million metric tonnes	M	tonnes
n	number of days	T	days
N	the number of households in the micro-catchment	No	-
n_m	days per month	-	days/month
n_r	rain-days for given duration	-	days/month
P	phosphorus flux as P	M/T	kg/month or kg/annum
$p.d$	per person per day	-/T	/p. day
$p.month$	per person per month	-/T	/p. month
P'_{bu}	potential biomass P-uptake corresponding to Y'_{bu}	M/T	kg/month
p_{01}	transition probability of occurrence of a rain-day after a dry day	-	-
p_{11}	transition probability of occurrence of a rain-day after a rain-day	-	-
P_b	brown water P-flux emanating from toilet flushing of human faecal material	M/T	kg/month
P_{br}	biomass residue P-flux which remains in the agricultural subsystem after the harvest and is subsequently ploughed back into the soil during land preparation	M/T	kg/month
P_{bs}	biomass storage P, with an initial value set at zero and then accumulates P during the plant growth period and is depleted to zero after the land preparation activity	M	kg
P_{bu}	biomass uptake P-flux from the soil storage occurring in the agricultural subsystem during the plant growth phase	M/T	kg/month
P_{csw}	composted organic solid waste P-flux emanating from household activities	M/T	kg/month
P_{fb}	food and beverage P-flux reaching the household subsystem i.e. imported or produced within the system boundary	M/T	kg/month
P_g	grey water P-flux emanating from activities related to nourishing and cleaning	M/T	kg/month
P_{gi}	grey water irrigation P-flux arising from grey water application on on-plot agricultural land	M/T	kg/month
P_{hb}	harvested biomass or economic yield P-flux of the crop also contains the inedible biomass advertently or inadvertently transported from the agricultural subsystem to the household subsystem	M/T	kg/month
P_{im}	imported manure P-flux from outside the system boundary	M/T	kg/month
P_{la}	land to atmosphere P-flux either in gaseous form or attached to fine particles of organic and inorganic matter.	M/T	kg/month
P_{le}	leaching P-flux due to percolation and groundwater flow	M/T	kg/month
P_{ma}	manure application P-flux, which is a combination of imported manure and the proportion of organic solid waste which is composted and deposited on agricultural land ($P_{ma} = P_{im} + P_{csw}$)	M/T	kg/month

Symbol	Description	Dimension	Unit
P_{mf}	mineral P-based fertiliser P-flux applied on agricultural land	M/T	kg/month
P_{ms}	municipal sewage P-flux, which is a combination of yellow, black, and proportion of grey and storm water P-fluxes conveyed through a pipe to a sewage treatment plant	M/T	kg/month
P_s	foul sewage or '*black water*' P-flux which is a combination of yellow, brown, and proportion of grey water P-fluxes	M/T	kg/month
P_{sd}	soap and detergent P-flux reaching the household subsystem	M/T	kg/month
P_{sl}	soil loss P-flux arising from soil erosion phenomena, representing the particulate P losses from the agricultural subsystem	M/T	kg/month
P_{sr}	surface runoff P-flux dissolved in storm water	M/T	kg/month
P_{sr}	surface runoff P-flux dissolved in storm water, representing the dissolved soluble P losses from the agricultural subsystem	M/T	kg/month
P_{ss}	soil storage P, which represents the plant available (*labile*) P fraction	M	kg
P_{sw}	organic solid waste P-flux derived from household activities and local vegetation growth and die-off derived from the agricultural subsystem	M/T	kg/month
P_{wd}	wet and dry deposition P-flux from the atmosphere to the land mass	M/T	kg/month
P_y	yellow water P-flux derived from urinary excretion	M/T	kg/month
Q	discharge	L/T	mm/month
q	constant in $p_{01}= q(R)^r$	$(T/L)^r$	$(month/mm)^r$
q'	proportion of organic solid waste which is either deliberately composted or is uncollected and end up being manure on agricultural land	-	-
Q_f	infiltration	L/T	mm/month
Q_g	groundwater flow or seepage	L/T	mm/month
Q_p	percolation or capillary rise	L/T	mm/month
Q_s	surface run-on or run-off	L/T	mm/month
R	rainfall	L/T	mm/month
r	power in $p_{01}= q(R)^r$	-	-
R_{eff}	effective rainfall	L/T	mm/month
S	storage of water or moisture	L	mm
S	available soil moisture content	L	mm
S_a	storage of moisture in the atmosphere	L	mm
S_b	available soil moisture content at the boundary between moisture constrained transpiration and potential transpiration	L	mm
S_g	storage of water in renewable groundwater	L	mm
S_{max}	maximum available soil moisture for certain soil type and crop	L	mm
S_s	storage of water on the ground surface	L	mm
S_{start}	available soil moisture at start of month	L	mm
S_u	storage of moisture in the unsaturated soil	L	mm
S_w	storage of water in water bodies	L	mm
t	time	T	month
T	transpiration due to rainfall	L/T	mm/month
T	Tera is a prefix to a unit of measure (10^{12})	-	-
T_i	transpiration due to garden irrigation using municipal water	L/T	mm/month
T_{max}	Maximum monthly transpiration, given certain	L/T	mm/month

Symbol	Description	Dimension	Unit
	initial and soil conditions		
T_{pot}	potential transpiration per day	L/T	mm/day
T_t	total transpiration ($T_t = T + T_i$)	L/T	mm/month
u	constant in $p_{11} = u(R)^v$	$(T/L)^v$	$(month/mm)^v$
V	volume	L^3	m^3
v	power in $p_{11} = u(R)^v$	-	-
V_e	effective volume for a specified soil horizon (i.e. total soil volume less >2 mm fraction)	L^3	m^3
W	municipal water supply normalised to the catchment area A	L/T	mm/month
W_b	brown water generated from household activity related to toilet flushing after defaecation	L/T	mm/month
W_c	municipal water consumed by population either directly or contained in ingested food products	L/T	mm/month
W_g	grey water generated from activities related to nourishing and cleaning (kitchen, bathroom and laundry)	L/T	mm/month
W_i	municipal water used for garden irrigation	L/T	mm/month
W_{ms}	municipal sewage water, which is a combination of yellow, black, and proportion of grey and storm water conveyed through a pipe to a sewage treatment plant	L/T	mm/month
W_s	foul sewage or '*black water*' which is a combination of yellow, brown, and a proportion of grey water. This corresponds to dry weather flow.	L/T	mm/month
W_y	volume of yellow water excreted by an equivalent adult population per month normalised to the micro-catchment area	L/T	mm/month
y	urinary volumetric rate of excretion for an adult person per month in winter or summer season	L^3/T	m^3/p. month
Y'_{bu}	combined water-limited dry biomass production potential (economic produce and straw)	M/L^2	kg/m^2
Y'_{hb}	water-limited dry mass economic yield	M/L^2	kg/m^2
Y_{hb}	economic yield or harvested biomass of maize	M/L^2	kg/m^2
Y_{sb}	yield of maize straw biomass	M/L^2	kg/m^2
θ	proportion of the recommended commercial fertiliser application rate applied by urban farmers as Compound D	-	-
ρ_b	bulk density of soil	M/L^3	kg/m^3
σ_{fb}	P-content in food and beverage material as P	M/M	kg/kg
σ_{hb}	P-content of the harvested maize crop biomass as P	M/M	kg/kg
σ_{im}	P-content in imported manure as P	M/M	kg/kg
σ_{mf}	phosphorus concentration in fertiliser as P_2O_5	M/M	kg/kg
σ_{ms}	P-concentration of municipal sewage as P	M/L^3	kg/m^3
σ_{sb}	P-content of the maize straw biomass as P	M/M	kg/kg
σ_{sd}	P-content in soap and detergent material as P	M/M	kg/kg
σ_{sr}	phosphorus concentration in runoff as ortho-P	M/L^3	mg/l
σ_{ss}	bio-available phosphorus concentration in soil storage as P (0.2 m deep plough layer)	M/M	kg/kg
σ_{sw}	P-content of the organic solid waste fraction as P	M/M	kg/kg
σ_y	P-concentration in urine as P	M/L^3	kg/m^3
Z_l	soil loss from agricultural land per unit area	M/L^2	kg/m^2

1 Introduction

1.1 Closing nutrient and water cycles

Paradigm shift

The world is experiencing rapid growth in urbanisation and industrialisation. The pace, depth, and magnitude of these changes, while bringing about benefits to local people, have exerted severe ecological stresses on both local human living conditions and regional life support ecosystem. Urban sustainability can only be assured with a human ecological understanding of the complex interactions among environmental, economic, political, and social-cultural factors and with careful design, planning and management grounded on ecological principles (WCED, 1987; Mansson, 1992; Van der Ryn & Cowan, 1996). The city, with its water, wastewater and other infrastructures, sits within an industrial-urbanised ecology of individual entities arranged one to another and bound together in a web of energy and material flows. In an ideal world the "waste" residue from the metabolism of one of these entities should become the feedstock of another[1] (Ausubel, 1992; Ekins & Cooper, 1993; Hallsmith, 2003).

It is now broadly agreed, as a principle that human activities on a global scale should be in harmony with the environment, for both present and future generations (WCED, 1987). "Sustainability" although a mantra for the 1990's and beyond is a way in which society utilises the environment (Mitcham, 1995). The environmental load that follows from social activities should be "ecologically suitable". This means that the functioning of regeneration systems, absorption capacities, and other parts of the ecosystems[2] is guaranteed both quantitatively and qualitatively. The environment is seen as a set of resources for society. Sustainability therefore refers to the continued existence of the socially functional components of ecosystems; it limits the use that is made of these components (NWO, 1992; Daly, 1996; Van der Ryn & Cowan, 1996).

Conventional economics emphasises the seemingly self-generating flows of goods and money between firms and households in the marketplace. However it is blind to the irreversible unidirectional material flows that sustain the economy (Barbier, 1990; Rees, 1992; Daly, 1996; Wackernagel & Rees, 1996). The circular economic flows are actually sustained by the unidirectional throughput of ecological goods and services from and to the ecosphere (the "natural income" stream). All the energy and much of the matter that passes through the economy is permanently dissipated into the environment never to be used again. Trade and technology have enabled human-kind progressively to exploit nature far beyond sustainable levels so that present

---

[1] Urban ecosystems, designed 'from scratch' to imitate nature by utilising the waste products of each component as raw material or input for another are an attractive theoretical idea, but as yet most projects are at an infancy stage (Odum; 1983; Ayres & Ayres, 1996).
[2] A natural ecosystem is a self-organising system consisting of interacting individuals and species, each programmed to maximise its own utility (survival and reproduction), each receiving and providing services to others, each therefore dependent on the system as a whole. The ecosystem normally maintains itself in a balanced condition, or evolves slowly along a developmental path. But such 'dissipative systems' remain far from (thermodynamic) equilibrium (Ayres, 1989).

consumption exceeds natural income (the "interest" on our capital). This leaves the next generation with depleted capital and less productive potential even as the population and material expectations increase (Wackernagel & Rees, 1996; Hawken *et al.*, 1999; Chambers *et al.*, 2000).

The world today is oriented towards "green", "clean" and "compact" solutions where urban planning is re-thought from an "urban ecology" perspective[3]. This brings local decentralised solutions (neighbourhoods) into focus as opposed to the conventional centralised water and wastewater systems (Lyle, 1985; Matthews, 1996; Kalbermatten & Middleton, 1998). It is not clear whether urban water management in its present form is sustainable or not (Cobb, 1995; Harremoes, 1997; Larsen & Gujer, 1997; Butler & Parkinson, 1997; Savenije, 2000). What is clear though is that too much water and non-renewable resources of high quality are taken from the eco-system and returned to the ecosphere as pollution (Daly, 1992; Hodges, 1993; Niemczynowicz, 1993). Overall it is argued that the present forms of agriculture, architecture, engineering and industry are derived from design epistemologies incompatible with nature's own (Van der Ryn & Cowan, 1996; Smit 1996).

How can cities perform better with regard to resource use and waste management? Urban expansion has caused loss of fertile land and other ecological changes (Fazal, 2000; Morello *et al.*, 2000; Hallsmith, 2003) despite the fact that urban areas occupy less than 2% of the total earth's land surface (Miller, 1988). At present there is almost an explosion of seemingly novel technologies of potential strategic interest. How should these future technologies be judged to be better or worse than current arrangements? Does the form of present technology commit us to further developments that are incompatible with the needs of a sustainable city?

Cities, like organisms, are associated with flows of material and energy. Within the broad context of the global cycles of certain principal materials, the ways in which futuristic urban water and drainage system might introduce minimal distortion to these "natural" material cycles is an important subject of exploration lately (Ayres & Ayres, 1996; Herrmann & Klaus, 1997; Grotteker, 1998). Specifically, the cycles of Carbon (C), Nitrogen (N), Phosphorus (P) and Sulphur (S) bearing materials, together with those of heavy metals, synthetic organic chemicals and pathogens, have to be examined carefully. These represent the principal categories of pollution associated with the activities of a city. Much of the analysis points towards the desirability of returning the non-aqueous output fluxes of the urban drainage system to the land, as opposed to the aquatic environment (Lijklema & Tyson, 1993; Beck *et al.*, 1994; Staudenmann *et al.*, 1996).

The principal nutrients (P and N) flow in a circular, closed loop system in nature, but human activities use and dispose nutrients in a linear, open-ended system (Walker, 1991; Vitousek *et al.*, 1998; Gijzen & Mulder, 2001; Smil, 2001). The danger is that once one closed loop system is opened, it may force open other closed loop systems elsewhere in the ecosystem (Esrey, 2000). Short-cutting or closing the P-cycle in the urban environment is closely related to closing of water cycles (Savenije, 2000).

---

[3] Urban Ecology is a neologism intended to call attention to a biological analogy; the fact that an ecosystem tends to recycle most essential nutrients, using only energy from the sun to 'drive the system'. This analogy to a complex web of potentially synergistic and symbiotic relationships is appealing (Ayres, 1989).

There is still much debate however, on whether robust municipal systems can achieve 100% source control treatment and reuse technologies as in industry, from an economic, engineering and ecological resilience point of view (Ayres & Ayres, 1996; Matthews, 1996; Jeffrey *et al.*, 1997; Otterpohl, 2001).

New solutions in terms of water management, sanitation and management of organic wastes for the 21st century cities (Cosgrove & Rijsberman, 2000) are perceived to be source orientated, local and small scale, non-mixing, ecologically sound closed loop systems approaching the life-support system of a spaceship[4] (Beck *et al.*, 1994; Roelofs, 1996; Van der Ryn & Cowan, 1996; Nelson, 1997; Otterpohl *et al.*, 1997; Crites & Tchobanoglous 1998; Ross *et al.*, 2000; Lens *et al.*, 2001). Since cities need to close the open loop of limited resources such as P, urban agriculture seems to be a viable option by reusing and transforming the by-products of human metabolism especially, which, usually are dumped as polluting waste into the bio-region (Smit & Nasr, 1992; Hodges, 1993).

There is a growing agreement among engineers, planners, and activists that: 'Large scale end-of-pipe wastewater treatment facilities create a spectacular example of technology that must change in order to provide a sustainable, global solution. Several non-conventional, ecologically sound wastewater treatment technologies do exist and they are ready for implementation. The strongly entrenched connection of urban residuals with pipes, culverts, conduits, concrete, and tanks has enabled the society to forget that ultimately the interest must lie in studying the relationship between all types of land use (which determines water use and vice versa) and receiving water quality (Lyle, 1985; Lijklema & Tyson; 1993; Gumbo, 2000a, 2003a).

The Bellagio principles

The pressures of humanity on a fragile water resource base, and the corresponding need for environmental and freshwater protection requires that human excreta and other societal wastes (solid and liquid) be recycled and used as a resource. The so-called Bellagio principles, following a Water Supply and Sanitation Collaborative Council (WSSCC) meeting of experts in Bellagio (Italy), underpin the basis for this new approach to environmental sanitation (SANDEC & WSSCC, 2000; Schertenleib, 2001).

There are two main concepts emanating from the Bellagio principles, which make the basis of this dissertation. Firstly, Household Centred Environmental Sanitation (HCES) puts the household at the focal point of environmental sanitation planning and; secondly, the Circular System of Resource Management (CSRM) that emphasises conservation, local recycling and reuse of resources (SANDEC & WSSCC, 2000; WSSCC, 2000). The house, the most familiar habitat has been used

---

[4] For space missions, high tech solutions such as vapour compression distillation, reverse osmosis, or multi-filtration are considered for the recovery of water from urine and faeces (Silverstone, 1993; NASA, 1994). In short-term missions, however, human wastes are usually stored for post flight treatment or disposed using an overboard-vacuum-system. On long-term space missions it is necessary to close not only the water cycle but also the food and nutrient cycle, for this purpose bio-regenerative life support systems have been designed (Alling *et al.*, 1993a; 1993b; NASA, 1994; Nelson, 1997; Allen, 2002). The only example of full-scale implementation is Biosphere 2 (Nelson, 1997; Alling *et al.*, 1993b).

as a good starting point in many ecological designs. The rural or village homestead is viewed as once the centre of a largely self-sufficient system that produced a family's livelihood, its food and fibre, and its tools and toys (Van der Ryn & Cowan, 1996). Therefore rethinking home metabolism has become the focus of ecological design (Olkowski *et al.*, 1979; Sowman & Urquhart, 1998).

Implementation of the HCES and CSRM approaches for environmental sanitation requires integration between excreta disposal, wastewater disposal, solid waste disposal, and storm water drainage (Cosgrove & Rijsberman, 2000; SANDEC & WSSCC, 2000; WSSCC, 2000). Firstly, the HCES makes the household the focal point of environmental sanitation planning, reversing the customary order of centralised top-down planning (Hodges, 1993; Zeeman & Lettinga, 1998; Schertenleib, 2001). The approach argues that only problems not manageable at the household level should be '*exported*' to the neighbourhood, town, and city and so on up to larger jurisdiction. Secondly, the CSRM, in contrast to the current linear system, emphasises conservation, recycling and reuse of resources as illustrated in Figure 1.1 (Schertenleib, 2001). Many water supply and sanitation problems would be resolved by this new paradigm, which places all aspects of water and waste within one integrated service delivery framework (Larsen & Gujer, 1997; Niemcynowicz, 1997; Esrey *et al.*, 1998; Schertenleib & Gujer, 2000; Schertenleib, 2001).

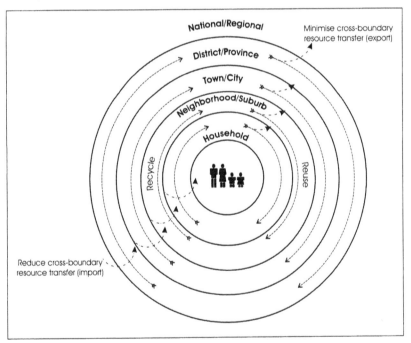

Figure 1-1 The concept of household centred environmental sanitation (HCES) and circular system of resource management (CSRM)
Source: Adapted from Schertenleib (2001)

Land and water demands by humanity

Sustainability requires living within the regenerative capacity of the biosphere. One approach is to identify spatial and physical dimensions of urban demands on the natural capital and also to explore the extent to which surrounding rural areas are affected by cities (Wackernagel & Rees, 1996; Fazal, 2000). In an attempt to measure the extent to which humanity satisfies this requirement, Wackernagel *et al.*, (2002a, 2002b) used existing data to translate human demand on the environment into an area required for the production of food and other goods, together with the assimilation of wastes generated, under the predominant management and production practices in any given year (see Box 1.A: Land as a finite resource). Not only human demand on nature, but also nature's supply changes over time because of innovations in technology and resource management, changes in land use, and cumulative damage of past impacts (Wackernagel *et al.*, 2002a).

According to the exploratory assessment by Wackernagel *et al.*, (2002a, 2002b) indications are that human demand on nature may well have exceeded the biosphere's regenerative capacity since the 1980's (Box 1.A). The accounts of biologically productive space[5] presented by Wackernagel *et al.*, (2002a) are based on six human activities that require biologically productive space. They are (i) growing crops for food, animal feed, fibre, oil, and rubber; (ii) grazing animals for meat, hides, wool, and milk; (iii) harvesting timber for wood, fibre, and fuel; (iv) marine and freshwater fishing; (v) accommodating infrastructure for housing, transportation, industrial production, and hydro-electric power; and (vi) burning fossil fuel. The calculated human demand and existing capacity in year 2000 for each category is shown in Table 1.1.

Table 1-1 Summary of humanity's area demands, and earth's biological capacity in 2000

Area	Equivalent factor [a]	Average equivalent global area demand	Equivalent existing bio-capacity
	Global m^2/m^2	m^2 per person	
Growing crops	2.1	5 300	5 300
Grazing animals	0.5	1 000	2 700
Harvesting timber	1.3	2 900	8 700
Fishing	0.4	1 400	1 400
Accommodating infrastructure	2.2	1 000	1 000
Fossil fuel and nuclear energy	1.3	11 600	0
Total	-	23 300	19 100

a) To make aggregation reflect differences in bio-productivity, areas are expressed in standardised global m^2, which correspond to m^2 with world average bio-productivity.
Source: Adapted from Wackernagel *et al.*, (2002a)

---

[5] Biologically productive space is the land and water area that is biologically productive. It is land or water with significant photosynthetic activity and biomass accumulation. Marginal areas with patchy vegetation and non-productive areas are not included. The total biologically productive space adds up to 0.10 billion Mm^2 and hosts over 95% of the planet's terrestrial biomass production.

Box 1-A Land as a finite resource
Adapted from Redefining Progress (2002) and
Wackernagel *et al.*, (2002a; 2002b)

The surface area of the planet is about 0.51 billion Mm^2. Of this, less than 0.13 billion Mm^2 is land. About two-thirds of this, at most, could be classed as productive in the sense that it can support 10% or more forest cover, is permanent pasture, or arable land. The remainder is inhospitable desert, ice-covered, and rock.

If an upper estimate of 0.10 billion Mm^2 usable land is assumed, and a year 2000 global population of 6 billion, then this gives an average 'Earth share' of about 17 000 m^2 per capita (an area 165 metres by 100 metres). A figure which will reduce to just over 10 000 m^2 by 2050 if the United Nations medium-term population predictions of around 9 billion are realised. Adding to this the productive regions of the oceans, takes the year 2000 average 'Earth share' to around 19 000 m^2, this includes a very modest 12% allowance for other species and conservation; WCED (1987) (see Table 1.2). This shrinking space must service all human needs (food, materials, recreation, living space, energy) and be capable of assimilating all wastes (arguably the most pressing being the sequestration of greenhouse gases).

Based on current global average yields, and ignoring the massive energy inputs common with modern agriculture, around half the year 2000 'Earth share' was used to provide meat and vegetables alone to the human population. Available data suggests that humanity is drawing down the planet's capital account rather than living within the interest. Whereas the available 'Earth share' is 19 000 m^2, the average global citizen has a footprint of 23 000 m^2 - an overshoot of 37%.

Table 1-2 The year 2000 global ecological footprint

Parameter	Value
Population (in billions)	6
Ecological footprint[6] (000's m^2)	23
Bio-capacity[7] (000's m^2)	19
Ecological deficit[8] (000's m^2)	4

The latest estimates from Wackernagel, *et al.*, (2002b), based on 1995 data suggests that consumers in developed countries have very large feet. The United States has a per capita footprint of 96 000 m^2, Australia 94 000 m^2 and Canada 72 000 m^2. Wackernagel *et al.*, (2002b) also calculate the 'Available Bio-capacity' within a country. Where this is higher than the Footprint then it is theoretically possible for that country to meet its own bio-productive needs. Of the three nations mentioned, Australia and Canada have theoretically 'spare' bio-capacity. The US uses around twice its nationally available bio-capacity with a deficit of 41 000 m^2/person. An average US citizen consumes at five times the total 'Earth share'.

There are two main ways in which deficits are resolved. First, a country trades with others that have a surplus of resources - as in the case of fossil fuel use (stored energy of '*Paleozoic summers*'), where use of ancient bio-productivity leads to present-day pollution (Sturm *et al.*, 2000). Second, and more significantly, ecological natural capital is diminished and/or renewable resources are 'used' at unsustainable rates (Chambers *et al.*, 2000; Sturm *et al.*, 2000). Examples include soil degradation, deforestation, over-fishing, and over-grazing

---

[6] Ecological footprint is a measure of how much productive land and water an individual, a city, a country, or humanity requires to produce all the resources it consumes and to absorb all the waste it generates, using prevailing technology. This land could be anywhere in the world. The Ecological Footprint is measured in 'global area units'.

[7] Biological capacity: the total biological production capacity per year of a biologically productive space, for example inside a country. It can be expressed in "global m^2", i.e. the equivalent area of space with world-average productivity.

Infrastructure for housing which include transportation, industry, and hydroelectric power results in built-up land (Sowman & Urquhart, 1998). The space occupied by this infrastructure is the least well documented, because low-resolution satellite images are not able to capture dispersed infrastructure and roads. Wackernagel *et al.*, (2002a) use an estimate of 0.003 billion Mm^2 (approximately 2% of global land area), a minimum estimate of the extent of infrastructure worldwide today, and assume that built-up land replaces arable land, as has been documented for the United States.

It is in cities that a considerable (and growing) proportion of the world's population lives and a much higher proportion of all resource use and waste generation is concentrated. The form and structure of any city influences it global impact (Hardoy *et al.*, 1992; Rees, 1992). Many cities and economies in the North achieve sustainable development goals within their own region by drawing heavily on the environmental capital of other regions or nations and on the global sink for their wastes (see Box 1.B: The manure problems in the Netherlands). Through trade and natural flows of ecological goods and services, all urban regions appropriate the carrying capacity of distant 'elsewheres', creating dependencies that may not be ecologically or geopolitically stable or secure (Rees, 1992; Chambers *et al.*, 2000; Sturm, 2000).

World Bank predictions indicate that over the next 25 years production of food must increase by a factor of at least three, yet present statistics show that total grain production per capita is decreasing, with no signs of change. In order to reverse this trend, huge amounts of water and nutrients will be required to increase food production. This will require expansion of agriculture, which can be achieved with increasing use of finite fossil fertilisers and pesticides, more artificial impounding reservoirs for irrigation, potentially bringing economic burden, environmental pollution and a further decrease of clean water resources (Niemczynowicz, 1997a; 1997b; Falkenmark *et al.*, 1998; Savenije, 1999). Meanwhile the global fertiliser use has soared from less than 14 Mt/a in 1950 to about 140 Mt/a as macro nutrients (N, P & K) in year 2000 (Steen, 1998; Johnston & Steen, 2000).

Ecological agriculture can be viewed as a component of sustainable agriculture which includes all agricultural systems that promote the environmentally, socially and economically sound production of food and fibres (Rees, 1992; Nasr, 2000). These systems take local soil fertility, culture and biodiversity as a key to successful production (Tilth, 1982; Hands *et al.*, 1995; IFOAM, 2000). By respecting the natural capacity of plants, animals and the landscape, it aims to optimise quality in all aspects of agriculture and the environment (Halweil, 2001). Sustainable agriculture dramatically reduces external inputs by refraining from the use of chemo-synthetic fertilisers, pesticides, and pharmaceuticals. Instead it allows the powerful laws of nature to increase both agricultural yields and disease resistance. Sustainable agriculture also calls for efficient use of '*green water*[9]'. Of all water resources, *green*

---

[8] Ecological deficit: the amount by which the Ecological Footprint of a population (e.g. a country or region) exceeds the biological capacity of the space available to that population. The national ecological deficit measures the amount by which the country's footprint (plus the country's share of biodiversity responsibility) exceeds the ecological capacity of that nation.

[9] The concept of *green water* was first introduced by Falkenmark (1995). The storage medium for green water is the unsaturated soil. The process through which green water is consumed is transpiration. Green water is a very important resource for global food production. About 60% of the

water is probably the most under-valued resource. Yet it is responsible for by far the largest part of the world's food and biomass production (Savenije, 1999).

Box 1-B The manure problem in the Netherlands
Adapted from Van Ruiten (1998)

Increasingly intensive methods of stock farming have resulted in a surplus of animal manure in the Netherlands (a developed country in western Europe with a population of about 15.9 million and land area of about 34 000 Mm^2), which primarily means a surplus of minerals (nitrogen and phosphorus). Above all, the emergence and growth of 'land-independent' (intensive) stock farming establishments in the sandy regions of the southern, eastern and central Netherlands has made a major contribution to this state of affairs. The reason is that the soil type in these regions is not suitable for a high level of production of 'land dependent' agricultural crops, and the cultivated area per farm is small. The import of cheap raw materials for livestock feeding and the good infrastructure have encouraged the development of intensive stock farming in these regions. As a counterweight to the import of mostly pig and cattle feed, there are substantial exports of products (meat, eggs, milk, cheese, etc.). However, the excreted minerals remain in the soil. The result is a net increase in the amount of minerals within the country's borders.

A surplus of minerals causes environmental pollution in the form of contaminated soil, groundwater and surface water. In addition, unpleasant odours are emitted. Although the signs of the negative effects were already recognised in the 1960's, it was not until the 1970's that measures were gradually introduced to limit the loss of minerals into the environment. Although the emission of nitrogen is just as harmful from an environmental point of view as the emission of phosphorus, the Dutch government has always geared its legislatory and regulatory measures to phosphate (P_2O_5). The significance of this is primarily practical. Nitrogen is more difficult to incorporate in an analysis of inputs and outputs, due to its presence in the form of volatile compounds such as ammonia (NH_3) and nitrogen gas (N_2). In the course of the years, the limiting measures taken by the government have been steadily tightened up (phased approach) and the amount of research work into proposed problem solutions has simultaneously increased. In its search for solutions the government has elected for a three-pronged policy of geographically spreading (from regions of surplus to regions of shortage), limiting minerals in cattle feed, and treating and processing manure.

At the end of the 1980's the Dutch government urged the stock farming sector to take 25 000 Mg P_2O_5 (equivalent to about 6 Mt of pig manure) out of the market by 1995 through processing. A large number of initiatives were started at the major ones involving: Processing pig slurry into fertiliser granules; Concentrating liquid fraction of sow and veal calf manure; Drying chicken manure into fertiliser granules. With the exception of the processing of veal calf manure (total processing capacity 660 000 Mg/a) and slurry (evaporation plants with a total processing capacity of 160 000 Mg/a), these large-scale initiatives ultimately led to nothing, despite the enormous amounts of time and money invested.

There were various causes for these large-scale failures, such as an excessively high processing price in relation to competing options such as geographically spreading; insufficient support from the sector; problems with the choice of location and licences; and uncertainty with regard to the market for the end product. In addition, in 1995 the EU prohibited the continuation of subsidies for large-scale manure processing and long-distance transportation, and this had an extremely disadvantageous effect on the competitive position of the various problem solutions.

---

world staple food production relies on rainfed irrigation, and hence green water. The entire meat production from grazing relies on green water, and so does the production of wood from forestry. In Sub-Saharan Africa almost the entire food production depends on green water (the relative importance of irrigation is minor) and most of the industrial products, such as cotton, tobacco, wood, etc (Savenije; 1999, 2000) (see also Chapter 4).

Box 1-B The manure problem in the Netherlands (conti)

Surpluses also occur in some regions of other countries, for example in Flanders, Brittany, northern Germany and the Po delta. Although this problem is gradually attracting more attention in these regions, partly under the influence of European legislation, far less effort has been devoted to solving it than in the Netherlands. On the basis of the most important manure flows, the emphasis is on the following categories of animals: cattle, veal calves, pigs and chickens. According to the data from the CBS (Dutch Central Bureau of Statistics) the number of animals in 1997 is shown in Table 1.3. It should be noted that the cattle, with the exception of veal calves, are held on the land (grazing animals). Veal calves, along with pigs and chickens, are raised in indoor establishments. This is important with regard to the form in which the manure becomes available (see Table 1.4). The type of stall plays an important role - for example, farmyard manure (solid) versus slurry (semi-liquid).

Table 1-3 Number of major livestock in the Netherlands (1997)

Livestock	Number (000's)
Cattle	4 410
Veal calves	740
Pigs	15 190
Chickens	93 110
Other[a]	2 160

a) Sheep, goat, fur bearing animals and rabbits

Table 1-4 Manure production (1997)

Livestock	Production (Mt)	
	Slurry	Solid
Cattle	56.6	1.0
Veal calves	2.5	-
Pigs	16.2	-
Chickens	0.9	1.2
Other	1.6	0.4
Total	77.8	2.6

Large amounts of cattle feed, and therefore of phosphate, are imported for the livestock industry. A considerable proportion of this phosphate is excreted by the animals. In principle, the amount of excreted minerals can be calculated from the produced amount of manure and its composition (Gerritse & Vriesma, 1984). In the case of the mineral phosphate, and on the basis of the data in Table 1.4, the phosphate as P_2O_5 excreted by the Dutch livestock in 1997 is shown in Table 1.5. The table shows that cattle make the biggest contribution to phosphate excretion (about 50%).

A considerable proportion of the phosphate excreted in the manure is used to fertilise grassland and arable land (placement in farm where produced and distribution; Farm surplus = Production - In-farm placement). A small proportion is exported or processed. In view of the fact that not all manure can be disposed of in this way, a national manure and phosphate surplus arises. This is illustrated in Table 1.6. The amount of manure that can be spread on cultivated land inside the Netherlands depends on the legal standards drafted by the government for this type of use. Since the early 1990's these standards have been tightened up. In 1997 the standards for spreading on grassland, maize silage and other arable land was 13.5, 11.0 and 11.0 g/m^2 as P_2O_5, respectively (Van Boheemen, 1987; Van Ruiten, 1998).

Table 1-5 Phosphate excretion by Dutch national livestock (1997)

Livestock	Mt as P_2O_5
Cattle	0.096
Veal calves	0.003
Pigs	0.058
Chickens	0.029
Other	0.006
Total	0.192

Table 1-6 Phosphate disposal options (1997)

Option	Capacity (Mt)
Production	0.192
In-farm placement	0.118
Farm surplus	0.088
Distribution	0.071
Export or processing	0.014
National surplus	0.003

From Table 1.6 it is clear that a national surplus (National surplus = Farm surplus – Distribution + Export or processing) was created in 1997. In future as a consequence progressive tightening up of the manure policy, more manure surplus would need to be taken out of the market or the number of animals reduced.

Agricultural management can successfully mimic natural ecosystems, with - for instance - the use of mulches, green manure crops, legume covers, alley cropping and appropriate crop rotations (Hands *et al.*, 1995; Gruhn *et al.*, 2000). Sustainable agriculture adheres to globally accepted principles, which are implemented within local social-economic, geo-climatical and cultural settings. Cumulative negative nutrient balances heighten the impact of climatic factors, insecure tenure arrangements, and land and demographic pressures on soil fertility (see Box 1.C: Human-induced soil degradation). In Sub-Saharan Africa net annual nutrient depletion was estimated at 2.20 g N/m^2; 0.25 g P/m^2 and 1.5 g K/m^2 during 1982 to 1984 (Stoorvogel *et al.*, 1993).

Box 1-C Human-induced soil degradation
Adapted from Gruhn *et al.*, (2000)

Soils in many countries suffer from declining fertility. Their physical and chemical structures are deteriorating and the vital nutrients for plant growth are slowly being depleted. By some estimates, the annual cost of environmental degradation in some countries ranges from 4 to 17% of gross national product (GDP). Three-quarters of the area degraded by inappropriate agricultural practices, overgrazing, and deforestation is in the developing world. Tables 1.7 and 1.8 illustrate the extent and human-induced causes of soil degradation in Africa, Asia, and South America (WRI, UNEP, & UNDP 1992; Oldeman, 1992).

Table 1-7 Extent of human-induced, nutrient-related soil degradation in selected regions

Region	Degree of degradation (land area in 10^3 Mm^2)		
	Light	Moderate	Severe
Africa	204	188	66
Asia	46	90	10
South America	245	311	126

Source: Oldeman *et al.*, (1992)

Table 1-8 Human-induced causes of soil-degradation

Region	Degrading activity (% of total)				
	Deforestation	Overexploitation	Overgrazing	Agriculture	Industry
Africa	14	13	49	24	-
Asia	40	6	26	28	1
S. America	41	5	28	26	-
World	29	7	34	28	2

Source: Oldeman (1992)

It is suggested that the lack of manure, and excessive exploitation of soil and water resources beyond their population-carrying capacity, have been major factors in the collapse of several ancient civilisations. Cases in point are the Roman Empire, the Sumerians in Mesopotamia and the Mayans in Central America, where soil disintegration was caused respectively by salinisation and deforestation – two of the major threats of today (Ponting, 1991; Johnston, 1995; IFA & UNEP, 1998).

The urban food system is so far not sufficiently reflected in the urban planning process in many countries (Jarlov, 2000; Nasr, 2000; Argenti, 2000; Pothukuchi & Kaufman, 2000). The urban food system connects to many other urban systems – notably the agricultural sector, the economy and ecological systems. Urban people are not passive food recipients; in many locations they are actively involved in food production (Argenti, 2001; Balbo *et al.*, 2001).

Agriculture in urban areas can mitigate negative impacts on surrounding and more distant biodiversity (the urban footprint). The lead feature of urban agriculture, which distinguishes it from rural agriculture, is *its integration into the urban economic and ecological system* (the urban 'ecosystem') (Smit, 1996; Mougeot, 2000). Urban agriculture produces food and energy crops close to the market demand, some within the neighbourhood. This proximity of production to consumption reduces traffic, storage, and packaging as sources of the pollution that erodes biodiversity (Smit, 2000; Hallsmith, 2003)

It is argued that urban agriculture is inherently more biodiversity-prone than modern rural agriculture, by being more sustainable, less chemically dependent and more biologically friendly. Urban agriculture occurs on smaller sites and typically has a more diverse or integrated crop mix (Smit, 2000). Urban agriculture may close open nutrient and energy loops. Perhaps the most effective example is modifying urban wetlands to food, fuel and recreation instead of filling them with waste and converting to built-up uses. Further research might be worthwhile on differences in biodiversity in different climate zones, associated with urban versus rural farming.

1.2 Urban agriculture

Significance

The poor throughout most of Africa have experienced increasing difficulties over recent years as a result of the imposition of structural adjustment programmes. One of the main coping mechanisms has been increased self-help in satisfying basic households needs (Matshalaga, 1997; Argenti, 2000; 2001). Food is one of these basic needs and urban agriculture[10], both legal and illegal, has grown as a consequence of the difficult economic climate. As yet relatively few studies have attempted to asses the role that urban agriculture plays or might play in social or environmental terms (Drakakis-Smith, 1992; UNDP, 1996; FAO, 1999c).

Urban farming is often minimised as being merely "kitchen gardening" or marginalised as a leftover of rural habits. Certain myths still predominate e.g. urban agriculture means household and community gardening, urban agriculture is a temporary activity, urban agriculture is a marginal activity or means of survival, urban agriculture pre-empts "higher" land uses and cannot pay full land rent, urban agriculture competes with and is less efficient than rural farming, urban agriculture is unhygienic, urban agriculture causes pollution and damages the environment, urban agriculture is unsightly and aesthetically inappropriate in the city and the "garden city" is an archaic, utopian concept that cannot be created today (UNDP, 1996; Nugent, 1997).

---

[10] Urban agriculture is defined as the production of crops and or livestock on land, which is administratively and legally zoned for urban uses. This activity is done within these zones or at the periphery of urban areas i.e. land likely to be rezoned from rural agriculture to urban land.

In 1996 the UNDP estimated that some 15% of food production in the world comes from urban agriculture (farming, horticulture, animal husbandry, fish ponds, etc.). Nearly 1 billion people are engaged in urban agriculture, 200 million producing food for markets (UNDP, 1996; Bakker *et al.*, 2000; IDRC, 2003; RUAF, 2003). In cities such as Lusaka and Dar es Salaam as much as 50% of the food is produced within the city. Shanghai, which has a population of 11 million, produces 100% of its fresh vegetables in community gardens (Yeung, 1993; Yi-Zhang, 1999). For city dwellers, community gardens are another option for farm-fresh food. It has been estimated that having market gardens located throughout suburbs and cities could cut the dollar cost of food by 70%. Given that half of the world population soon will live in urban areas, it could be expected that re-circulation of nutrients in urban areas will be featured high on the agenda in the near future (Hayward, 1997). However despite many benefits, urban agriculture is still an ill-understood industry (Mbiba, 1995; Ruel *et al.*, 1999; Bakker *et al.*, 2000).

Many questions arise (Drescher, 2000; FAO, 1999c): Where are urban agricultural activities concentrated and why? Who is involved? Is it for psychological or cultural reasons (hobby, leisure, growing your own cultural food, growing organic food)? What kinds of crops are grown and by which groups of city dwellers? What contribution does the product make to nutrition and food security? What type of land tenure system has to be adopted to ensure sustainability? What kinds of soils do urban agriculturalists prefer? How available is water and what is its quality? How far does the producer have to travel to market the products grown? What are the risks to human health? What are the possible environmental impacts from urban agriculture? How can harmful health and environmental impacts be mitigated? What are the possibilities and limitations for integration of urban agriculture in urban planning and zoning?

Urban agriculture constitutes an important component of sustainability in terms of ensuring that people have sufficient income and food, both of which sustain other aspects of household life such as health and education (Maxwell *et al.*, 1998; Martin *et al.*, 2000; Nugent, 2000). The World Food Summit held in Rome in November 1996 gave priority to the development of urban and peri-urban agriculture as well as improving the efficiency of food supply and distribution systems and linkages between production and consumption areas, with the aim of facilitating access to food by low-income households and hence improving food security in developing countries and countries in transition (Armar-Klemesu, 2000). The Habitat II Conference in 1996 drew attention to the high urbanisation levels and the rapid urban growth rates in developing countries and countries in transition, and stressed their direct relationship with urban poverty.

Historically, public support for access by the poor, to urban land for food production has arisen for economic and cultural reasons (*schreber* gardens, allotment gardens). In the post-World War II period, urban farming examples include the Gorbachev reforms in Russia (Moldakov, 2000), Mozambique's *zonas verdes*, Cuban *hydroponicos*, (Bourque & Canizares, 2000; Cruz & Medina, 2003) Mongolian school gardens (Yeung, 1993), South Africa's provincial urban small-scale farms (Jarlov, 2000), and 'community gardens' in France and the USA.

Urban agriculture draws on the often unmanaged and 'un-recovered' urban waste stream inherent to a majority of cities in the developing world and attempts to re-direct these resources toward the production of food and fibre in an economically and environmentally sound fashion. Food production schemes can be augmented and enhanced by recycling human and solid waste (Furedy, 1996) if low-cost and reliable waste recovery technologies and approaches can be demonstrated and proven feasible[11] (Tilth, 1982; Smit & Nasr, 1992; Gardner, 1998; Rose, 1999).

Feasibility

In many countries, an important and so far unsolved problem is how to devise appropriate methodologies to integrate agricultural activities in cities into urban planning processes. To what extent are various functions of city life - such as farming - included in the urban planning process (Drescher, 2000; Quon, 1999; Jacobi *et al.*, 2000)? The ultimate objective of an urban plan is to create a liveable city – relatively free of conflicts among dwellers and uses, providing for the needs of its citizens, and maintaining its natural resources. The role of urban and peri-urban agriculture in a city plan is to contribute to those ends (Mougeot, 2000). In most of the world's cities, little is known about the actual extent to which inner city areas are used for agricultural purposes (Drescher & Iaquinta, 1999).

In many urban centres, open space for agriculture is limited and the faster the urbanisation, the more limited that space becomes. For this reason, the efficient protection of open spaces in inner cities is an important issue. However, despite the lack of open spaces for agriculture in urban areas, there is huge, unused potential for this activity in many cities of the world[12] (Fernades & Varley, 1998; Balbo *et al.*; 2001). Strips of land at the sides of roads, railways and power lines; along river banks and on seasonally flooded land; and on other partially unproductive areas could be legally used for agriculture, in many cities of the world, providing it is carried out in a proper, environmentally sound way (Dowall & Giles, 1996; Drescher, 2000).

Despite traditions in some countries, access to land must be distinguished from availability of land; land may be available or present in a city but not accessible to farmers because of political or social constraints to its use or redistribution (Mbiba, 1995; UNDP, 1996; Meadows, 2000). Generally, agriculture in urban areas suffers greater ecological and economic pressures than rural agriculture, requiring more intensive and better controlled production to stay competitive and secure. When land availability is restricted, urban farmers tend to be opportunistic, and find creative ways to use the smallest plots or strips of land and water. This leads to farming on land originally set aside for other purposes, on land that is hazardous and therefore unusable, or land that has been abandoned or contaminated by past uses, sometimes

---

[11] The International Federation of Organic Agriculture Movements (IFOAM) specifies that manures containing human excrement (faeces and urine) shall not be used on vegetation for human consumption, except where all sanitation requirements are met. Certification bodies or standardising organisation should establish sanitation requirements and procedures, which prevent transmission of pests, parasites and infectious agents (IFOAM, 2000)

[12] There are lots of interesting ideas for intensive gardening. One of the most inspiring ideas is "The Productive Homestead" by Dr Gus Nilsson in Gaborone in Botswana. He claims that a 1000 m^2 plot including a house can feed a family and pay for the building of the home over a 20-year period. The effective adoption of this system, however, depends on the level of education attained by the people (Esrey *et al.*, 1998).

without the farmer even being aware of the hazard (UNDP, 1996; Smit, 1996; Dubbeling, 2000). Such opportunistic use may result in unregulated production and processing that may be hazardous to consumers. Planners can assist poor families (De Zeuw *et al.*, 2000; Balbo *et al.*, 2001).

Negative impacts of urban agriculture

Today, as in earlier times, agriculture in the city poses a range of possible negative impacts. Irrigation with polluted water, animal waste in the streets, or spraying chemical insecticides can be injurious to man and the community biosphere (Birley & Lock, 1999; Flynn, 1999; Smit, 2000). Converting park-like open space to mono-cropping can diminish biodiversity of the site. The management of an ecologically sustainable or 'biogenic' city, which conserves biodiversity, will require a much higher level of environmentally sophisticated management than current practices (Van der Ryn & Cowan, 1996).

Several reviews or studies of health issues have tended to highlight the health risks of urban agriculture (Birley & Lock, 1999). This has served to reinforce the perceptions of many governments and municipal authorities that urban agriculture is a (marginal) activity that has substantial health risks and should not be supported (Lock & De Zeeuw, 2001; Lock & Van Veenhuizen, 2001). According to Lock & Van Veenhuizen (2001), the main health risks associated with urban agriculture are summarised in Box 1.D.

Box 1-D Overview of major health risks in urban agriculture
Lock & De Zeeuw (2001)

	Communicable diseases	Non-communicable diseases
Crop production	• Crops irrigated with untreated (or inadequately treated) domestic wastewater or fertilised with improperly produced compost may be infected with bacteria (shigella, typhoid, cholera), worms (like tape and hookworms), protozoa, enteric viruses or helminths (ascaris, trichuris) • In Africa, mosquitoes that are the vector for malaria may breed in clean, shallow irrigation water and crop land with serious water-logging. Incidence of malaria mainly relates to wet rice and ridge cultivation of yams and sweet potatoes • Mosquitoes that are the vector for filariasis and dengue may breed in standing water heavily polluted with organic materials (drains blocked by organic refuse, latrines, septic tanks) • Food may be contaminated with bacteria due to poor hygienic conditions in informal food preparation and marketing, causing diseases such as salmonella and E-coli	• Crops may take up heavy metals and other hazardous chemicals from soils, irrigation water or sewage sludge polluted by industry • Crops grown close to main roads or industry, and food purchased from street vendors may be contaminated by air-borne lead and cadmium • If waste materials are not separated at source, the resulting compost may contain heavy metals, which can be taken up by crops • Occupational injury of agricultural workers is an important source of disability including musculoskeletal disorders or poisoning by agrochemicals

Box 1-D Overview of major health risks in urban agriculture (cont)

	Communicable diseases	Non-communicable diseases
Animal husbandry	• Closeness of animals and humans may lead to occurrence of zoonotic diseases like bovine tuberculosis (cattle) and tapeworms especially when animals are scavenging waste tips • Animal products can become contaminated with pathogens due to contamination of animal feed with infected faeces (salmonella, campylobacter)	• Animal products (like red meat, poultry meat and eggs) may be contaminated with pesticides (especially organo-phosphates) and or antibiotics, if animals are kept intensively • Freely wandering animals can injure people and may cause traffic accidents • Allergens from livestock wastes or dust (especially. poultry) can cause occupational diseases in farm workers (asthma, allergic pneumosis) • Tanneries may discharge hazardous chemicals in their wastes (tannum, chromium, aluminium)
Aquaculture	• If fish (especially shellfish) are fed with wastewater and or human and animal excreta, there are potential risks of : a. passive transfer of pathogens (hepatitis A) by fish and aquatic macrophytes; and b. transmission of trematodes whose life cycles involve fish and aquatic macrophytes. This is only a problem where trematodes are endemic and fish is consumed raw • Contamination of fish with human or animal faecal bacteria may occur during post-harvest operations (e.g. salmonella) • Poorly managed fish ponds may become a breeding ground for malaria mosquitoes • Use of antibiotics in fish feed may lead to development of antibiotic-resistant bacteria in the food chain	• Fish products may be contaminated with heavy metals if fed with wastewater or organic wastes contaminated by industry • Fish products may be contaminated with agrochemicals, if produced in an input-intensive way

Environmental reasons have often been used to criticise urban agriculture and in many instances to prohibit it. In Zimbabwe the main criticism is that it causes soil erosion and subsequent siltation of water supply sources (ENDA, 1995; Mbiba, 1995; Bowyer-Bower *et al.*, 1996; Bowyer-Bower & Tengbeh, 1997). A review of available literature reveals that very little field research has, in fact, been undertaken to provide empirical substantiation of these claims. Without such evidence, potentially damaging environmental impacts may be ignored, or actions (often costly and unpopular) taken that if investigated could be found to be unnecessary (Drakakis-Smith, 1992; ENDA, 1995).

Bowyer-Bower & Tengbeh (1997) identified four main categories of potential environmental effects of widespread urban agriculture on public land in Harare. These are listed in Box 1.E together with their implications on the quality of urban life and cost of urban management.

Box 1-E Potential environmental impacts of urban agriculture

Adapted from Bowyer-Bower & Tengbeh (1997)

Category	Examples	Primary effects	Quality of life and environmental costs
1. Change in the hydrological regime of the area	More runoff; more flooding; less infiltration	Flood damage to property, transport routes etc.; apparent drying of wetter areas	Costs of maintaining urban infrastructure are affected
2. Soil erosion	Lowering of the land surface; deposition of eroded sediments; particulate pollution of the air	Clogging of city drains; nuisance to transport; health problems	Increasing costs of maintaining urban infrastructure; loss of aesthetic quality of urban space; increasing the health-hazard of urban living
3. Chemical pollution	Eutrophication; vegetation or crop toxicity; water quality	Algal blooms; potential health hazards; threat to wildlife	Increasing costs of water treatment for safe urban water supply; loss of aesthetic quality of urban space
4. Vegetation change	Reduced plant species diversity; change in dominant plant type e.g. from open grassland to tall maize and weeds; loss of ground cover	Loss of species habitat; loss of biodiversity; loss of aesthetic quality of open land, loss of land for recreation; increased incidence of crime; cause of soil erosion	Aesthetic quality of urban space affected; loss of an urban amenity (recreational space); gain of an urban amenity (cultivation); increasing the crime-hazard of urban living by providing hiding habitats for muggers

1.3 The eco-city concept

Introduction

The eco-city-building concept is aimed at improving the city's structural coupling, metabolism process and functional sustainability through cultivating an ecologically vivid landscape (eco-scape), totally functioning production (eco-industry) and systematically responsible culture (eco-culture) (Van der Ryn & Calthorpe, 1991; Girardet, 1992; Smit, 1996). It is a healthy human ecological process towards sustainable development within the carrying capacity of the local ecosystem through changing production mode, consumption behaviour and decision instruments based on ecological economics and system engineering (Mansson, 1992). Institutional integration, scientific incubation, entrepreneur's investment and citizen's incentives as provided by government are the key in eco-city development (Van der Ryn & Calthorpe, 1991; Todd & Todd, 1994, Tjallingii, 1995).

The eco-city network is inspired by transformative environmental movements through their political wings (exemplified by the Greens) almost in every part of the world. The "Greens" share the age-old longing for the genuine community, where face-to-face interactions create a web of political, economic, social, and cultural life (Roulofs, 1996). Green theory, which emphasises on ecology, owes much to the Club of Rome's (1972) *The Limits to Growth*, Schumacher's (1973) *Small is Beautiful*, Ebenezer Howard's (1965) *Garden cities of tomorrow* and Daly's (1977) *Steady-State Economics*. The sustainable city concept is meant to remedy most of the modern metropolis and suburb's ills through re-creation of locally or bio-regionally based economies (Hallsmith, 2003). Today, many planners and related professionals have joined the environmentalists in the eco-city movement, and advocate both environmental protection and social justice (Roulofs, 1996). Local governments world-wide have enacted formal environmental policies, which are seriously enforced (Satterthwaite, 1999, UNEP, 2002).

Several models influence eco-city designers of today, some utopian and bordering on fantasy villages proposed by a Frenchman Charles Fourier (1772-1837). Fourier believed that 1 620 people should live in communities (called phalansteries) to share the work, the wealth, the fun, and the enormous variety of tastes and passions in the combined order. Growing food (in gardens and orchards), cooking, and eating would be major activities in his new world. Sex, in all the non-violent varieties imaginable by this diverse crowd, was also an important feature. Manufacturing would be reduced to bare necessities, such as fine glassware for *le vin* and tureens for *potage*. His compact scheme for a totally full life would also encourage resource conservation, reduce pollution, and facilitate the utmost in conviviality (Roulofs, 1996). Fourier's vision was reincarnated in the 1960's and 1970's in Collenbach's (1975) novel, *Ecotopia*, which in turn influenced bio-regionalism and the USA green political movement. In Ecotopia all materials are recycled, including sewage sludge used as fertiliser. Towns are decentralised and built around public transit hubs. Energy production and the economy are bio-regional, with only a tad of imports and exports.

One of the earliest attempts to create a model eco-city dwelling is described in Olkowski *et al.*, (1979) *The integral urban house: Self reliant living in the city*. The integral urban house project in Berkeley, USA involved transforming a typical urban house into a self-reliant life support system involving food production, wind and solar energy, composting, grey water recycling, and a host of other "revolutionary" integrated systems concepts.

Urban eco-villages have been created in various parts of the world e.g., in Denmark; Vancouver, Canada; Munich, Germany; Davis in California and Ithaca in New York, USA; The Halifax eco-city project in Adelaide, Australia; Stockholm, Sweden; The Green roof (*Het Groene Dak*) in Utrecht, the Netherlands; and many others are either in planning or construction stages. Most of them have adopted a radical approach by abandoning the private household in favour of communal living another legacy of Fourier (The Catalyst, 1993; Tjallingii, 1995; Roulofs, 1996).

The joint UNCHS (Habitat) and UNEP Sustainable Cities Programme (SCP) initiative is presently working in more than 40 cities around the world (UNEP, 2002). The primary focus of the Sustainable Cities Programme is at the city level, with country, regional and global levels being recognised. According to UNEP, (2002), a sustainable city is a city where achievements in social, economic, and physical development are made to last. It has a lasting supply of the natural resources on which its development depends (using them only at a level of sustainable yield). A Sustainable City also maintains a lasting security from environmental hazards which may threaten development achievements (allowing only for acceptable risk) (Satterthwaite, 1999; Hallsmith, 2003).

The "Melbourne Principles on Sustainable Cities" governing the design and function of sustainable cities were formulated in Melbourne Australia in 2002 by a group of experts and launched in the same year during the World Summit on Sustainable Development in Johannesburg. The Principles provide cities with a framework to develop a consensus around a sustainable development policy and programs (UNEP, 2002).

Feasibility of eco-city concept

Eco-city concepts are theoretically interesting, however in practice the model is fraught with controversies and contradictions and even the advocates differ about the goals. There are disagreements and dilemmas on self-reliance of cities and the actual urban form (Rees, 1992; Roulofs, 1996). Are all resource imports and waste exports necessarily exploitative? Some believe that ultimately there should not be a clear separation between town and country (Harremoes, 1996; Chambers *et al.*, 2000). Nonetheless there are some common widely shared objectives which include resource conservation, waste reduction, toxic reduction, social justice, participatory process, health and cultural vitality. These objectives are espoused in the 1987 United Nations "Brundtland Commission" report *Our Common Future* (WCED, 1987; Mansson, 1992).

According to Kalbermatten & Middleton (1998), the "City of the future" would be a city of neighbourhoods. This urban renaissance will include the provision of water supply, sanitation, solid wastes and storm drainage services at the lowest economically efficient level, by neighbourhood or combination of neighbourhoods. The emphasis will be on resource conservation and recycling, synergism between services, commercialisation of services (whether through public or private ownership), and multiple use of facilities for maximum public benefit (Beck & Cummings, 1996).

Even when considering single services, there is evidence that economies of scale limits may have been reached - bigger is not necessarily better. As services become more integrated or inter-related, and especially if they are increasingly turned over to communities to manage, decentralised smaller packages of services will probably be found to be more appropriate (Schumacher, 1973; Lens *et al.*, 2001).

Urban ecosystem management should be based on a network of small-scale, "closed" water and nutrient cycles (Beck & Cummings, 1996; Lens *et al.*, 2001) associated with a variety of patches, spatially and temporally connected in the landscape by water and nutrient flows. Control and prediction of water and nutrient movement may depend on the number, size, shape, sequence and configuration of patches in a landscape. This is a level of complexity well beyond most current landscape-, erosion- and water and nutrient cycling- models (Tiessen, 1995).

In as yet no city in the world has demonstrated that it can sustain itself by drawing only on the resources within its boundaries. What is sought in sustainable development is not "cities that sustain themselves" but cities (and rural areas) where the inhabitants' development needs are met without imposing unsustainable demands on local or global natural resources and systems (Hardoy *et al.*, 1992). In the context of this study, the most essential concept of eco-city development is the closing of nutrient cycles, the short-cutting of cycles and the implementation of this process at the household and neighbourhood scale, both technically and in terms of environmental feasibility of such systems.

The question is how to achieve mastery and control with the attribute of "engineering resilience" and at the same time "ecological resilience" (Somlyody, 1995; Beck, 1997). The complexity of ecological systems is daunting and so too is the complexity of urban social systems. There are overwhelming calls globally – in the slogans of eco-cities (green or sustainable cities), clean technology, clean households, source control of pollution, recycling and recovery of valuable materials, consumption emission, closing material cycles, cross-sectoral and other, improved integration etc. - to consider revolutionising the urban water and sanitation systems in particular. A number of questions still emerge. Is it more than one of the many buzzwords, which can not really be defined? Can it be used operationally? How is it done? It is also becoming a frequent belief that "low-tech" approaches are "naturally" better than "high-tech". That may not be the case at all. Low-tech solutions maybe technically simple, but they can be "ecologically" complicated and frequently difficult to operate (Beck, 1997; Chen & Beck, 1997; Harremoes, 1997).

1.4 Ecological sanitation and organic waste recycling

Global sanitation problem

In spite of the efforts during the International Drinking Water Supply and Sanitation Decade (1981-1990)[13] (Cairncross, 1992; Evans *et al.*, 1990) still 1.2 billion of the world's population lack access to safe drinking water and about 2.4 billion lack adequate sanitation (UNDP, 2001, 2003). WHO's Collaborative Council Working Group on Sanitation (WHO, 1996a, 1996b) concluded that the progress of sanitation in the developing countries is hindered by basic misconceptions and myths stating that, water supply is always needed for good sanitation. Another myth originates from the assumption that water is need for to flush the toilets (Niemczynowicz, 1997a; WMO, 1997; Savenije, 2000).

Conventional sanitation in the form of waterborne sanitation has offered limited solution to this global sanitation crisis particularly in dry and soil-nutrient deficient environments (Hardoy *et al.*, 1990) suggesting a need for a new paradigm (Niemczynowicz, 1993, 1997a; 1997b; Savenije, 2000).

---

[13] So declared by the General Assembly of the United Nations; thereinafter, referred to as 'the Decade.'

Over the period 1990-2000, access to improved sanitation increased globally from 51 to 61%, resulting in 1 billion additional people with access to sanitation. Despite these gains, in year 2000 about 2.4 billion people, 80% of them in Asia, still lacked access (Cosgrove & Rijsberman, 2000; MDG, 2000; UNDP, 2003). The gap between rural and urban areas still remains extremely wide, especially in Eastern and South-Central Asia, where coverage in rural areas is only about one quarter of the population, while urban coverage is 70% (WHO, 1996). The Millennium Development Goals[14] (MDG Goal 7 Target No. 10[15]) championed by the United Nations, World Bank, International Monetary Fund (IMF) and the Organisation for Economic Development (OECD) has set up an ambitious target of halving the proportion of the world's population without improved sanitation[16] by 2015. This will require reaching an additional 1.7 billion people, a challenge for greater financing and more effective sanitation programs (MDG, 2000, United Nations, 2002).

The Millennium Development Goals set targets[17] for reductions in poverty, improvements in health and education and protection of the environment. They commit the international community to an expanded vision of development that vigorously promotes human development as the key to sustaining social and economic progress in all countries, and recognises the importance of creating a global partnership for development (Cosgrove & Rijsberman, 2000). The goals have been commonly accepted as a framework for measuring development progress (MDG, 2000; UNDP, 2003).

The sanitation practices that are promoted today fall into one of two broad types: '*flush-and-discharge* and forget' or '*drop-and-store*' (Franceys *et al.*, 1992; Pickford, 1995; Drangert *et al.*, 1997; Winblad, 1997; Esrey *et al.*, 1998; 2001). Over the past hundred and fifty years flush-and-discharge has been regarded as the ideal technology, particularly for urban areas (Winblad, 1997). Many municipalities in developing countries, often with the help of international financing, have copied this model (Niemczynowicz, 1993, 1996; Drangert, 1997). For those without access to flush-and-discharge the conventional alternative has been a 'drop-and-store' device, usually a pit toilet, based on containment and indefinite storage of human excreta. Drop-and-store is often regarded as an inferior; temporary solution compared with flush-and-discharge (Winblad, 1997, 2000).

---

[14] The Millennium Development Goals and targets come from the Millennium Declaration signed by 189 countries, including 147 Heads of State, in September 2000. The goals and targets are inter-related and should be seen as a whole. They represent a partnership between the developed countries and the developing countries determined, as the Declaration states, "to create an environment – at the national and global levels alike – which is conducive to development and the elimination of poverty."

[15] Goal 7 of the MDG's refers to ensuring environmental sustainability and Target 10 reads: "Halve, by 2015, the proportion of people without sustainable access to safe drinking water and basic sanitation".

[16] The definition of *access to improved sanitation facilities* and methods for assessing it are more contentious than those for water, with national definitions of "acceptable" sanitation varying widely.

[17] *Proportion of the population with access to improved sanitation* refers to the percentage of the population with access to facilities that hygienically separate human excreta from human, animal and insect contact. Facilities such as sewers or septic tanks, pour-flush latrines and simple pit or ventilated improved pit latrines are assumed to be adequate, provided that they are not public, according to the World Health Organization (WHO) and United Nations Children's Fund's (UNICEF) *Global Water Supply and Sanitation Assessment 2000 Report*. To be effective, facilities must be correctly constructed and properly maintained (MDG, 2000).

The major question in sanitation today is? How can a rapidly growing city short of money and water and with limited institutional capabilities achieve safe, non-polluting sanitation for all its inhabitants (or even recover its resources)? Conveniences like flush toilets are totally dependent upon the electrical grid and completely reliant on a constant water supply. When the electricity is out and water is unavailable, how does one flush the toilet? Supplying the clean water, treating the sewage, and providing all the delivery and collection requires sophisticated systems whose cost strains the resources even in wealthy countries (Niemczynowicz, 1993, 1996).

The historical background for the waterborne sanitation in cities needs to be outlined in order to give the rationale for the technical solutions that have been inherited from last centuries (Niemczynowicz, 1993; Hayward, 1997; Larsen & Gujer, 1997; Niemcynowicz, 1997a; Otterpohl *et al.*, 1997). The key element is maintaining the hygienic conditions in the cities. The success is illustrated by the absence of water borne diseases in the modern developed city (Butler & Parkinson, 1997; Harremoes, 1997).

History of conventional waterborne sanitation

Methods of waste disposal date from ancient times, and sanitary sewers have been found in the ruins of the prehistoric cities of Crete and the ancient Assyrian cities. Storm-water sewers built by the Romans are still in service today (Herschel, 1973; Goubert, 1989; Drangert, 1997). Although the primary function of these was drainage, the Roman practice of dumping refuse in the streets caused significant quantities of organic matter to be carried along with the rainwater runoff through the *Cloaca Maxima* to the river Tiber emptying into the Mediterranean Sea (Herschel, 1973). The one million citizens of Rome imported much of their food from neighbouring countries and did not have to worry about the reuse of nutrients in the effluent. After the decline of the Roman Empire, advanced sewer systems initially fell into oblivion.

Toward the end of the Middle Ages, below-ground privy vaults and, later, cesspits were developed. When these containers became full, sanitation workers removed the deposit at the owner's expense. The wastes were used as fertiliser at nearby farms or were dumped into watercourses or onto vacant land (Gotaas, 1956; Del Porto & Steinfeld, 1999).

In England, Chadwick's[18] report in 1842 had advocated for combined hygiene and recycling of nutrients by application of wastewater on agricultural fields through the use of the 'pail system'. However the unique conditions in London made it possible to flush all human waste via the perennial River Thames into the North Sea (Gayman, 2000). As with the Romans, with cheap imported fertilisers, the English then were not concerned about a decline in soil fertility (Asano & Levine, 1996; Drangert, 1997).

---

[18] Chadwick's Report on Sanitary Conditions: Edwin Chadwick (1803-1890) had taken an active part in the reform of the Poor Law and in-factory legislation before he became secretary to a commission investigating sanitary conditions and means of improving them. 'Report from the Poor Law Commissioners on an Inquiry into the Sanitary Conditions of the Labouring Population of Great Britain. London, 1842, pp 369-372.

In the first half of the 19th century Europe experienced epidemic Cholera, called the Asian disease, because it migrated in waves from Asia to Europe e.g. in 1853 the epidemic reached Copenhagen, Denmark, with the result that 7% of the population died in two month (Doudoroff & Adelburg, 1970; Hunter, 1997). Nobody knew the real cause. Not until after the discovery of bacteria by Pasteur, in 1880 did real understanding of the mechanisms of transmission of water borne diseases have an impact on the application of technology. Before that, the "experts" were divided into two camps. The two concepts were *Miasma* versus *Contagonism*. Miasma was the belief that diseases are the result of a foul environment (mostly air). Contagonism was the belief that some (as yet unknown agent) cause a contamination that could be transferred by infection through contact (Gayman, 2000). John Snow during the Cholera epidemic in Soho, London 1855, made one of the first real epidemiological investigations that indicated transmission via water (from the Broad Street pump); however his approach was not recognised until much later. This era can be referred to as the '*great sanitary awakening*' era (Taras, 1981; Goubert, 1989; Winblad & Kilama, 1985).

In the 19th Century the concept of miasma prevailed: the cities had to be cleaned up. Clean water, air and food had to be provided and waste had to be carried out of the cities in an organised fashion. Therefore the introduction of water supply and sewerage was based on a false concept, a misunderstanding of the issue (Gayman, 2000. Development of municipal water-supply systems and household plumbing brought about the modern sewer systems. Despite some reservations that sanitary sewer systems wasted resources, posed health hazards, and were expensive, many cities built them in Europe and in the USA (Taras, 1981; Goubert, 1989; Niemczynowicz, 1996).

At the beginning of the 20th century, a few cities and industries began to recognise that the discharge of sewage directly into the streams caused health and environmental problems, and this led to the construction of sewage-treatment facilities (Asano & Levine, 1996). At about the same time, the septic tank was introduced as a means of treating domestic sewage from individual households both in suburban and rural areas (Gayman, 2000). Because of the abundance of diluting water and the presence of sizable social and economic problems during the first half of the 20th century, few municipalities and industries provided wastewater treatment '*The sewer overfloweth*' (Del Porto & Steinfeld, 1999).

During the 1950's and 1960's, the USA and most European governments encouraged the prevention of pollution by providing funds for the construction of municipal wastewater treatment plants, water-pollution research, and technical training and assistance. New processes were developed to treat sewage, analyse wastewater, and evaluate the effects of pollution on the environment (Metcalf & Eddy, 1995). In spite of these efforts, however, expanding population and industrial and economic growth continued to cause pollution and health difficulties in both the developed and developing countries (Young, 1985; Del Porto & Steinfeld, 1999).

Up until the 1990's emphasis has been on reclamation and reuse of water in wastewater and not the contained nutrients (Harremoes *et al.*, 1991; Asano & Levine, 1996; Nakazato, 1997). From the 1970's technology and research has been advanced for removing nutrients from wastewater (notably biological nutrient removal in activated sludge treatment works) so as to mitigate eutrophication of fresh water lakes. In arid areas like Windhoek in Namibia reclamation of wastewater for potable reuse has been achieved with minimal potential for health risks (Harhoff & Van der Merwe, 1998).

Human excreta have traditionally been used for crop fertilisation in many countries (*Night Soil Fertiliser Industry*; Del Porto & Steinfeld, 1999). Medical science in the 19th Century posed no barriers to the use of human wastes directly on the soil. The Massachusetts sanitary commission of 1850 argued that, when applied 'in the open country', the wastes where "diluted, scattered by winds, oxidised in the sun: vegetation incorporated its elements". Miasmas and contagions would be dealt with by nature (Del Porto & Steinfeld, 1999). In China human and animal excreta have been composted for millenia and is still widely practiced (King, 1973; Matsui, 1997). In 1952 an estimated 70% of all human excreta produced in China was collected and used as fertiliser and this represented about a third of all fertiliser use in the country (Gotaas, 1956; Winblad & Kilama, 1985; Matsui, 1997). Whilst in Japan it is recorded that recycling of urine was introduced in the 12th century (Matsui, 1997; Shiming, 2001).

In the 1970's alternative dry sanitation systems exemplified by the *Clivus Multrum* were developed and marketed in Sweden and the USA in particular (Winblad & Kilama, 1985; Del Porto & Steinfeld, 1999, Jenkins, 1999). In Sweden initially, these were intended for use in summer cottages rather than in apartments, but this earlier interest among ecologically minded individuals has now broadened into a public concern. The Swedish Environmental Protection Authority (SEPA) has approved a number of these dry systems and the current regulations make the user responsible for maintaining the system (Drangert, 1997).

Wastewater and sludge[19] has also been widely used in agriculture in the last two centuries (Smith, 1996). However, when raw sewage was used in Berlin in 1949, it was blamed for the spread of worm-related diseases. In the 1980s, it was said to be the cause of typhoid fever in Santiago, and in 1970 and 1991, it was blamed for cholera outbreaks in Jerusalem and South America, respectively (Metcalf & Eddy, 1995). The economic value of human excreta has gradually been outweighed by the demands for hygienic and aesthetic conditions. Incidentally, the production of synthetic fertilisers began at the turn of the last century and accelerated after the Second World War (IFA, 1999; IFA & UNEP, 2002).

It is realised that sanitation is not just a 'technical fix' but an intricate interplay of norms and attitudes among professionals as well as users. The reasons for installing an improved collection of excreta may vary, and often include status, convenience, hygiene and improved health (Lens *et al.*, 2001). Rarely is improved nutrition mentioned, since re-circulation of nutrients is hardly practised or contemplated (Staudenmann *et al.*, 1996; Esrey *et al.*, 2001). Sanitation in its global sense should include collection, sanitisation and beneficial use of human excreta, sullage or grey water and solid waste. In broad terms sanitation calls for the holistic management of

---

[19] Phosphorus bound to the sludge makes the sludge interesting as phosphorus fertiliser. Unfortunately not only P is bound to the sludge, but also toxic compounds that derive from the wastewater (Ekvall, 1995).

both the water cycle and the food cycle to avoid health and environmental repercussions.

Ecological sanitation

According to many water professionals, researchers, environmentalists and public health experts a new paradigm in sanitation has emerged based on three fundamental aspects: 1. rendering human excreta safe; 2. preventing pollution rather than attempting to control it after polluting and; 3. using the safe products of sanitised human excreta for agricultural purposes. This approach can be characterised as 'sanitise-and-recycle' and can be called 'ecological sanitation' or 'eco-san' for short. It is a cycle - a sustainable, closed-loop system. It treats human excreta as a resource. Human excreta are processed on site or off-site (commonly applying dehydration and decomposition techniques) and then, if necessary, further processed off site until they are completely free of disease organisms. The nutrients[20] contained in the excreta are then recycled by using them in agriculture (Winblad, 1996; Drangert *et al.*, 1997; Otterpohl *et al.*, 1997; Esrey *et al.*, 1998; Winblad, 2000; Esrey *et al.*, 2001; Werner; 2001).

The principles underlying eco-san are not novel. The Swedish ecosan programme is based on "collection, containment, sanitisation and use of excreta" (Jonsson *et al.*, 2004). In different cultures sanitation systems based on ecological principles have been used for hundreds of years (Matsui, 1997; Del Porto & Steinfeld, 1999). Ecological sanitation systems are still widely used in parts of East and South-East Asia (mainly China, Japan and Korea) (Polprasert *et al.*, 1981). In Western countries this option was largely abandoned as flush-and-discharge became the norm but in recent years there has been a revival of interest in ecological sanitation. With ecological sanitation it is preferable to avoid mixing (Winblad, 1996; Winblad, 2000; Esrey *et al.*, 1998; Otterpohl, 2001):

- Human urine (*yellow water*) and faeces (*brown water*)
- Human excreta and water (dry-sanitation systems are preferable)
- *Black water* and *grey water*
- Household organic wastes and industrial wastes
- Wastewater and rainwater

According to this concept, by not mixing human excreta and flushing water the sanitation problem is limited to managing a comparatively small volume of urine and faeces. This leads to savings in water, savings on pipe networks and treatment plants, employment creation and preservation of the environment (Winblad, 1996; Winblad, 2000; Esrey *et al.*, 1998; 2001; Werner, 2001). Domestic water use can also be reduced by recycling grey water for washing cars, flushing toilets, and watering gardens i.e. non-potable water uses (Whelans, Maunsell & Palmer, 1993).

The purpose of ecological sanitation systems is the closing of the water and nutrients cycles, taking into account that the main task of sanitation is to assure highest

---

[20] In this dissertation plant nutrients refer to all types of nutrients, whether organic or inorganic, that combines with energy from the sun to result in plant growth. The word "fertiliser" will usually refer to chemical or inorganic fertiliser unless it is explicitly qualified with the adjective 'organic'. Therefore, plant nutrients include both organic and inorganic fertilisers.

hygienic standards in a cost- effective, environmental sustainable way (Esrey *et al.*, 2001; Simpson-Herbert, 2001), saving both water and energy and keeping soils fertile. There is a fundamental connection between agricultural development and actions to be taken in the sanitation sector and organic waste management (UNDP, 1996; Savenije, 1998).

Feasibility of ecological sanitation

The plant nutrients stemming from human metabolism contained mainly in urine is of particular interest in ecological sanitation (Olsson; 1995; Larsen & Gujer, 1996; Herrmann & Klaus, 1997; Jonsson *et al.*, 1998; Johansson *et al.*, 2000). Ways of recovering the resources in urine, which include – diversion, separation, absorption and combined processing with faeces, vacuum toilet systems, waterless urinals (some toilet pedestals are shown in Figure 1.2) – are currently under intensive investigation and piloting in numerous settlements around the world (Del Porto & Steinfeld, 1999; Jenkins, 1999; Morgan, 2000; Clark, 2001; Hoglund, 2001; Vinneras, 2002; Werner *et al.*, 2003). There are different possibilities to design sanitation systems, in accordance with the Bellagio Principles. One option is certainly the installation of on-site dry sanitation systems using ecological toilets with or without urine diversion (Esrey *et al.*, 1998; Morgan, 1999; Jenkins, 1999, Del Porto & Steinfeld, 1999). Figure 1.2 (b) shows a non-flush urine diverting toilet seat from Zimbabwe (note the ash container on the right: ash is used as a desiccant and to facilitate composting, an optional air freshener container can also be seen to the left).

The focus on urine is because it contains the bulk of the plant nutrients in domestic wastewater, approximately 80% of nitrogen, 55% of phosphorus and 60% of the potassium (see Table 1.9 and Figure 1.3), furthermore, this is provided in the correct forms for uptake by crops - nitrogen as urea, phosphorus as super-phosphate and ionic potassium, with urine also being free from heavy metals and usually sterile and less objectionable to handle. The plant availability of the nutrients Nitrogen (N), Phosphorus (P) and Potassium (K) in source separated urine is high (Lentner *et al.*, 1981; Kirchmann & Pettersson, 1995; Larsen & Gujer, 1996) and the concentrations of different heavy metals are low (Jonsson *et al.*, 1997; 2004). Furthermore, after being stored the hygienic quality of the source separated urine improves considerably (Blumenthal *et al.*, 1989; Cairncross *et al.*, 1995; Hoglund *et al.*, 1998; Hoglund, 2001).

The arguments favouring the ecological sanitation approach are convincing at least from an agricultural point of view. The average global cereal output is about 0.1 kg/m^2 (FAO, 1986; 1999a; 2001). This implies that the land area needed to produce an average adult person's annual intake[21], of say, 250 kg of cereals would be 2 500 m^2 (Drangert, 1997). This varies substantially between different agricultural zones and whether irrigation or dry-land farming is contemplated: from some 500 m^2 in irrigation agriculture to as much as 5 000 m^2 in dry-land farming on marginal land (Drangert, 1997). In a year, each person excrete the fertilisers needed to grow the 250 kg of cereal they require over the same period of time, with urine accounting for around 90% of this fertiliser value (Drangert, 1996; 1997; Lienert & Larsen, 2003).

---

[21] There are many different suggestions of the area needed to feed a person. The span is generally between 100 and 800 m^2. The requisite area differs with the climate, the soil, the access to water and fertiliser and the diet of the family – vegetarians need less area.

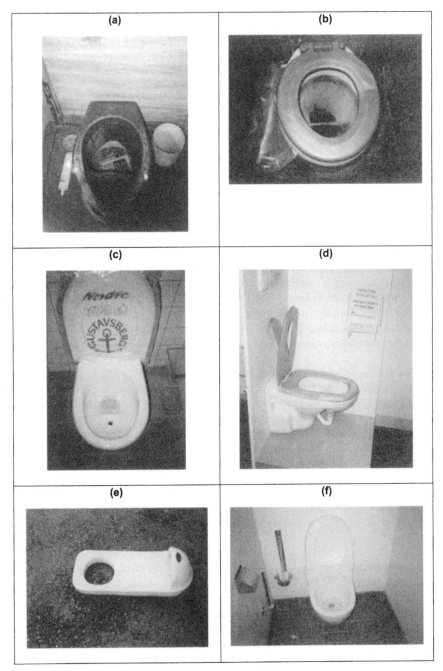

Figure 1-2 Some types of ecological sanitation toilet receptacles
(a) Urine diverting non-flush dry toilet concrete used and manufactured by Mvuramanzi Trust in informal settlements in Harare; (b) Porcelain non-flush urine diverting dry-toilet seat designed by Aquamor, Harare, Zimbabwe; (c) and (d) Double-flush urine-diverting toilet seat, the Nordic 393U from Gustavsberg and Dubbletten from BB Innovation and Co AB; (e) Chinese urine-diverting squat pan; (f) Swedish vacuum toilet system.
Photographs: Gumbo (1998 to 2002)

Table 1-9 Elemental composition and volume of domestic wastewater expressed per adult person per day (p.d)[22]

Parameter	Unit	Urine		Faeces [a]		Grey water	
		Value	Range	Value	Range	Value	Range
Volume	litres/p.d	1.2	0.6-1.5	0.15	0.07-0.4	150	50-300
Weight [b]	g/p.d	1 200	600-1 500	150	70-400	15 000	5 000-30 000
Total solids	g/p.d	60	20-150	45	30-60	80	40-150
Total nitrogen	g/p.d	11	4-16	2	1-4	1	0.2-1.5
Total phosphorus	g/p.d	1	0.5-2.5	0.5	0.1-1.5	0.2	0.1-0.4
Potassium	g/p.d	2.5	1-5	0.5	0.2-1.2	2	1-4
BOD$_5$	g/p.d	7.5	2-14	14	6-18	28	10-40
COD	g/p.d	15	4-28	35	20-55	60	30-90

a) Values exclude flush water. Flush water volume ranges from 15 to 80 l/p.d with an average value of 30 l/p.d

b) Density assumed to be 1.0 kg/dm^3

Source: From various sources including, Lenter *et al.*, (1981); Metcalf & Eddy, (1991); Feachem *et al.*, (1993); Butler *et al.*, (1995); Kirchmann & Pettersson, (1995); Hellstom & Karrman, (1996); Jonsson *et al.*, (1997; 2004); Del Porto & Steinfeld, (1999); Johansson *et al.*, (2000); Hoglund, (2001); Vinneras, (2002)

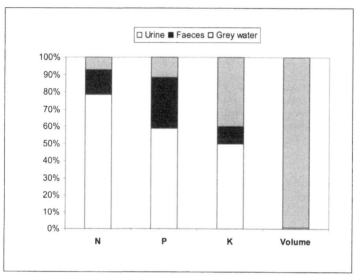

Figure 1-3 Content of major plant nutrients and volume in typical domestic wastewater

To grow 250 kg of maize requires 5.6 kg of nitrogen, 0.7 kg of phosphorus and 1.2 kg of potassium and urine provides 5.6, 0.4 and 1.0 kg respectively (*The urine equation*, Drangert, 1996; Hellstrom & Karrman, 1996; Jonsson, 1997; Niemczynowicz, 1997a; Niemczynowicz, 1997c).

Other points in urine's favour are that (see Box 1.F: Concise facts on human urine as fertiliser), when collected in 'no-mix' toilets that keep urine and faeces separate, flushing can be carried out with only 100 millilitres (Herrmann & Klaus, 1997;

---

[22] There is a wide range of variability in the content of excrement and water use from person to person and place to place. Factors include nutrition, climate, health, age, lifestyle and levels of service for water and sanitation. For example, vegetarians produce higher quantities of faeces with higher water content than those who eat meat (Lentner *et al.*, 1981; Feachem *et al.*, 1983; Del Porto & Steinfeld, 1999).

Vinneras, 2002). As a result, the yearly volume of urine and flush water will only be about 0.7 m^3 per person, so that transportation costs are likely to be low.

Box 1-F Concise facts on human urine as fertiliser: Separation, collection, storage, transportation and application
Adapted from Johansson *et al.*, (2000)

- Human urine is a quick acting fertiliser that can replace mineral fertiliser in crop production.
- The relationship between nitrogen, phosphorus, potassium and sulphur is well-balanced and, with appropriate doses, broadly corresponds to the needs of most crops.
- The odour problems in connection with urine separation toilets do not appear to be greater than with other toilets.
- Transport systems and technology for the storage of urine are currently available on the market
- Nitrogen losses during transportation and storage can be kept very low if the urine is stored in non-ventilated tanks[23].
- A high temperature, low dilution and high pH levels promote sanitisation of the urine mixture. Recommendations are now available for storage times and suitable crops.
- Urine in itself presents a negligible hygienic risk.
- Faeces that enter the urine bowl can contaminate the urine.
- Many pathogens are killed during a storage period of about 6 months.
- Nitrogen loses in the form of ammonia in connection with application are normally less than 10% (usually 5%) (Jonsson *et al.*, 2004).
- Conventional techniques for applying liquid manure work well for human urine too. New technology is also of interest e.g. umbilical hose systems.
- There is a noticeable odour while urine is being applied, but this subsides within 24 hours. At a short distance from the field the odour is not a problem
- The risk of nitrogen leaching into water is no greater than when mineral fertiliser is used.

In general it can be said that pathogen survival depends mainly on time-temperature combinations. When the temperature is high enough for a certain period of time, pathogens will die (see Figure 1.4; Feachem *et al.*, 1983; Mara & Cairncross, 1989). It is bacteria in faeces, which produce urease, which is responsible for conversion of urine to ammonia gas thereby creating odour problems (Esrey *et al.*, 1998).

It is obvious that the open space available in densely populated urban areas my not allow 100% in-situ recirculation of all human excreta (mainly urine and including the organic fraction of solid waste), even if all open space was allotted to agriculture (Drangert, 1997). Drangert, (1997) proposed a relationship between outdoor space and plant uptake of nutrients (see Figure 1.5). There is a biochemical limit to what plants can absorb and soils can retain nutrients and another limit to what is administratively allowable. In between these limits there is a 'feasibility gap' where interesting combinations of local recirculation and export to distant sites can be found. The capacity for 100% local recirculation depends on the population density. In very densely populated areas like Khayelitsha in Cape Town, South Africa or Dzivarasekwa Extension in Harare, Zimbabwe where lees than 10 m^2 of open space per person are available, it would require strong efforts by skilful and keen horticulturalists (Drangert, 1997).

---

[23] One method to reduce this risk is to prevent the decomposition or urea to ammoniacal nitrogen by adding small amounts of acid to the fresh urine before storage (Vinneras, 2002).

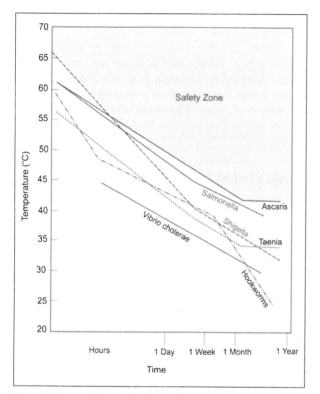

Figure 1-4 Survival times of various pathogens versus temperature
Source: Adapted from Feachem *et al.*, (1983)

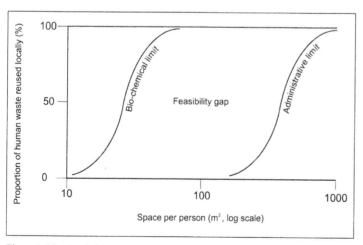

Figure 1-5 Recirculation potential of human waste in urban agriculture in relation to population density
Source: Adapted from Drangert (1997).

In China the old tradition of human manure (humanure) use in agriculture still continues (see Box 1.G: Overview of some ecological sanitation projects) and a range of new toilet receptacles have been developed (King, 1973; Jiang, 2001; Shiming,

2001). In Mexico around 100 000 urine separation toilets have been distributed (Clark, 2001; Esrey *et al.*, 2001; Cordova & Knuth; 2003), and in Africa a number of projects are being implemented in other countries notably Mvula Trust and CSIR in South Africa (Austin, 2003; Holden, 2003), WaterAid and ESTAMOS in Mozambique (Breslin & Dos Santos, 2001), SUDEA in Ethiopia (Faul-Doyle, 1999; Terrefe & Edstrom, 1999; 2001), OSIENALA in Kenya (Munyirwa & Onganga, 1999) and a number of NGO's in Uganda a country which has adopted ecological sanitation on a wider scale (Schattauer *et al.*, 2001; Nyiraneza, 2001). In Zimbabwe and many other similar case studies full utilisation of separated urine is yet to be achieved (Faul-Doyle, 1999; Morgan, 1999; Guzha, 2001).

Box 1-G Overview of some ecological sanitation projects

Sweden

The Understenshöjden eco-village in Sweden is situated in the suburb of Björkhagen, south of Stockholm. The village meets high ecological standards with regard to waste management, construction materials, energy systems and the outdoor environment. Ecological sanitation was introduced in 1995 to 44 apartments with 160 residents. Urine-separating toilets (wall hung Dubletten models) were installed. The urine is collected in two series connected tanks of 40 m^3 each. When the first tank is full, the urine mixture overflows into the other one. About once a year the urine is transported to holding tanks at Lake Bornsjön in Salem where it is stored for about 6 months. The remaining toilet waste and grey water is treated in a local biological treatment plant before being relesed for tertiary treatment in a systems o ponds and ditches (Johansson *et al.*, 2000).

The Palsternackan housing estate situated in Enskede south of Stockholm has 51 rental apartments with 160 residents was commissioned in 1995. The Dubletten urine-separating toilets are used. The urine is collected in three subsystems of roughly equal size, each connected of 30 m^3 capacity tanks. The tanks are emptied once a year and the urine is transported to holding tanks at Lake Bornsjön where it is stored for about 6 months. The remaining wastewater (faeces, flush water and grey water is discharged into the Stockholm wastewater system (Johansson *et al.*, 2000; Vinneras, 2002).

At Lake Bornsjön three 150 m^3 'balloon' tanks made out of rubber are used for storage. The tanks are airtight (minimal nitrogen losses during storage), require little construction work (can be easily re-located), and are reasonably priced. A farm close to the tanks uses the urine solution by means of a feeder hose to grow spring barley for forage.

In the Understenshöjden and Palsternackan there were some starting problems: odour problems through improper connection. Groundwater was leaking into pipes, because they were not watertight. Blockage of the urine collection separation system by hair and other objects and precipitation of urine. However the toilets were found not to be smelly compared to conventional flush toilets. Although a little more work was required for cleaning. Well motivated and informed residents contribute to the efficient functioning of the system and hence the recovery rate of the nutrients in urine (Fittschen & Niemczynowicz, 1997; Johansson *et al.*, 2000).

Zimbabwe

Mvuramanzi Trust a Zimbabwean NGO has been making ecological sanitation trials since 1998. About four different technologies have been developed to date and are being used in a number of community trial projects. Before introducing ecological sanitation practice among communities the Trust carried out a basic base line knowledge attitude and practice survey. The survey showed that people in Zimbabwe and the region were using urine and faeces as fertiliser or for medicinal purposes. Trials on crops showed that urine and sanitised faecal matter are useful ingredients to plant growth and improved harvest (Morgan, 1999; Esrey et al., 2001; Guzha, 2001).

Box 1-G Overview of some ecological sanitation projects (conti)

During the experimental trials Mvuramanzi Trust came up with the following ecological sanitation technology designs which are being used in rural and peri-urban situations (Guzha, 2001):

1. *Compost latrine*: The compost latrine consists of two vaults 1 m × 1.2 m × 3 m long these vaults are separated by a dividing wall. The vault is fitted with an access opening through which the digested excreta can be empty. A compost toilet normally has two non- urine diverting squat or and pedestal leading to each vault. One such squat and vault is used at a time whilst the other one is closed. Organic materials such as leaves, ash, soil, saw dust and grass is added into the toilet as the people uses the toilet. Earthworms may also be added to the mix, these together with fly-maggots help to digest the matter. The contents are left for six months to mature before being used as manure.

2. *Abhor-loo "Tree toilet"*: The Abhor-loo consists of a movable bottom slab and a portable upper structure; the movable components are placed on a 0.7 m shallow pit lined with bricks to prevent the pit from collapsing. Users are encouraged to add ash, leaves and other organic material as they use the toilet. The mixture of soil, ash, urine and faecal matter forms a rich mix of compost. When the pit becomes ¾ full a layer of top-soil is added and a choice fruit tree is planted. The toilet is used to furnish orchards and have been used extensively to address sanitation and deforestation problems in some peri-urban informal settlements. Communities are planting bananas, guava, marlberry and mango trees.

3. *Fossa Alterna:* This operates like an Abhor- loo, it consists of movable upper structure and slab which is placed on top of 0.7 m pit, in the case of the Fossa Alterna twin pits are used. The materials used to make the supper structure ranges from plastic sheets, hessain and wood. The family is encouraged to add organic matter, ash and soil to the human excreta as they use the toilet. When the first pit is ¾ full it is covered top soil is added, short term plants like maize and flowers may be planted into the pit the plant roots helps in breaking down the faecal matter facilitating decomposition. The matter can be left for six months or so to allow for composting to take place, after the 6 months the rich organic manure is recovered and used in agriculture.

4. *Sky-loo Urine Diverting toilet:* Sky- loo refers to a step up toilet built with a vault above the ground to minimise the possibility of ground water contamination especially in areas where the water table is very high. A wooden or brick super structure is placed on top of the vault. The faecal matter drops directly into the vault or into the plastic dish place in the vault. The Sky-loo urine diverting toilet is fitted with a urine-separating pedestal that diverts urine and ensures that urine and faeces do not mix. The separation of urine and the addition of soil and ash accelerate drying of faeces and create an environment that hinders multiplication of pathogenic bacteria. The diverted urine go through a network of pipes into a small soak away where a tree is planted to absorb some of the nitrates from the urine to prevent contamination of underground water.

Germany
In the ecological housing estate Flintenbreite in Lübeck vacuum toilets are installed in combination with anaerobic digestion with co-treatment of organic waste in a semi-centralised biogass-plant. Grey water is treated in a decentralised vertical flown constructed wetland. Storm water is retained in an infiltration constructed wetland. The project began in 1999. Until 2003 28 houses with 95 inhabitants had been constructed. The vacuum system has been running for 2 years without any technical problems. The flushing system is optimised and uses 0.7 litres for flushing. The average drinking water consumption is around 77 litres per capita per day. Problems and their reasons can be identified very easily. The vacuum toilets have been well accepted by the inhabitants and are now viewed as more hygienic than conventional flushing toilets (Otterpohl *et al.*, 1997; Otterpohl 2001; Wendland & Oldenburg, 2003).

Box 1-G Overview of some ecological sanitation projects (conti)

In the ecological village 'Braamwisch' in Hamburg composting toilets were installed in 1999. The Biolett system was installed in 15 houses whilst the TerraNova composting toilet was installed in 18 houses with. An additional 7 houses had a normal flush toilet with rainwater used for flushing. Co-composting of kitchen refuse together with the faeces and urine was selected. The major motivation of the inhabitants or users was contribution to the environment sustainability though the use of composting toilets. The composting process was rather complex and needed much attention. The users had to take the initiative themselves and solved some of the problems through 'learning by doing'. The grey water system composed of mechanical pre-treatment in a septic tank and biological treatment in reed bed soil filters of about 2 m^2 per person (Bijleveld, 2003).

China
Under the influence of the long tradition (dating back almost 5 000 years; King, 1973), human excreta is always used as fertiliser for crops in China (Shiming, 2001; Wenhua & Rusong, 2001). The main application methods are (1) direct usage for crops and fruits as basal or top application after fermentation in a ditch for a certain period, (2) composting with crop stalk for basal application, (3) direct usage as feed for fish in ponds. Even human waste generated in the cities and towns has been very valuable for farmers. Before 1949, there were private companies in Wuhan, Beijing, and other cities to control the commercial selling of human excreta (Shiming, 2001). Although the tradition of using human waste still continues, the percentage of human excreta used is decreasing due to a host of reasons, among them: increased urbanisation, reduced capacity to use the waste in peri-urban farms, transportation logistics, unsanitary methods of treatment, handling and application of excreta. It is estimated that in year 2000 an average of 31% of the absolute amount of human excreta generated in Beijing, Xian, Shanghai, and Changchun was collected and applied on farm land (Shiming, 2001).

With financial assistance from the Sida-funded EcoSanRes programme a pilot ecological sanitation project was introduced in 1998 in a large number of villages. The project was well received by villagers as well as by the government. The original pilot project covered 70 households, the following year 2 000 households were provided with ecosan systems, in year 2000 another 8 000 households. By the end of 2001 a total of 30 000 households in Guangxi are had ecosan toilets with urine diversion. The urine-diversion squatting pan was designed during his programme and patented in China. Many of the toilets were built inside the dwelling, and often upstairs. Apart from toilets for individual households a total of 7 ecosan school toilets have been built (Jiang, 2001).

Netherlands
In 1993 the housing estate 'Groene Dak' in Utrecht in The Netherlands installed two Clivus Multrum composting toilets. Grey water was treated in oxidation beds or reed beds. Despite a lot of effort and motivation the composting toilets never worked. There was too much moisture and an anaerobic environment, probably due to little aeration. This lead to problems of flies, spiders and smell nuisance. A very strong energy consuming ventilator was installed to mitigate the smell problem. In 2000 the toilets were removed because the composting process had failed. The removal of the black cake got national attention because of a suspected danger of explosion. The composting toilets were replaced by water saving toilets connected to the sewer system (Bijveld, 2003; translated from URL: www.groenedak.nl).

At the Twaalf ambachten in the province Noord Brabant in The Netherlands around 20 users have used the Paper Leaf Toilet, a sort of composting toilet. Composting is not done in or under the toilet. The bucket with faeces and paper is mostly emptied on the compost pile in the garden. The urine flows into the sewer or into a constructed wetland. Most of the users used the toilet because of environmental concerns, some because they had no connection to the sewer system. In the beginning problems occurred due to a construction failure in the first model. After some improvement most of the users where satisfied. However some felt emptying the container was not very pleasant and hygienic (Bijveld, 2003).

In Sweden, recycling of nutrients produced in sanitation systems to agriculture without increasing concentrations of conservative pollutants in soils is being given special attention (Johansson *et al.*, 2000). This requires non-flush dry, dubble-flush, micro-flush, foam flush, vacuum flush and urine-diverting solutions, and changes in, for example, construction of houses and layout of cities (Fittschen & Niemczynowicz, 1997). In Sweden more than 50 000 non-flush dry systems have been sold in 42 models from 22 manufacturers (Drangert, 1997; Esrey *et al.*, 1998). They cost scarcely more to buy and can cost less to install than a non-diverting toilet plus its sewer connection. Many other countries around the world have come up with a variety of proprietary urine separating and composting toilets. In Zimbabwe, two NGO's, Mvuramanzi Trust and Aquamor have developed a number of such toilet seats and about 7 000 have been installed mostly in rural and urban squatter settlements (Morgan; 1999; Esrey *et al.*, 2001; Guzha, 2001).

The popular Swedish WM-Ekologen (now Wost Man Ecology AB) dehydrating toilet costs about US\$ 360. The total on-site installation cost including a seat-riser; fan, processing vault, transport container and a 1 m^3 urine tank is about US\$ 750 (Johansson *et al.*, 2000). The Mvuramanzi Trust non-flush dry toilet seat version (which is a hybrid of South African and Mexican design) costs about US\$ 10 and the full installation including the superstructure is about US\$ 70 (Morgan, 1999; Guzha, 2001).

Urban organic waste recycling

Urban solid waste reduction and reuse involves, among other things, composting of urban organic wastes (especially in cities of developing countries where the organic fraction of municipal solid waste is high) (Furedy & Chowdhury, 1996; Polprasert, 1996; Lardinois & Furedy, 2000). Discussions of urban agriculture frequently point out that city farming often absorbs urban solid waste, thus reducing the volume of waste and the need to collect and transport wastes to distant dumps. In practice, urban farmers in many cities acquire municipal wastes as resources (Lewcock, 1994; Rosenberg & Furedy, 1996; UNDP, 1996).

The combination of urban organic wastes and urban agriculture creates particular issues in the modern urban setting (Gardner, 1998; Rose, 1999; Ojeda-Benitez *et al.*, 2000). On the one hand, the interests of urban waste reduction mesh well with the promotion of urban agriculture, since urban and peri-urban farmers are in need of organic matter as soil conditioner or fertiliser and animal feed, and cities and towns wish to conserve disposal space and reduce the costs of municipal solid waste management (UNCHS, 1989; WHO, 1991). At the same time, some tensions occur between public health officials (with their concerns about diseases affecting both humans and animals and accidents associated with the reuse of municipal solid wastes, (see Box 1.H: Waste recycling potential in India) on the one hand, and the proponents of urban agriculture (who emphasise job creation and increased food production, especially for the urban poor) on the other (Gotaas, 1956; Flintoff, 1976; Furedy & Chowdhury, 1996).

Composting solid waste for use as a soil amendment, fertiliser, or growth medium is important in many countries (Tchobanoglous *et al.*, 1993; Furedy & Chowdhury, 1996; Gardner, 1998; Rose, 1999). Asian countries in particular have a long tradition of making and using compost (Flintoff, 1976; Polprasert, 1996). In Western Europe, a range of modern technologies is used to produce compost (Suess, 1985).

Box 1-H Waste recycling potential in India
Adapted from Harender & Bhardwaj (2001)

India's current manurial potential of livestock and human excreta is estimated at 14 Mt of nitrogen, phosphorus and potassium nutrients, which is close to the present fertiliser consumption of the country (13 Mt). A few more million tonnes can be made available from crop residue, municipal solid waste, agro-industrial waste, and industrial waste. Careful collection, bio-conversion, conservation and recycling of all these available organic wastes or manures would enable India to meet its nutrient requirement of the crops to a considerable extent and thus its agriculture on a sustainable basis.

India generates around 25 Mt of municipal solid waste every year and about 60% of the waste comprises of biodegradable organic wastes originating from kitchen and markets. In addition, 273 Mt of crop residue and 6 Mt of fruit waste is also generated every year. This waste material can be processed with the help of earthworms in order to produce organic fertiliser at site without any extra cost. Earthworms play a key role in soil biology by serving as versatile natural bioreactor, converting organic wastes into valuable organic manure. The benefits are now being globally realised that earthworms can do wonderful job in the management of different pedo-ecosystems.

They are useful in land reclamation, soil improvement and organic waste management. Earthworms are also efficient environmental monitoring tool because worms can accumulate certain heavy metals, industrial effluents, various biocides, pesticides and their residues. Earthworms improve the soil texture, soil aeration, enrich the soil with nutrients and promote useful soil micro flora required for plant growth. Earthworms eat and mix a large amount of soil and organic matter, then deposit their castings (vermicompost) either on the surface of the soil or in burrow, depending on species. The vermicompost contains high concentration of organic material, silt, clay and is rich in many soil nutrients such as nitrogen, sulphur, potash, phosphorus, calcium, magnesium, etc. In soil, much of the phosphorus is bound in organic matter in a form that is not available to plants. Earthworms change the phosphorus into a form that the plant roots can easily absorb. The mixing action of the earthworms can also make slow-release forms of phosphorus fertilisers more readily available.

At the same time, composting has the distinction of being the waste management system with the largest number of failed facilities worldwide. In cities of developing countries, most large mixed-waste compost plants, often designed by foreign consultants and paid for by aid from their home countries, have failed or operate at less than 30% of capacity (Holmes, 1984; Furedy, 1992; Rosenberg & Furedy, 1996; Lardinois & Furedy, 2000). The problems most often cited for the failures of composting include: high operation and management costs, high transportation costs, poor quality product as a result of poor pre-sorting (especially of plastic and glass fragments), poor understanding of the composting process, and competition from chemical fertilisers (which are often subsidised). In many urban places, collection systems are too unreliable for urban authorities to consider running composting facilities efficiently (Tchobanoglous *et al.*, 1993; Lardinois & Furedy, 2000).

Industrialised and transition countries have more mature urban infrastructure, and a more clear separation between urban and rural food production practices. Developing countries tend to have more agriculture and horticulture within urban limits, providing a ready market for compost, depending on its organic and nutrient content (UNCHS, 1989; WHO, 1991; Furedy, 1992). Developing and transition countries tend to have a higher proportion of vegetable and animal wastes[24], sometimes as high as 90% (Tchobanoglous *et al.*, 1993; Lardinois & Furedy, 2000). People in industrialised and transition countries are also more likely to keep their yard wastes separate, while in developing countries these are likely to be mixed with other household wastes (Lardinois & Furedy, 2000).

Backyard and neighbourhood scale offers an attractive option to the many failed centralised systems in developing countries. Generally backyard composting consists of household-level aerobic decomposition of household organic garden and kitchen wastes, with the resulting compost being used in the yard itself. Such facilities can provide a waste management opportunity to a small group of people at a relatively low cost. Close proximity of yards to each other in many neighbourhoods in both industrialised and developing countries implies a need for management of the compost for vector and odour control, including periodic aeration or turning. In neighbourhoods with gardens or urban agricultural activities ready markets exists and compost can be sold at a price adequate to meet costs (Poerbo, 1991; Rosenberg & Furedy 1996; Lardinois & Furedy, 2000; Ojeda-Benitez *et al.*, 2000).

In recent years, a number of governments in industrialised countries have treated backyard composting as a means of waste reduction, since the materials which are composted remain at home and do not enter the municipal waste stream. Such backyard composting and mulching programmes, which have operated successfully in Northern Europe, North America, Australia, and New Zealand, are much less costly to a community than centralised compostable collection programmes. They have participation rates approaching 30%, with significant results in terms of wastes diverted from the municipal waste stream (Suess, 1985; Poerbo, 1991; Furedy, 1992; Rosenberg & Furedy, 1996; Lardinois & Furedy, 2000).

However, there are also negative impacts on the environment associated with making and using compost. These impacts depend both on the technical approach used and the waste composition of the input streams. These may include unpleasant odours from gases released from improperly maintained compost piles; leachate production; and the potential to convey heavy metals to the soil (Tchobanoglous *et al.*, 1993; Lardinois & Furedy, 2000). Control of bad odours and rodents can be carefully controlled if composting is performed within household backyards.

---

[24] As-delivered (wet basis) municipal solid waste from Accra, Ibadan, Dakar, Abidjan, and Lusaka shows a range of per-capita generation rates of 0.5-0.8 kg/pd (compared to 1-2 kg/pd in the OECD countries); putrescible organic content ranging from 35-80% (generally toward the higher end of this range); plastic, glass, and metals at less than 10%; and paper with a percentage in the low teens. Densities in the range of 90-180 kg/m^3 for un-compacted municipal solid waste are common in OECD case studies, whilst in Africa the range is thought to be between 180-540 kg/m^3 (Suess, 1985; UNCHS, 1989; WHO, 1991; Tchobanoglous *et al.*, 1993; Lardinois & Furedy, 2000).

Comparison of ecological sanitation and traditional waterborne sanitation

Various studies have been conducted to compare ecological sanitation with traditional waterborne sanitation systems (Jonsson, 1997; Jonsson *et al.*, 1998; Loetscher, 1999; Vinneras, 2002; Bijleveld, 2003; Drangert, 2003). Aspects compared range from social, environmental, health, economical and financial and technological. Box 1.1 summarises some of the broad aspects characterising ecological sanitation and traditional waterborne sanitation. Emissions of nutrients, heavy metals, and pathogens into aquatic systems have been compared. The absence of large scale ecological sanitation projects as compared to waterborne sanitation has necessitated the reliance on limited available data and assessment by use of models the possible effects of ecological sanitation (Lundin *et al.*, 1999; Johansson *et al.*, 2000, Karrman, 2000; Van der Vleuten-Balkema, 2003).

Although simulations are not measurements of actual systems, they can provide useful information which cannot be obtained in real life. Models like ORWARE have been developed to simulate the production, treatment and handling of solid and liquid organic wastes (Jonsson *et al.*, 1998; Dalemo, 1999; Karrman, 2000) on various system solutions on the basis of equivalent conditions.

Paradigm shift in sanitation

Organic matter and nutrients present in human excreta and the organic part of urban solid waste constitutes an important source of plant nutrients and renewable bio-energy (Niemczynowicz, 1993; 1997a). It could be suggested that a distinct path of evolution is occurring: one that is progressively engaging in the low-technology of naturally occurring processes, such as the ability of wetlands, reed beds, crop products and other biomass products present in terrestrial and aquatic eco-systems, in assimilating urban residues (Beck *et al.*, 1994; Matthews, 1996; Zeeman & Lettinga, 1998). This path has a certain philosophical satisfaction about it - a return to the spirit of the sewage farm, redolent of the 'good old days' of a rural society. Yet this is not an especially sound basis on which to prefer one form of technological development over another. However, the barrier against waterborne diseases is still a basic feature of the present urban water system that new technology must not jeopardise (Harremoes, 1997; Drangert, 2003).

Assuming that this path will result in waste load reduction and minimal if not 'zero' transfer of the waste problem to land, water and air contamination, then there is sense to propose a "*Paradigm shift*". This evolution of water supply and sanitation systems can be represented graphically as in Figure 1.6. Four eras are distinguished based on the major driving forces. Within these time frames certain types of technologies have been developed or dictated by the desired goal or driver. Some important events are also presented in Figure 1.6 as milestones (Niemczynowicz, 1993; 1997a; Gumbo, 2003a).

Box 1-I Summary features of traditional waterborne sanitation and ecological sanitation

	Traditional waterborne sanitation	Ecological sanitation
Process	• Mixing system • Dilution • Linear "end-of-pipe" • Separation at "end-of-pipe", containment, bio-digestion and disposal • Promotes discharge of nutrients into aquatic systems • Point source pollution copious and endemic • Tends to work at one scale at a time • Promotes additional water pollution arising from organics, pharmaceuticals and hormones	• Source separating system • Minimum dilution or dry • Circular "closed loop" • Sanitisation, bio-digestion, recycling and reuse • Enhances return of nutrients to the soil • Can result in dispersed pollution if scale and composition of wastes do not conform to the ability of local ecosystems to absorb them • In principle it should integrate multiple scales, reflecting the influence of larger scales on smaller scales and vice versa
Technology	• Easy to design and control, standardised templates are readily available and can easily be replicated all over the world • Design criteria usually based on economics, custom and convenience. • Narrow disciplinary focus • High to low-tech at "end-of-pipe" • Extends the pathogen cycle away from the user	• Complex natural processes, difficult to design and control, design responds to the bioregion i.e. integrated with local soils, vegetation, culture, climate, and topography • Design criteria based on human and ecosystem health, and ecological economics • Integrates multiple design disciplines • Low to high-tech at user level • Pathogen cycle closer to the user
Resources	• Energy demanding • Energy sources usually non-renewable and destructive, relying on fossil fuels or nuclear power • Can require chemicals e.g. coagulants and disinfectants • Complex by-products e.g. sludge, effluent	• Energy efficient, as it relies mostly on natural processes and renewable sources, solar, wind, biomass and small scale hydro • Minimal or no synthetic additives required • Simple by-products e.g. methane gas, ammonia and manure
Management	• Centralised and can benefit from economies of scale • Specialised agencies or institutions involved in operation and maintenance limited scope for community participation • Limited user education and awareness • Proven history to guarantee public health	• Decentralised • Users need to be mobilised and motivated in the management • Education and awareness of users is crucial and a clear commitment to discussion and debate of possible solutions • Proven history to provide the much needed plant nutrients • Other spin-offs include food production, poverty reduction

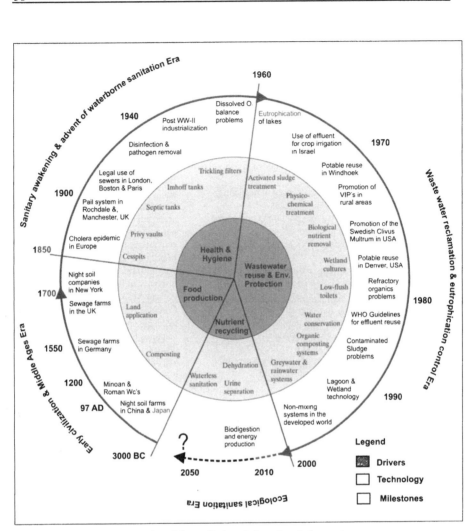

Figure 1-6 Graphical representation of the cyclical evolution of sanitation systems
Source: Gumbo (2003a)

1.5 Systems analysis and material flow and stock analysis

Integrative analysis of city systems helps to see beyond their current environmental and social problems to underlying causes, and it suggests different opportunities for possible interventions. Cities can be viewed as systems involving people's (social and economic institutions) interactions with one another and with the built environment they have created (Meadows *et al.*, 1972, 1992; Ross *et al.*, 2000). The built environments interact with the natural ecological processes of their sites.

Urban planning and design is usually accompanied by budgets, spreadsheets, bills of quantities, parts lists and so forth. Rarely does the parallel set of accounts that link the designs to the health of ecosystems considered. These accounts cover large square metres of misused land, kilowatt-hours of energy, cubic metres of water, tonnes of eroded soil and many other environmental impacts of the designs. The built-environment can be developed in the future to cooperate with natural functions and preserve their health (Beck, 1997; Hallsmith, 2003). This entails, for example, recognising the nature of the flood plain system and its drainage requirements and configuring the built environment where possible to complement rather than to resist it (Lyle, 1985; Van der Ryn & Cowan, 1996; Ross *et al.*, 2000). Systems analysis and material flow and stock analysis (MFSA) present attractive tools in desegregating the complex web of cycles, stocks and flows. The two provide a basis of tracking the flow of materials and products through society and the environment, an activity of increasing prominence and consequence throughout the world.

Systems analysis

A system is an entity, which maintains its existence through the mutual interaction of its parts (Von Bertalanffy, 1975; Coyle, 1996). A system is a human construct, an analytical artefact. Systems are abstracted for study from much more complex sets of interactions which occur in the 'real world'. In setting the boundaries of a system, the scale selected (river basin, city or neighbourhood), or which parts of the system to focus on is a question of choice. There is no problem in focusing on a particular aspect of a system, or on a small scale, as long as there is an understanding on how this aspect fits into the rest of its system (or puzzle) or indeed the web of potential systems. Land use and built environment configuration can extensively impair the functioning of the natural ecosystem. What is done at one scale has subtle impacts, both negative and positive, at many other scales (Senge; 1990; HPS, 1992; Senge *et al.*, 1994).

The characteristics of world problems and the functioning of the socio-ecological systems are changing. The cause-effect chains tend to change from local to global levels (e.g. global warming), from specific to diffuse (e.g. air and water pollution), from short delay to long delay (e.g. CFC's and the ozone layer) and from low complexity to high complexity (e.g. land use changes) (Winograd, 1997). The existence of vertical and horizontal linkages between scales and components and the relationship between variables imply the need to use tools, such as systems analysis, systems dynamics and geographic information systems.

The broad picture in systems analysis implies looking at global and transcending down to local scales; literally when considering natural resource use, but also figuratively in terms of the wider social, economic and political context. Nature's processes are inherently scale linking, for they intimately depend on the flow of energy and materials across scales (Ayres & Simonis, 1992;Van der Ryn & Cowan, 1996; Ford, 1999). Global cycles link organisms together in a highly effective recycling system crossing about seventeen tenfold jumps in scale, from a ten-billionth of a metre (the scale of photosynthesis) to ten thousand kilometres (the scale of spaceship Earth itself). In other words, a systems analysis approach requires a "bi-focal perspective": i.e. one eye examines the broad picture, seeing the whole, and the second eye focuses on the detail at a micro scale (Figure 1.7).

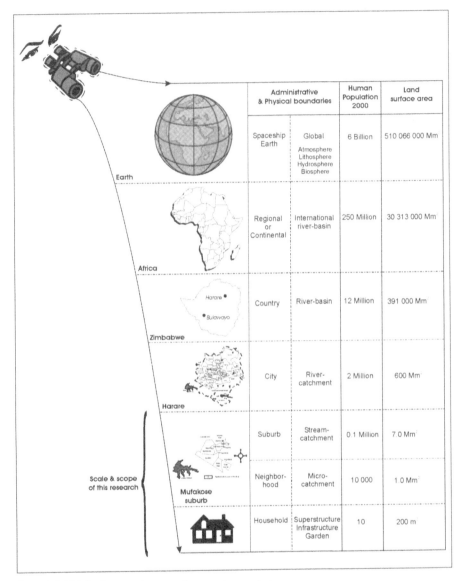

	Administrative & Physical boundaries		Human Population 2000	Land surface area
Earth	Spaceship Earth	Global Atmosphere Lithosphere Hydrosphere Biosphere	6 Billion	510 066 000 Mm
Africa	Regional or Continental	International river-basin	250 Million	30 313 000 Mm
Zimbabwe	Country	River-basin	12 Million	391 000 Mm
Harare	City	River-catchment	2 Million	600 Mm
Suburb	Suburb	Stream-catchment	0.1 Million	7.0 Mm
Mufakose suburb	Neighbor-hood	Micro-catchment	10 000	1.0 Mm
	Household	Superstructure Infrastructure Garden	10	200 m

Scale & scope of this research

Figure 1-7 A bi-focal perspective towards eco-system analysis

The development of sustainability frameworks and indicators should be developed and used at different levels (administrative and ecological), with different scales (local, national, regional and global) and components (economic, social and environmental) (Moldan & Billharz, 1997). The indicator tools should be able to produce a range of information from local to global, from detailed to aggregated and from scientific to policymaking. In the case of land use, for example, it is necessary to have indicators reflecting other variables, in addition to indicators about the pressure, state, impact and response on land (Figure 1.8).

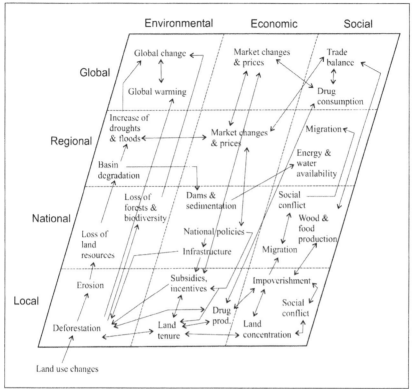

Figure 1-8 Vertical and horizontal linkages: the case of land use
Source: Adapted from Winograd (1997)

Material flow and stock analysis

Material Flow and Stock Analysis (MFSA) is the investigation of the physical flows of materials, typically on a geographic basis. MFSA can help in understand how changes in land use, industrialisation, consumption and population affect the cycles of elements or chemicals of concern in a watershed. It provides a means of taking a comprehensive rather than an ad hoc view of the drivers and source of substances (Ayres, 1978; Baccini & Brunner, 1991; Wackernagel & Rees, 1996; Ayres & Ayres 1998).

MFSA is an activity of compiling numerical measures of things or actions, in a form that makes comparison and analysis easy, or at least, possible. Although monetary accounts are certainly necessary for modern life, there is no good way to assign monetary values to most of the essential services provided by the environment, ranging from benign climate, breathable air and fresh water to biodiversity, nutrient recycling and waste assimilation (Daly & Cobb, 1989; Ayres & Ayres 1998).

Direct measurement is not enough and never can be enough. The environment is too complex and heterogeneous to understand in terms of direct measurement alone. The essence of real science, therefore, is selection and simplification. Vastly simplified world models are needed, at first. The trick is to build understanding step-by-step by starting with simple models that explain the most fundamental phenomenon, adding complexity (or making changes in the assumptions) only, when the existing models are clearly inconsistent with the reality (Anderberg *et al.*, 1993; Daly, 1996; Ayres & Ayres 1998; Brunner *et al.*, 1998). For MFSA, a consistent and meaningful choice of spatial and time boundaries helps to focus the accounting procedure and lend greater value to the problem at hand (Brunner & Baccini, 1992; Van der Ryn & Cowan, 1996).

Broadly there are three different types of data sources which can be used for MFSA purposes as illustrated in Figure 1.9 (Obernosterer *et al.*, 1998):
1. Primary data - generally raw data (e.g. direct experiment results or results from questionnaires, surveys etc.)
2. Secondary data – raw data that has been processed, collated and interpreted (e.g. statistics, publications, proceedings and reports, internet, models)
3. Tertiary data – informal or non-traditional data sources (estimations, interviews, non-published data, local or hands-on knowledge, individually performed estimates, the media etc.)

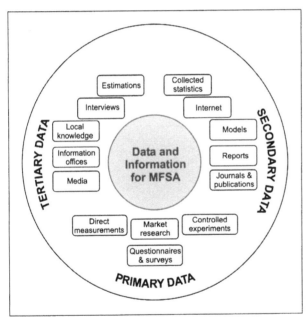

Figure 1-9 Overview of sources and types of data and information for MFSA
Source: Adapted from Obernosterer *et al.*, (1998)

There is a level of uncertainty associated with all forms of data. To use MFSA as a tool in the decision making process, the uncertainties of the presented results must be known. Since MFSA requires numerous data sources, this means, different levels of data uncertainty must be combined and taken into account. As a consequence, the system is usually not "balanced" i.e. input minus output does not equal stock variation (Baccini & Brunner, 1991; Ayres & Ayres 1998; Obernosterer *et al.*, 1998).

If different values are available for a flow or a stock, the minimum-maximum values are taken into account (data range). This data range could be misleading, if one supposes that the true value is within this range. It is sometimes the case that this range exists as a result of two different literature values, each with their own level of uncertainty. In addition, sometimes data on material flows and stocks or on a substance concentration of goods is limited or unknown. When no values are available, estimations may need to be made (Baccini & Brunner, 1991; Obernosterer *et al.*, 1998).

System dynamic models

The field of system dynamics originated in the 1960's with the work of Jay Forrester and his colleagues at the Sloan School of Management at the Massachusetts Institute of Technology. Forrester and his colleagues developed the initial ideas by applying concepts from feedback control theory to the study of industrial systems (Ford, 1999). One of the best-known applications of the new ideas during the 1960's was Forrester's *Urban Dynamics*, (1969). It explained the pattern of rapid population growth and subsequent decline that had been observed in a number of cities in the USA (Schroeder & Strongman, 1974). *Urban Dynamics* highlighted the field's expansion outside the industrial area. Its approach came to be known as system dynamics (Meadows *et al.*, 1972).

System dynamics maybe used to simulate material flow through a system (Ford, 1999). Stocks and flows are building blocks of systems dynamics models. The stocks are key variables in the model, they represent were accumulation or storage takes place in the system. Stocks tend to change less rapidly than other variables in the system, so they are responsible for the 'momentum' or 'sluggishness' in the system (HPS, 1992, 1993; Ford, 1999). Flows are the actions changing the system and they directly influence the stocks. Flow variables are measured in the same units as the stock variable, divided by the appropriate unit of time. The right combination of stocks and flows provides a good foundation for a model.

The third model builder in systems dynamic is the converter, which helps to describe the flows. Frequently converters are used to provide model inputs and those variables that don't logically meet the description of stocks and flows. The other purpose of converters is to calculate additional measures for the system performance (Ford, 1999). The converters serve a utilitarian role in the software. it holds values for constants, defines external inputs to the model, calculates algebraic relationships, and serves as a repository for graphical functions (HPS, 1996).

Figure 1.10 shows the generic symbols used to represents stocks, flows and converters in computer system dynamics models (HPS, 1993). Rectangles are used to represent accumulations, stocks, reservoirs or inventories. The pipes, with spigot or valve flow regulator and circle attached, represent flows. The connecters of wires indicate relationships. They link stocks back to flows, and in some cases flows to flows. Circles represent converters.

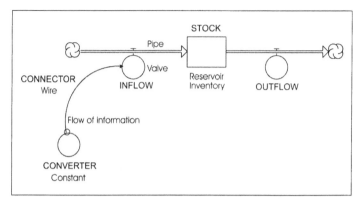

Figure 1-10 The four conceptual model building elements in system dynamics modelling

Almost all the concepts basic concepts and techniques needed for constructing and running System Dynamics (compartment-flow) modelling are shared with amongst the most common visual System Dynamics modelling software, such as Stella and ithink (HPS, 1992, 1993), Simile, Model-Maker, Dynamo (Forrester, 1961; Richardson & Pugh, 1981), Power-Sim (Powersim Corp, 1996) and Vensim (Kirkwood, 1995; Ventana, 1995; 1996). Computer simulation models can help in developing instincts for managing ecosystems.

1.6 Scope of research

Problem statement, the case of the Harare Metropolis, Zimbabwe

The Harare metropolis in Zimbabwe, extending upstream of Lake Chivero in the Upper Manyame River Basin, consists of the City of Harare and its satellite towns: Chitungwiza, Norton, Epworth and Ruwa (Figure 1.11). The existing urban water and drainage system which is the subject of study in this dissertation is typically a single-use-mixing system: water is used and discharged to "waste", excreta are flushed to sewers and eventually, after some "treatment", the effluent is discharged to the main drinking water supply source, Lake Chivero (see Box 1.J: Lake Chivero: A polluted Lake). Polluted urban storm water is evacuated as fast as possible without retention or any opportunity for environmental assimilation, therefore a multitude of other materials form an important source of non-point pollution. This system not only ignores the substantial value in "waste" materials, but it also exports problems to downstream communities and to vulnerable fresh water sources (Bailey *et al.* 1996; Moyo; 1997a; Gumbo, 2000a).

The main research question in this dissertation is how can the Harare metropolis system, which is complex and has evolved over time be rearranged to achieve sustainability (i.e. water conservation, pollution prevention at source, protection of the vulnerable drinking water sources and recovery of valuable materials like nutrients) (Gumbo, 2000a; 2003a).

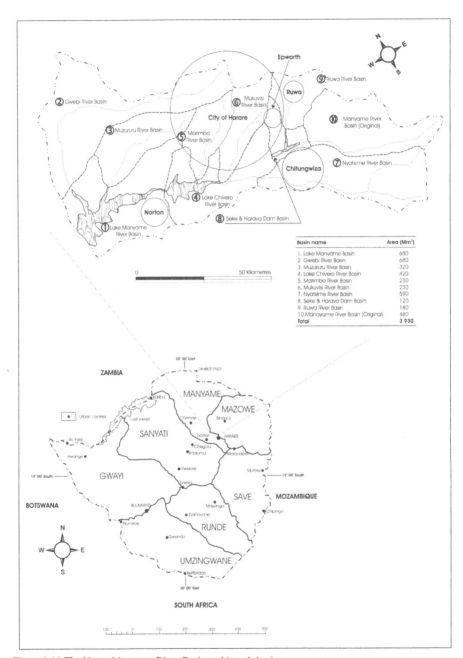

Figure 1-11 The Upper Manyame River Basin and its sub-basins

In the year 2000, approximately 2.4 million people, 22% of the population of Zimbabwe lived within the Harare metropolis. By 2020 the population of the catchment (2 200 Mm^2) is estimated to rise to 3.4 million people which will be 25% of the total population (Zanamwe, 1997). Being on the watershed, the city of Harare and its satellite towns lie within the catchment area of the main sources of water supply.

As a result most of the drainage from the City and the towns flows into its own water supply lakes (Thornton & Nduku, 1990; Moyo, 1997a). Settlement densities are among the highest in the country, with the Harare-Chitungwiza urban areas accounting for about 50% of the national urban population (CSO, 1992). Harare urban has a settlement density of 3 000 people/Mm^2. Density in the rural parts of the catchment are a lower average of 30 people/Mm^2 (JICA, 1996; Zanamwe, 1997).

Box 1-J Lake Chivero: A polluted Lake
Adapted from Nhapi *et al.*, (2002)

Numerous hydro-biological investigations carried out in Lake Chivero in the 1970's showed the Lake to be eutrophic, with the eutrophication of the impoundment being caused by the rapid chemical changes from the input of treated sewage effluent to the lake by the City of Harare (Thornton, 1980). Since the 1980's, no comprehensive research programme has taken place and thus there is a death of information on vital limnological parameters which are useful in the determination of the lake's present trophic status (Moyo, 1997a). Thornton (1980) made a comparative study of the phosphorous loadings in three reservoirs of varying trophic states. The summer phosphorous loads to Lake Chivero accounted for 80% of the annual phosphorous loading to the lake and much of the phosphorous entering the lake was lost to the sediments. The study by Thornton, (1980) showed the following annual phosphorus loadings to the lake: 74 600 kg/a, or 2.8 g/m^2 and 0.3 g/m^3. Before the diversion of municipal wastewater from the lake catchment to pasture irrigation schemes between 1970 and 1975 the loadings were respectively 288 000 kg/a or 11.0 g/m^2 and 1.2 g/m^3. Magadza (1997) also attributes this increased phosphorous loading to the sewage works in the Lake's catchment. The high phosphorous content of the effluent is a combination of inadequate treatment, due to overloading, as well as breakdown in the treatment works.

Lake Chivero is already eutrophic (Robarts & Southall, 1977; Thornton, 1980) and a number of fish kills have been reported (Moyo, 1997b). The major cause for the fish deaths in 1996 was de-oxygenation of water compounded by ammonia toxicity (Moyo & Mtetwa, 1999; Magadza, 1997). Most fish are sensitive to DO levels of below 3 mg/l (Welch & Lindell, 1980). Ammonia-N is toxic to fish at concentrations above 0.5 mg/l (Abesinghe *et al.*, 1996).

Both the lake and the inflow rivers are heavily infested with the water hyacinth *(Eichhornia crassipes)* and blue-green algae, principally *Microcystis aeruginosa* and *Anabaena sp.* and this is attributable to the nutrient loadings from sewage treatment works (Jarvis *et al.*, 1982; Mathuthu *et al.*, 1997). Excessive amounts of algae and other organic matter have seriously impacted on raw water abstraction and water treatment (McKendrick, 1982; Moyo & Mtetwa, 1999) and led to clogging of downstream commercial farming irrigation pipes (Bailey *et al.*, 1996). The lake is also losing its value as a recreational area (e.g. yatching, skiing, angling). In addition, the aesthetic qualities of the water has also deteriorated (Mbiba, 1995).

Lake Chivero created in 1952 (full supply capacity of 250 Mm^3) is the major raw water source and receives sewage effluent in excess of 120 000 m^3/day from an industrialised and densely populated area via Firle and Crowborough sewage treatment works (Gumbo, 2000a; Nhapi *et al.*, 2002). A pollution analysis of rivers was conducted in 1995, 1997 and 2001 in terms of Chemical Oxygen Demand (COD), Total Nitrogen (TN) and Total Phosphorus (TP), under dry and wet season conditions (JICA, 1996; Marshall, 1997; Nhapi *et al.*, 2002; Nhapi, 2004). The research indicates that nutrient inflows into the Lake have increase dramatically since the impoundment formed forty five years ago. Total phosphorus concentrations in the water column of the lake have increased from 0.5 mg/l in 1995 to about 0.8 mg/l as TP in 2001, suggesting a steady build-up of phosphorus in the lake.

The mass fish death witnessed in April 1996 on Lake Chivero is a graphic representation of the consequences of excessive pollution of water sources within the catchment (Moyo, 1997b; Gumbo, 1997a). Other ongoing effects of pollution of Lake Chivero include water treatment difficulties, water hyacinth infestations (Figure 1.12), toxic algae blooms and the clogging of downstream commercial farming irrigation pipes (JICA, 1996; Moyo, 1997b; Marshall, 1997; Magadza, 1997).

Figure 1-12 Water hyacinth infestation on a weir upstream of lake Chivero
Photograph: Gumbo (2000)

Sewage works are the main and most easily identifiable source of pollution contributing about 40% of the nutrient input (Gumbo, 2000a; Nhapi *et al.*, 2002; Hranova *et al.*, 2002). The remainder is from non-point sources emanating from flows within the catchment area which include urban storm runoff and runoff from commercial farming areas. Virtually all domestic and industrial wastewater in the urban centres goes into the sewerage system because of high sewerage service coverage ratio and ideally all wastewater reaches the treatment plants at the end-of-the-pipe and its "treated". In some locations however rapid development has clearly outstripped urban sewerage infrastructure. In some localities, overflowing sewage from manholes is encountered where domestic sewage is directly discharged into sewer lines. This environmental problem is associated with deficient solid waste management and habitual behaviour of residents, aside from limited absorbing capacity of sewer lines and lack of appropriate maintenance. Due to overload and lack of maintenance of infrastructure specific sections experience sewer blockages on a frequent basis (Taylor & Mudege, 1997).

A study in the 1970's, on storm water drainage in the catchment showed a clear relationship of increased nutrient loading with urbanisation (Thornton & Nduku, 1982). From estimations it was found that diffuse source storm water runoff can potentially supply sufficient nutrients to Lake Chivero to maintain a 'eutrophic state'. Commercial farming is a source of pollution from runoff containing nutrients and various noxious substances from the use of fertiliser, herbicides and pesticides. According to studies into pesticide residues in Lake Chivero (Mhlanga & Madziva, 1990; Zaranyika, 1997) traces of BHC's, aldrin, dieldrin, DDE, DDD, and DDT were detected in water, soil, fish and sediment samples.

In Zimbabwe indirect reuse of wastewater to augment potable water supplies is encouraged through a ministerial policy in view of the impending water stress situations in this semi-arid country (Lock, 1994). Even before the advent of planned water reuse, Harare's citizens had for many years drunk their "own bath water". However reduced water flows in feeding rivers during droughts have resulted in effluent flows increasing with respect to natural flows (JICA, 1996). If the lake, due to droughts, does not spill then it acts as a 100% *sink* of all the effluent flowing into it (Fullstone, 1980). This is a form of indirect reuse, which entails the incorporation of reclaimed wastewater into raw water supply *source*. This concept is based on *mixing* and assimilation with natural discharge into an impoundment of water, such as a domestic water supply reservoir, where *dilution* becomes very important (Stewart Scott, 1982; Gumbo; 1995; Asano & Levine, 1996). The wisdom for indirect reuse of wastewater for potable supplies in this basin and in Zimbabwe in general is questionable, "Is it a nuisance or a resource?" The old concept that the solution to pollution is dilution is proving to be unsuitable under these circumstances. The state of Lake Chivero is a manifestation of the '*dangerous dilution disposal method*' of wastewater (Gumbo, 1995; 1997b).

The Lake Chivero case downstream of the Harare metropolis demonstrates that societies can no longer rely on end-of-pipe sewage treatment works, as kidneys and liver to separate and remove nutrients and other toxic materials in a concentrated form to prevent them from entering the aquatic environment. The existing "conventional" urban water and sanitation system (end-of-pipe) is not efficient and compatible with the needs of an eco-city; the long end-of-pipe recycling loop of wastewater, storm water and improper application or disposal of sludges and solid organic residuals needs some rethinking (Tilth, 1982; Todd & Todd, 1994; Esrey, 2000).

In principle short "circular" or closed systems at neighbourhood scale fit better in the concept of sustainability: these seek to minimise the net use of resources and encourage local recycling, water conservation and source protection inline with the needs of a sustainable urban environment. The main question is however the feasibility such a closed system and particularly of ecological urban agriculture within the Harare metropolis through the use of nutrients derived from human metabolism (excreta and organic residues). What land use changes or infrastructural arrangements have to be developed to promote urban agriculture? What are the social, economic and environmental impacts, costs and bnefits of incorporating urban agriculture as part of the urban fabric?

The quantification and establishment of a comprehensive material flux and storage analysis for an urban eco-system in the Lake Chivero basin is the starting point in identifying where short "circular" systems or material recovery techniques could be introduced without transferring the problem from one sphere (hydrosphere, lithosphere, or atmosphere) to the other (Gumbo, 2000a).

Sustainable use of resources has to satisfy several conditions. One can distinguish physical, economic, social, financial, institutional and environmental or ecological sustainability (Savenije, 1998). Physical sustainability means closing the resource cycles and considering the cycles in their integrity (water and nutrient cycles). The closing of cycles implies that resources are transported back to where they came from. Obviously, in the case of phosphorus (P) this does not imply that P needs to be returned to the mine (lithosphere) it came from, but that food and human wastes need to be returned to the farm as fertiliser.

Closing cycles maybe costly or wasteful in terms of energy (transport mainly). As a result, closing of cycles implies that they should be kept as short as possible. The economic sustainability relates to the efficiency of the system. If all the societal costs and benefits are properly accounted for, and cycles are closed, then economic sustainability implies a reduction of scale by short-cutting the cycles. Examples of short cycles, in the context of this research include; local or on-site recycling, nutrient conservation and recovery; water conservation, making maximum benefit of rainfall where it falls (and not to capture it downstream to pump it up for irrigation or water supply).

Any attempt to close water and nutrient cycles, would mean closing of energy cycles and so would be food supply lines. The distance between "civilised humans" -both urban and rural- and their food source is part of the problem which manifests itself as water pollution problems in the case of Lake Chivero, among others. For instance the energy costs of the food system are enormous. Machinery, fuel, fertilisers, pesticides, water supply, long distance transport (generally requires more energy than is available from the food produced), drives to the supermarket, and home refrigerators. Other resource impacts in the food and agriculture system are related to soil, water and forests (erosion, nutrient loss, water storage for arid land agriculture, irrigation and return flows, deforestation as more and more acreage is needed for food and fodder production or grazing) (see Boxes 1.B and 1.C). These long food-supply lines common in a modern society are also a security risk. If war does not disrupt them, the collapse of economies might. The shortest food supply line is therefore in need of investigation as one of the most interesting alternatives for a sustainable future.

History, however, is not to be ignored. There are sewer networks in Harare metropolis and their existence sets the constraints of the initial conditions for any developments into the future. Similarly, an infrastructure of wastewater collection, conveyance, treatment, reuse and disposal is already in place so that, retrofitting it or "unhooking" certain components from the public urban drainage system network, may not make the best sense for the future. There is still not enough knowledge about what the environmental, economic and social effects of the more general use of source oriented, local and small-scale methods and technologies will be. This is leading to several questions. Should such methods be used only in new developing housing areas or should all systems be gradually replaced? How will the general use of source control options and reuse of nutrients for urban agriculture impact on the entire river basin? A view to the totality of the urban water and drainage system with a proper analysis of flow of water and matter through society is required. Among the tools are input-output analysis, cradle-to-grave analysis, mass balancing, material flow and stock accounting and systems analysis (Gumbo, 1999a; 1999b, 2001).

Aims and objectives

This research intends to analyse the '*paradigm shift*' in the water and sanitation sector in urban areas, by providing a computational and evaluation framework for implementing ecological sanitation concepts based on natural cycles of plant nutrients The finite resource phosphorus is the main subject of analysis. Recycling of phosphorus (P) in urban or peri-urban ecological agriculture (without synthetic fertilisers) is used to assess the feasibility of concepts such as the Bellagio principles for environmental sanitation (section 1.1). The focus is on the origins, sources and sinks, the different mechanisms which, transform and export phosphorus through an urban-shed of a selected micro-study catchment lying within the Lake Chivero basin in Zimbabwe. These processes are mirrored against the global P-cycle, transformation and transfers. The thesis investigates the flows and stocks of phosphorus in a typical high-density residential suburb in Harare.

The hypotheses is that short-cutting or closing open-ended water and nutrient cycles at the lowest appropriate level would ensure sustainable use of limited resources like phosphorus at the same time protecting the same resources from possible contamination. The reasons for the choice of P as an assessment parameter are outlined in Box 1.K.

Box 1-K Choice of P as an assessment parameter

1. Phosphorus is a major constraint on food and fibre production in many parts of the world especially in developing regions. Therefore, an economic supply of P is a necessity for a secure production in agriculture and forestry (Tiessen, 1995). Since P is an important nutrient in ecosystem productivity in general, its recovery or reuse is attractive and both economically and ecologically justifiable.
2. An inefficient way to satisfy the agricultural P demand is by the exclusive use of inorganic fertilisers, which normally are expensive imports, and which often have relatively low use efficiencies in the field.
3. Management and research should aim at nutrient and organic matter cycling (this includes human and animal waste) in combination with sufficient fertiliser use to avoid 'nutrient mining' in agriculture and forestry.
4. Phosphorus originates from a mined rock, therefore its presence in the urban-shed demonstrates the impact of urbanisation and anthropogenic influences on natural cycles of material flow.
5. In passing through the urban system P is mobilised from particulate to soluble forms. Therefore, in the case of P-bearing material fluxes can it be said that the activities of the urban population on the aquatic environment are globally a problem.
6. P is an important limiting nutrient in fresh water quality assessment and its role in eutrophication and its control requires further investigation. Eutrophic inland waters and their watersheds should be managed to limit P inputs, because P is easier to control than N for example, which can be entrained from the atmosphere.
7. Its resistance to be degraded into soluble forms and affinity for binding into particulate matter is representative for other elements (e.g. Si, K, N, Fe).
8. P bearing materials are most easily removed from storm and sewage water in particulate form, therefore making P separation, recovery and reuse less energy intensive.

At the core the real concerns are about food security and environmental integrity. The specific objectives of the investigation are:

- To test the feasibility of ecological sanitation and eco-city concepts in urban centres at household or neighbourhood scale by critically assessing knowledge of the nature, sources, sinks and fluxes of phosphorus at a global scale. Urban agriculture is investigated as the main option for nutrient recycling.

- To establish an inventory of phosphorus bearing materials (fluxes and stocks) within the micro-study catchment where agriculture is already a major activity, by studying; the material use and disposal at the household level; and the urban agriculture dynamics in terms of phosphorus inputs, storage and outputs. Using a system's thinking approach and MFSA two compartments or subsystems are defined to enable accounting and analysis of P-bearing materials namely the "household" (consumption or use and excretion or waste) and "agriculture" (soil-plant interaction) (Figure 1.13). The identified flows and stocks are used to draw up the P-balance for the micro-study catchment.

- To develop mathematical and conceptual relationships which describe the flows and stocks of phosphorus (origins, sources, pathways and fate) on a monthly time-step in the household and agricultural subsystems of the micro-study catchment.

- To develop a suitable model which can be used to demonstrate the possibility of short-cutting the phosphorus cycle through synchronisation of the activities at household level and those of the urban farm or garden. And to analyse the sensitivity and reliability of the model to input variables.

- To investigate the feasibility of recycling at neighbourhood level in terms of environmental criteria.

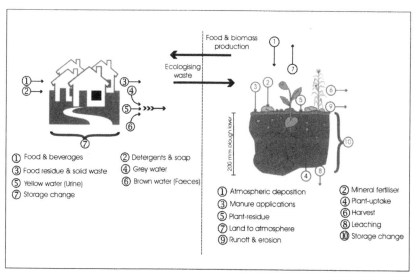

Figure 1-13 P-fluxes analysed in the household and agricultural subsystems
Source: Gumbo *et al.*, (2002a)

Limitations of methodology

The various natural cycles are interconnected and depend on one another to a great extent. For example, the burning of fossil fuels not only puts large amounts of carbon into the atmosphere, but also increases the amount of atmospheric nitrogen, phosphorus and sulphur. This interdependence is also obvious when one considers nutrient cycling through organisms. When a herbivore eats a plant or a carnivore an animal, it ingests at one go all the elements, which are found in organisms. Clearly, biogeochemical cycles involve complex interactions between the abiotic environment and living organisms and any one cycle is composed of numerous loops and steps.

In this dissertation in as far as it is an oversimplification to consider the P-flux in isolation, so it will be a limitation to assess the merits of each technological option in isolation. The key to sustainable urban eco-system will be a judgement on the best strand of unit process technologies that transfers wastes from source to sink (in an environmentally efficient manner). The issue, however, is that urban systems should be designed so as not to exacerbate any of the principal, global problems but rather alleviate them. The convenient simplification of analysing material cycles (using material accounting and system's analysis) in isolation must therefore be acknowledged as merely a device for starting this assessment of the technologies and management of the urban water, sanitation and organic waste system.

The "ecological rucksack[25]" of P imports is not considered in approximating the global and regional P fluxes and stocks. Waste emanating from use or consumption is actually a small percentage of material wasted. The world over resource wastage is ten times more than use (Von Weizsackers *et al.*, 1997). The calculations of material flow and stocks are based on the assumption that there is no difference in the ecological rucksack between imports and exports of products.

The P-flux calculations in this dissertation are based on monthly or annual time series. This is of practical significance since most of the urban agricultural activities are seasonal and usually occur between the months of November and March. Data used in the analysis is either aggregated daily data or disaggregated annual data. This approach introduces a number of errors as an average year is normally assumed.

1.7 A guide through the dissertation and its innovations

Chapter one presents an overview of '*paradigm shift*' with regard to sanitation and management of organic residues from urban centres. The eco-city concept is presented and it is suggested that the problems of water pollution experienced in the Harare metropolis in Zimbabwe can be solved by employing a decentralised system of waste management based on the Bellagio principles. The chapter introduces the methodology used in this dissertation i.e. material flow accounting and systems thinking and its limitations. The choice of P as a parameter of assessment is elaborated, and so is the focus on the urban P-flows (tracking its source to sink).

---

[25] The ecological rucksack of a product is the portion of the material input used for its production that is not incorporated within the product. This represents inefficiencies in for example in P mining, extraction and conversion to P-based fertilisers.

Ecological sanitation and organic waste recycling are a piece in the global cycling of materials (Figure 1.14) and, as such, it may be useful to postulate that '*desirable*' technologies for its implementation will be those that introduce minimal distortion of this cycling relative to the global material cycles of the pre-industrial era.

Chapter two provides details on bio-geochemistry of P, its history and exploitation by human beings especially in the agricultural sector. The Chapter quantifies the impacts of open-ended P-flows and the need to close its cycle at every possible scale. This chapter provides useful background information on P-cycling at a macro-scale i.e. global and regional. It is especially essential to the reader with little knowledge on soil-plant nutrient relationships and transfer of P from land to riverine ecosystems. Regional P-trade balances in crop, animal and fertiliser commodities is presented for year 2000 based on FAO (2001) data to illustrate the distortions on the natural cycling of P as a result of human activities.

Chapter three describes the area of study, within the context of Zimbabwe and the Lake Chivero Basin. The Chapter provides background information on the pollution problems and makes linkages between the various activities in the catchment including urban agriculture and waste management. The micro-study catchment is presented in detail in preparation for Chapter four.

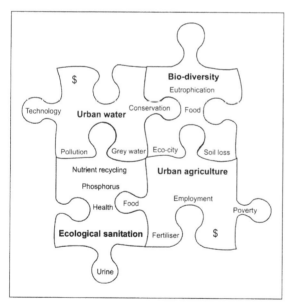

Figure 1-14 Ecological sanitation and organic waste recycling as part of a puzzle in nutrient recycling and sustainable urban development

Chapter four is the computational heart of this dissertation as it organises data and information collected into a monthly P-flux calculator using STELLA, a systems analysis software developed by High Performance Systems Inc. Mathematical algorithms are developed. The calculator can be used as a planning and decision-making tool for closing the P-cycle within urban ecosystems. It also provides the means to simulate and evaluate different options in linking household waste P-fluxes to agricultural P-requirements. The calculator is used in **Chapter five** in assessing promising options based on environmental sustainability criteria.

Chapter six summarises the methodology and presents recommendations and conclusions.

Chapter seven is a closing chapter outlining the feelings of the researcher during the six years of research on this topic and various interactions made with the promoter, supervisor, and resource persons and important of all family members. This Chapter is dedicated to those who might want to follow a similar mode of study and is presented in such a relaxed fashion reflecting that there is fun, frustrations, toil and sweat during this uncertain journey in attaining a PhD.

Figure 1.15 illustrates the linkages between the Chapters, the flow of ideas and thoughts from the initial stages of the research work to its conclusions.

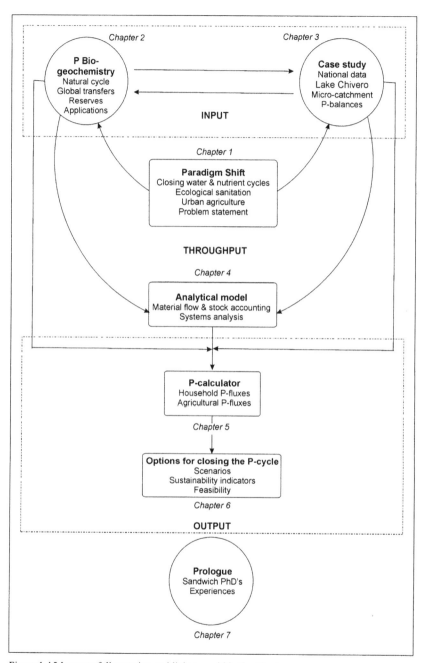

Figure 1-15 Layout of dissertation and linkages within the Chapters

2 Biogeochemistry of phosphorus global transfers and cycles

2.1 Origin and forms of phosphorus

Discovery and history of phosphorus

Elemental phosphorus may have been known to Arabian alchemists as early as the 12th century, but it was its discovery by a secretive alchemist, the self styled Herr Doktor Brandt around 1669 in Hamburg, Germany, that initiated its carrier in modern chemical developments. While making some chemical test with human urine, Brandt saw a strange, wax-like substance distil over from his retort, and as it dripped from the spout, it sputtered and glowed in his dingy dark room with an eerie, yellowish light. It was white phosphorus (Sauchelli, 1965). One imagines that Brandt envisioned a golden future as he toiled over his vats of urine, following the "golden stream" (turning base metals into gold), in pursuit of the philosopher's stone (Emsley, 2000). Until 1750's phosphorus was rare and expensive and its main use was medicinal.

The story of phosphorus and its multiple uses in agriculture and industry is most fascinating. The 19th century saw a great increase in the manufacture of phosphorus, which, until a mineral source was discovered in the 1860s, was derived from bones guano, and fish to improve the fertility of the soil with Albright and Wilson Plc, being the first large scale commercial venture using bones as a source of P (Emsley, 2000; Driver *et al.*, 1999). The new science of chemistry, focusing on the element which shone in the dark, established that it was an essential component of the bones of animals and man, a constituent of every living cell of both plants and animals, and essential to the nutrition of all living organisms. The name, phosphorus, was derived from two Greek words, *phos*, meaning light, and *phore*, meaning to bear.

For many years after the discovery of P, the use of bones and guano continued as the main source of phosphates for improving soil fertility. A revolutionary step in its agricultural history was the realisation that the treatment of bones with sulphuric acid would render the contained phosphate more assimilable to plants. The next significant, and for mankind's benefit the most important step, was the discovery that the phosphate in mineral deposits in many parts of the world was identical to that in bones, and, when acid-treated, released a highly water-soluble phosphate. This discovery was destined to become the foundation of the modern fertiliser industry, an industry dedicated to the maintenance and improvement of soil fertility, and the base that sustains man's hope that hunger and starvation can be kept from his door (see Figure 2.1 for the progressive history of the discovery of world phosphate resources).

The blessing in the discovery of P in terms of its vital role as an agronomic nutrient has been dented by its potential as a killer in the manufacture of phosphorus bombs and nerve gas with the former, ironically, used for the first time in Hamburg by the allied forces against Germany in the second world war (Emsley, 2000).

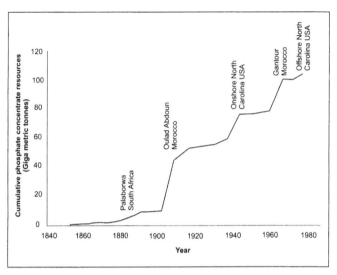

Figure 2-1 History of the dicovery of world phosphate resources
Source: UNIDO (1998)

The element phosphorus, never occurs in nature. It combines spontaneously and vigorously with oxygen and even its pentoxide, P_2O_5, combines eagerly with water to form ortho-phosphoric acid. There are two main types of phosphate rock deposits, igneous and sedimentary, which have widely differing mineralogical, textural and chemical characteristics. Phosphorus ranks eleventh in the order of abundance of elements in the igneous rocks of the earth's crust (estimated at 0.1% of the earth's crust). Originally all the phosphorus in the lithosphere must have been present in igneous rocks.

The most prevalent phosphate minerals in these rocks are species of apatite $(Ca_{10}X_2[PO_4]_6$, where $X = OH^-$ or F^-, Ca may also be substituted with Na, Mg and PO_4 with CO_3) i.e. a complex form of calcium phosphate with quartz, calcite, dolomite, clay and iron oxide components. It is the main phosphate mineral in the earth's crust and is very stable in calcareous environments, can be sand or silt-sized (and therefore easily identified) and can be occluded in other minerals such as quartz (Syers *et al.*, 1967; Lindsay *et al.*, 1989).

Igneous rock is often associated with carbonatites and or alkalic intrusions and is generally low in grade i.e. low concentration of phosphate. The abundance ratio of igneous versus sedimentary rock is 13 to 87. Some 80% of the world phosphate production is derived from sedimentary phosphate deposits. Nearly all the phosphorus in sedimentary rocks is present in the form of apatites, primarily carbonate fluoro-apatite (Sauchelli, 1965; Steen, 1998).

More than 75% of the globally commercially exploited phosphate rock is surface mined, which can take many forms from manual methods to the employment of highly mechanised technologies, with the remainder recovered by underground mining (Steen, 1998).

Phosphorus forms

There is now an accepted English jargon of abbreviation where D, P, and T stand for dissolved, particulate and total, respectively, and O and I for organic and inorganic. Thus PIP means particulate inorganic phosphorus. In addition H represents hydrolysable phosphorus. In most studies of rivers and standing waters, the forms of P are operationally defined based on analytical practicality. The distinction between particulate and dissolved P is made by the porosity of the filter used to separate the two fractions; filters with porosities around 0.5 μm are commonly used (see Box 2.A: Phosphorus forms in water and wastewater). Considerable effort has been devoted to improve analytical capabilities and understanding of the physical, chemical and biological meaning of the measurable P fractions; much of this work is reviewed in Bostrom *et al.,* (1988), Broberg & Pettersson (1988), Pettersson *et al.,* (1988), Froelich (1988) and Engle & Sarnelle (1990).

The phosphorous forms [1] or complexes usually considered are dissolved orthophosphate (including H_3PO_4, $H_2PO_4^-$ and HPO_4^{2-}) expressed as $DP-PO_4$ and total dissolved phosphorous, TDP, which includes the latter plus the polyphosphate originate from detergents, and dissolved organic phosphorous, (DOP). Particulate organic phosphorus, (POP), and particulate inorganic phosphorous, (PIP), are rarely measured or considered in P budgets (Meybeck, 1982; Crites & Tchobanoglous, 1998).

Analysis of water and wastewater

The total phosphorous (TP) is a mixture of orthophosphates and in some polluted waters, of polyphosphates, and of organic phosphorous on either dissolved or dissolved plus particulate material and inorganic particulate P. However, TP depends on the chemical attack made. Usually TP analysis is carried out on unfiltered waters after heating at 110 °C with H_2SO_4 as oxidant. The bulk organic P is thus mineralized into phosphate. Stronger acid or alkaline attacks may completely solubilise the inorganic particulate phosphorus (PIP) that includes the apatitic P and the non-apatitic P. Selective chemical attack of particulate matter may solubilise mixtures of various phosphorous forms. The AWWA (1975) methods thus separate: organic particulate P (POP), adsorbed P onto particles, apatitic P, and non-apatitic P (Suess, 1995).

---

[1] In this dissertation, the word phosphorus abbreviated as P is used in the text rather than identifying a particular phosphate

Box 2-A Phosphorus forms in water and wastewater
Modified from USEPA (1983); AWWA (1975)

USEPA (1983) provides methods for the determination of specified forms of phosphorus in drinking, surface and saline waters, domestic and industrial wastes. The methods are based on reactions that are specific for the orthophosphate ion. Thus, depending on the prescribed pre-treatment of the sample, the various forms of phosphorus defined below may be determined (see Figure 2.2). Except for in-depth and detailed studies, the most commonly measured forms are phosphorus and dissolved phosphorus, and orthophosphate and dissolved orthophosphate. Hydrolysable phosphorus is normally found only in sewage-type samples. Particulate forms of phosphorus are determined by calculation. All phosphorus forms are normally reported as P mg/l, to the third decimal place.

Total Phosphorus (TP): all of the phosphorus present in the sample regardless of form, as measured by the persulfate digestion procedure.

Total Orthophosphate (TP-PO$_4$): inorganic phosphorus [(PO$_4$)$^{-3}$] in the sample as measured by the direct colorimetric analysis procedure.

Total Hydrolysable Phosphorus (THP): phosphorus in the sample as measured by the sulphuric acid hydrolysis procedure, and minus predetermined orthophosphates. This hydrolysable phosphorus includes polyphosphates plus some organic phosphorus.

Total Organic Phosphorus (TOP): phosphorus (inorganic plus oxidisable organic) in the sample as measured by the per-sulphate digestion procedure, and minus hydrolysable phosphorus and orthophosphate.

Total Dissolved Phosphorus (TDP): all of the phosphorus present in the filtrate of a sample filtered through a phosphorus-free filter of 0.45 µm pore size and measured by the per-sulphate digestion procedure.

Dissolved Orthophosphate (DP-PO$_4$): as measured by the direct calorimetric analysis procedure.

Dissolved Hydrolysable Phosphorus (DHP): as measured by the sulphuric acid hydrolysis procedure and minus predetermined dissolved orthophosphates.

Dissolved Organic Phosphorus (DOP): as measured by the per-sulphate digestion procedure, and minus dissolved hydrolysable phosphorus and orthophosphate.

Total Particulate Phosphorus (TPP): (TP) = (TPP) + (TDP)

Particulate Orthophosphate (PP-PO$_4$): (TP-PO$_4$) = (PP-PO$_4$) + (DP- PO$_4$)

Particulate Hydrolysable Phosphorus (PHP): (THP) = (PHP) + (DHP)

Particulate Organic Phosphorus (POP): (TOP) = (POP) + (DOP)

Total phosphorus (TP) is sometimes measured on unfiltered waters. In this case the total phosphorous reported refers to the total dissolved phosphorous (TDP) plus an unknown part of particulate phosphorous – mostly the particulate organic phosphorous released to solution by acid attack. This type of measure is reported routinely in the United States rivers survey (USEPA, 1983). Orthophosphates are more commonly analysed and usually serve as a basis for P budgets. Orthophosphate levels in natural waters are low, between 1 and 24 mg P/m^3. The most commonly encountered values are around 8 mg/m^3 (see Table 2.2 and 2.3 in this Chapter[2]) (White, 1979, Meybeck, 1982). It is striking to note that particulate P is almost never surveyed, although it constitutes a major form of nutrient transport.

---

[2] When considering available data for the unpolluted rivers (Table 2.2), the ratio DP-PO$_4$/TDP is generally between 0.2 and 0.7 with a median value of 0.4. The same ratio is obtained for contaminated river waters (Table 2.3). However, these results must be considered with caution as many authors do not specify whether water is filtered or not prior to TDP analysis (Meybeck, 1982; Meybeck *et al.,* 1989).

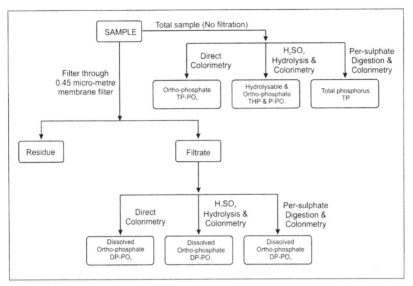

Figure 2-2 Analytical scheme for differentiation of various phosphorus forms in water and wastewater
Source: USEPA (1983)

Plant and soil tests

Plant and soil analysis are used extensively to diagnose the P status of farming systems. The methods for determining phosphorus in soil and plant tissues are many and varied. These can be grouped depending on the type of extractant used (Sauchelli, 1965; Frossard *et al.*, 1995). Pre-plant soil tests offer a better method of predicting P requirements for establishing crops whilst plant analysis is more suited to permanent crops or monitoring the effectiveness of a fertiliser programme (Hedley *et al.*, 1995).

Various tests have been developed in different countries to suit the forms of P present in their agricultural soils (Table 2.1) (Nye & Tinker, 1977; Anon, 1985). The form of soil P extracted by each test is determined by its solution pH and the reaction of the ions present in the extractant with sorbed or mineral P. For instance, the HCO_3^- and OH^- in the bicarbonate extract promote the desorption of P from $CaCO_3$ and Fe and Al hydrous oxide surfaces. In the strongly weathered soils, dominated by Fe-P, Al-P and organic P with little Ca-P, bicarbonate is an effective extractant of P but the amount is correlated poorly with identifiable soil P fractions. Bray 1 and Truog-extractable P are highly correlated to Al-P and Fe-P in such soils.

The Olsen extractant is suitable for calcareous and weakly weathered acid soils and less suitable on strongly weathered soils where Bray 1 and Mehlich tests are more appropriate (Kamprath, 1991). Water extraction techniques (Sissingh, 1971) are mostly suited to well fertilised or calcareous soils because unfertilised acidic soils are likely to yield P concentrations in the extract below detection limits (Frossard *et al.*, 1995). Anion exchange resin has been used as the preferred method to classify tropical soils into groups of P sufficiency or deficiency (Frossard *et al.*, 1995). Phosphorus extracted by the anion resin method of Sharpley *et al.*, (1984) and

Sharpley (1991) is strongly correlated to P extracted by Fe and Al oxide impregnated filter paper, and compares favourably as a soil P availability index (Sibbesen, 1983).

Table 2-1 Some extractive soil testing methods for assessing soil P status

Soil test (reference)	pH [a]	Soil: Solution ratio	Extractants	Time
Bray 1 (Bray & Kurtz, 1945)	3	1:7	0.03 M NH_4F + 0.025 M HCl	1 min
Bray 2 (Bray & Kurtz, 1945)	1	1:7	0.03 M NH_4F + 0.1 M HCl	40 sec
Colwell (Colwell, 1963)	8.5	1:20	0.5 M $NaHCO_3$	16 hr
Egner-Riehm (Grigg, 1965)	-	1:20	0.1M NH_4-lactate + 0.4 M Acetic acid	4 hr
Fe paper (Menon *et al.*, 1991)	nat	1:40	Fe hydroxide impregnated filter paper	16 hr
Lactate (Holford *et al.*, 1985)	3.7	1:50	0.2 M Ca-lactate + 0.01M HCl	1.5 hr
Mehlich (Holford *et al.*, 1985)	1.0	1:4	0.025 M H_2SO_4 + 0.05 M HCl	5 min
Truog (Truog, 1930)	3.0	1:200	0.002 M H_2SO_4 + 0.3% $(NH_4)SO_4$	30 min
Olsen (Olsen *et al.*, 1954)	8.5	1:20	0.5 M $NaHCO_3$	30 min
P_w test (Sissingh, 1971)	nat	1:2 then 1:60	Distilled water standing then shaking	22 + 1 hr
Resin s (Amer *et al.*, 1955; Sibbesen, 1983)	nat	-	Anion exchange resin beads	variable
Resin s (Van Raij, 1986)	nat	1:100	Anion + cation exchange resin beads	16 hr
Resin s (Saggar *et al.*, 1990a)	nat	-	Anion + cation exchange membrane	-

a)   nat = natural pH of soil
Source: Adapted from Hedley *et al.*, (1995)

Correlation and calibration of soil tests against crop yield pose a number of problems (Dahnke & Olsen, 1990; James & Wells, 1990; Fixen & Grove, 1990). The largest errors in soil test values are associated with soil variability and sampling technique. Designing appropriate soil sampling techniques requires a detailed knowledge of the effects of soil parent materials, pedogenesis, landscape and farming system on the distribution of labile soil P with respect to the crop rooting zone, timing of the previous fertiliser application and placement of the future application (James & Wells, 1990; Saggar *et al.*, 1990b). A way of grouping similar land areas and or coping with soil heterogeneity is required to reduce (or measure) the spatial error involved in estimating a representative soil test value (James & Wells, 1990).

The ideal nature of crop yield response to increasing soil P test value is curvilinear, rising to a maximum yield plateau where P no longer limits plant growth (see Figure 2.3). Differences in climate, other nutrient availability and soil characteristics cause variations in the maximum yield between years and sites. Yield data at each site can be transformed into percentages of maximum yield at each site to allow comparison of data between sites. Curvilinear or linear response-plateau models must be fitted to the relative yield data to test the predictive power of the soil test (Nelson & Anderson, 1977; Sanchez & Salinas, 1981; Hedley *et al.*, 1995)

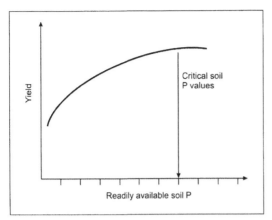

Figure 2-3 Schematic relationship between yield and readily available soil P
Source: Johnston & Steen (2000)

Intensity and buffering capacity methods can be used for assessing the P fertiliser requirements. In P-fertilised soils, Fox (1981) argues that variations in the relative yield of crops can best be explained by the concentration of inorganic P in the soil solution rather than the quantity of *labile* P[3] (extractive method) on the soil surface. Two factors control relative yield: (a) the initial concentration of P in soil solution that will support maximum plant yield, i.e. the critical solution P concentration or external P requirement, and (b) the ability of the soil to buffer the solution inorganic P during plant growth (Thompson & Troeh, 1973). By conducting plant growth experiments and P sorption studies with a range of fertiliser levels it is possible to describe the relationships between added fertiliser P, soil solution inorganic P concentration and relative yield, and thus estimate fertiliser requirements. While the method is useful for research purposes, the labour and time involved in constructing P sorption curves make it too expensive for routine soil testing laboratories (Hedley *et al.,* 1995).

---

[3] Phosphate on surfaces that can be readily be desorbed, plus phosphate in solution, is called *labile* P to distinguish it from the P held in insoluble compounds, or in organic matter, which is *non-labile* (White, 1979; Tiessen, 1995; Tunney *et al.,* 1997).

2.2 Biogeochemistry of phosphorus

Terrestrial nutrient cycles

Terrestrial ecosystems exist at the interface between four spheres: atmosphere, biosphere, hydrosphere, and lithosphere. The soil is also a dynamic system that has developed in response to the interactions between the four spheres. Soils are a critical component of terrestrial ecosystems providing storage pools for nutrients, water, and air, and a medium of support (i.e. root anchoring) for terrestrial plant (Lovelock, 1979; Schlesinger, 1997). Many important processes related to nutrient cycling occur in the soil and the productivity of a particular ecosystem is intimately linked to the quality of the soil (see Figure 2.4).

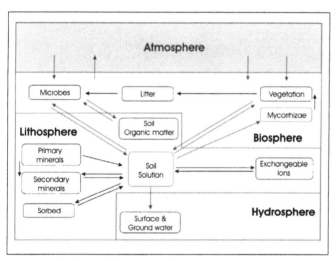

Figure 2-4 The flow and interchange of nutrients between the four spheres

Water represents the greatest flow of any material substance through an ecosystem and plays a vital role as a transporting agent and chemical solvent. With regard to nutrient cycling, water provides a dynamic pathway for redistribution of elements within different trophic levels and within the soil zone and lithosphere.

The energy required to drive nutrient cycling is obtained by processes occurring in the biosphere. The primary source of energy for ecosystem processes is photosynthesis. Directly or indirectly, photosynthesis provides the energy for all forms of life in the biosphere (Schlesinger, 1997). To maintain all the biochemical reactions necessary for plant growth, at least 14 essential mineral nutrients are required from the lithosphere (IFDC, 1992, Schlesinger, 1997). These nutrients are extracted from the soil by plant uptake, incorporated into the biomass store and returned to the soil directly or indirectly as organic matter.

The organic matter (organic store) in turn is a source of energy for heterotrophs which further oxidise the organic compounds by decomposition processes (i.e. by respiration), simultaneously releasing the mineral nutrients into the soil inorganic store (White, 1965; Thompson & Troeh, 1973). Thus, the biosphere processes of nutrient uptake, incorporation of mineral nutrients into biological tissues, litter fall and harvest residues, and the decomposition of organic matter with the concomitant release of nutrients by microorganisms are essential to the cyclic flow of nutrients in terrestrial ecosystems[4] (see Figure 2.5). Biosphere processes are intimately linked with processes in the atmosphere, hydrosphere, and lithosphere (Kupchella & Hyland, 1993; Schlesinger, 1997).

Herbivory is the consumption of living tissue by herbivores. Herbivory short-circuits the nutrient cycle by consuming living tissues prior to senescence and litter fall. Many of the nutrients may be returned directly to the ecosystem by herbivore defecation and death or the nutrients may be transferred to other ecosystems by herbivore migration (White, 1965; Thompson & Troeh, 1973).

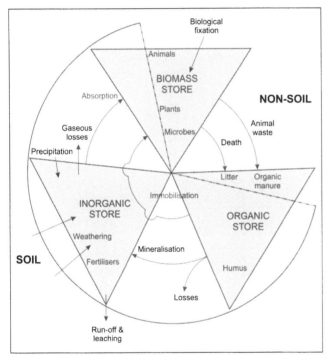

Figure 2-5 The fundamentals of nutrient cycling in the soil-plant system
Source: White (1979)

---

[4] In reality, the biosphere does not recycle all the important nutrient elements – notably phosphorus and calcium – without help from geological processes.

Microorganisms play an essential role in completing the intra-system nutrient cycle by releasing nutrient elements for plant uptake through the processes of decomposition and mineralization (Tisdale & Nelson, 1975).

There are differences in the way the nutrients are distributed within the soil. Nitrogen (N) always accumulates in the organic-rich A horizon, the content declining gradually with depth. Phosphorus (P) is similar, the decline with depth being more abrupt because of immobility of phosphate ions in the soil (see Figure 2.6). Sulphur (S), like N, accumulates in the surface of temperate soils, but not necessarily in tropical soils, owing to the mineralization of organic-S and the downward displacement SO_4^-. Many of the latter soils typically show a bimodal S-curve distribution, with organic-S in the surface and an accumulation of S-SO_4^- in the subsoil (White, 1979).

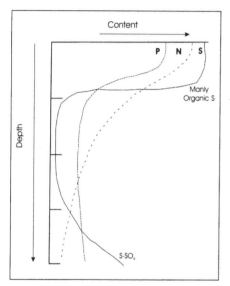

Figure 2-6 The distribution of N, P and S in a soil profile
Source: Modified from White (1979)

The phosphorus cycle

Phosphorus is a very important nutrient in reactions that store and release energy for use in biological cells. The first step of P cycling is the dissolution of apatite. In the natural environment, P is supplied through the weathering and dissolution of rocks and minerals with very low solubility. The dissolution of apatite has been extensively studied with phosphate rock (Olsen, 1975) but more seldom in natural systems where it is the first step of the P cycle (Walker & Syers, 1976). Apatite dissolution requires a source of H^+, which can originate from the soil itself or from roots or microbes, and sinks for Ca and P (Mackay *et al.*, 1986). Therefore phosphate rocks have lower agronomic value in soils above pH 6.2 (Fardeau *et al.*, 1988).

P is usually the critical limiting element for plant and animal production, and throughout the history of natural production and human agriculture, P has been largely in short supply (Sanchez & Uehara, 1980; Tiessen, 1995). Therefore in general availability of P is often a limiting factor in ecosystem productivity. Through erosion and weathering of rock[5], inorganic phosphate is made available to plants through uptake from the soil or, in the case of aquatic plants, from the aqueous environment (Figure 2.7). In spite of efficient recycling of phosphorus, some of it will be lost to deep-sea sediments. From there it can only be retrieved again through major geological events over millions of years (mountain formation) followed by renewed weathering (Ayres & Ayres, 1996).

A portion of the earth's phosphorus is continually passing out of the mineral reserve into living substance, and similarly, phosphorus is continually passing out of living matter to re-enter the mineral reserve. This movement of the element has been pictured as taking place within two cycles, a land cycle and a marine cycle, or one general cycle with its complicated circulations of the element since the two cycles are definitely interrelated (Kupchella & Hyland, 1993; Schlesinger, 1997).

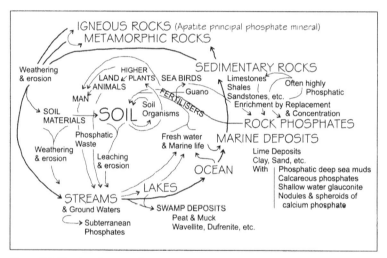

Figure 2-7 The phosphorus cycle
Source: Adapted from Sauchelli (1965)

Phosphorus in soil-plant systems

Studies of P cycling and availability have posed a challenge to agronomists and ecologists for many years because P exists in soils and sediments in many different physico-chemical forms, and it is involved in a myriad of biological processes. The cycling of P can be controlled by inorganic chemical reactions, but in many systems, the turnover of organic P controls the availability of P to plants through organic matter decomposition, or the release of P from microbial stores (Chapin *et al.,* 1982; Harrison, 1985; 1987; Walbridge, 1991; Lajtha & Harrison, 1995). Phosphate sorption by various soil constituents is a dominant process maintaining soil solution P at very low levels, and it has proven difficult to measure the amount of P in a soil that is available for plant uptake. Whereas important pools and fluxes of nitrogen are

---

[5] The soil parent material is the sole source of P for plant growth unless fertilisers or manures are applied. Few unfertilised soils release plant available P at rates sufficient to meet P requirements for continuous crop production, and P is commonly deficient (Sanchez & Uehara, 1980). Exceptions are younger soils formed from alluvium, glacial till and basic volcanic rocks of high P content (Wild, 1988).

readily measured in soils (e.g. 'soluble' inorganic N, rate of N mineralisation), the different physico-chemical forms of P are difficult to categorise or to define operationally, thus limiting the ability to develop a simple definition of bio-available P (Lajtha & Harrison, 1995).

The immobilisation, mineralisation, and redistribution of P is best discussed with reference to P cycling in soils (Cole *et al.*, 1977; White, 1979; Cole & Heil, 1981; Chauhan *et al.*, 1979, 1981) as represented by Figure 2.8. In uncultivated soils, the availability of P to plants is a function of the amount and form of soil P present and the rate at which it can be mobilised and transported to plant roots.

Agricultural systems influence soil P status by the nature, intensity and frequency of: 1) cultivation and tillage, 2) crop and product removals, 3) erosion and leaching, 4) manure and fertiliser application. Arable agriculture depletes soil P through removal of P in the crop and soil erosion (Tiessen *et al.*, 1983; O'Halloran *et al.*, 1987) and smaller leaching losses (Sharpley *et al.*, 1995). Crop removals account for 0.1 to 4.0 g P_2O_5/m^2 per crop (FAO, 1987a; 1987b; Withers, 1999).

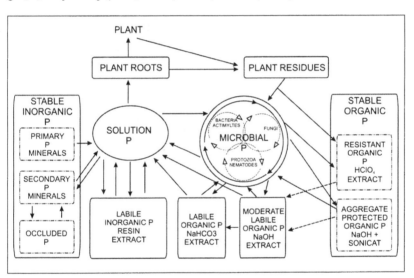

Figure 2-8 Compartments and flows of P in the soil-plant system of the P cycle
Source: Chauhan *et al.*, (1981)

The main source of P in soils is from primary P minerals (mainly hydroxyl-apatite). This material is sparingly soluble and releases small amounts of phosphate to solution, where it enters in equilibrium with labile inorganic P (i.e. surface-adsorbed or precipitated as soluble calcium phosphates, such as octo-calcium phosphate, that are exchangeable to ^{32}P and extractable by anion-exchange resins) (Sauchelli, 1965; Van Wazer, 1973; Richey, 1983). Secondary inorganic P compounds of reduced solubility and occluded forms of inorganic P may also be present, especially in more weathered soils (Kupchella & Hyland, 1993). P has a low concentration and low solubility in soils with typical values for topsoil of 0.1-3.0 P_2O_5 g/kg of soil and <0.0003 mol/m³, respectively (Frossard *et al.*, 1995; Hedley *et al.*, 1995). If the critical soil solution P concentration required to produce an appropriate crop yield in a soil is known, P sorption isotherms can be used to estimate P requirements (Fox & Kamprath, 1970; Sanchez & Uehara, 1980).

Soil solution P is usually quite low due to complex interactions of phosphate with various soil components. Because of low P concentrations in soil solutions and competition from micro-organisms, many plants have evolved mechanisms to enhance the absorption of P when it is in short supply; one of the most of these is *mycorrizhia*, a symbiotic association of fungus and root (White, 1965; Nye & Kirk, 1987; Harrison *et al.*, 1988; Frossard *et al.*, 1995; Lajtha & Harrison, 1995).

The topsoil is often identified as the *plough layer*, i.e. the 0.20 to 0.30 m depth of soil which is turned over before seedbed preparation. The crop takes up the majority of the nutrients it requires from the topsoil. The volume of a square metre of topsoil is around 0.25 m^3 and weighs approximately 350 kg (assuming a bulk density of 1 400 kg/m^3). The topsoil can contain between 20 and 700 g P_2O_5/m^2 (Johnston & Steen, 2000).

The organically bound phosphorus mineralises, i.e. becomes available through the activity of soil microbes and cultivation, but only slowly (Frossard *et al.*, 1995; Barrow, 1961; Dalal, 1977; Harrison, 1985). In general, organic P mineralisation rates are more rapid in tropical soils where organic P is an important source of available P (Adeptu & Corey, 1976, 1977; Morris *et al.*, 1992). In tropical soils initial net organic P mineralisation rates may range from 2.7 to 5.0 g P/m^2.a for the first year of cultivation after scrub or grass fallow (Johnston & Steen, 2000), which is sufficient to provide P for two crops per annum. In cooler climates where organic P mineralisation rates are slower, not enough P may be mineralised during one cultivation and growing season, and a cultivated fallow may be used to provide enough mineral P for the crop and to conserve moisture (Hedley *et al.*, 1995).

Organic P constitutes 10 to 65% of topsoil P (Harrison, 1987), most in low molecular weight compounds (Frossard *et al.*, 1995). Organic P is synthesised in plants and may constitutes approximately 20-80% of P in plant vegetative tissue, depending upon plant part, age and P supply, and 70-80% of P in seeds. The more important P compounds synthesised are phospholipids (membrane structure), nucleic acids (genetic code) and inositol phosphates (P storage). Their respective tissue concentrations in leaves are in the range 0.01%, 0.01% and <0.0001% and in seeds 0.002%, 0.02% and 0.05% (Frossard *et al.*, 1995). In general, P concentration is higher in meristematic tissue. It is common for P in older plant parts to be remobilised and transported to meristematic tissue or to seeds (Koide, 1991).

In general, neutral to slightly alkaline soils have shown higher soil solution P levels than acidic soils dominated by oxide-rich clays, suggesting that sorption processes may not have such a marked effect on P bio-availability in these systems. Soil pH, clay and sesquioxide content and exchangeable Al^{3+} all influence P availability. Where the effects of the free Iron (Fe) and Aluminium (Al) oxides are predominant, or in fertilised acid soils where compounds of composition $FePO_4.nH_2O$ and $AlPO_4.nH_2O$ exist and dissolve incongruently to release more P than Fe or Al, P in solution increases with pH rise (White, 1979; Lajtha & Harrison, 1995). By contrast, in phosphate-rich alkaline and calcareous soils, insoluble soils calcium phosphates such as hydroxy-apatite and octa-calcium phosphate comprise the bulk of the non-labile P, the solubility of which improves as pH falls. Consequently the, the availability of P in such soils is greatest between pH 6 and 7 as illustrated in Figure 2.9.

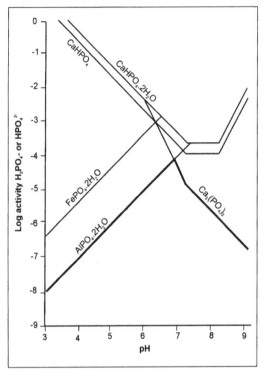

Figure 2-9 The solubility of phosphorus in the soil solution as a function of pH[6]
Source: Schlesinger (1997)

However acid soils of low P status which retain a surplus of clay minerals relative to sesquioxides show a contrary trend in P availability with pH change. Between pH 4.5 and 6, exchangeable Al^{3+} hydrolyses to form hydroxy-aluminium ions which act as sites for P adsorption because of their residual positive charge. The ions also polymerise readily so that hydroxyl-aluminium complex with $H_2PO_4^-$ ions *occluded* in its structure forms on the surface of the clay mineral, rendering a minimum in P concentration over the pH range 5-6.5. Above pH 7, Al begins to dissolve as the aluminate anion and P is released (Cole *et al.*, 1977, 1978; White, 1979).

Soils with the highest P requirements, *viz.* Andosols, Ultisols and Oxisols make up 43% of the land area of the tropics (Sanchez & Salinas, 1981; FAO-UNESCO, 1988a; 1988b; USDA, 1989; ITC & FAO, 1995) and represent areas that probably have to be brought into production to meet the future food requirements of developing countries (see section 2.6). With low P reserves and inherent P deficiency they will require careful management, but have the potential to produce high yields if the main chemical constraints to plant growth are alleviated (Sanchez & Salinas, 1981).

---

[6] Precipitation with Al sets the upper limit on dissolved phosphate at low pH (bold line); precipitation with Ca sets a limit at high pH. Phosphorus is most available at a pH of about 7.

Andosols and strongly weathered soils have high P sorption capacity and may require over 20 g P_2O_5/m^2 to raise the soil solution concentration of the plough layer to 0.2 g/m^3, at which level P limitation of crop yield is alleviated (Sanchez & Uehara, 1980)

Modelling the P-cycle

Most models of P dynamics have concentrated on a specific set of soil and plant processes operating over time scales of hours to months. While annual P budgets have been constructed for a variety of ecosystems, there have been few attempts to develop dynamic models of P cycling at ecosystem level over a number of years (Blair *et al.*, (1977; Frossard *et al.*, 1995).

An example of such models is the CENTURY model (Parton *et al.*, 1988; Figure 2.10) which simulates long-term P dynamics in a variety of ecosystems with an emphasis on soil organic matter. Plant residues and animal excreta are partitioned into structural and metabolic components, while soil organic matter is divided into active, slow and passive pools.

Given the monthly time-step of the model, the labile pool is assumed to be in equilibrium with a sorbed pool equivalent to 0.1M NaOH extractable P, in a curvilinear relationship described by sorption affinity and sorption maximum parameters. This pool is in slow equilibrium with a strongly sorbed pool. P can enter the cycling pool by weathering of a parent material and can be slowly fixed in an occluded pool.

There is still considerable scope for the development of mechanistical P cycling models which can be applied over long time scales in order to study the transformations of P between various inorganic and organic soil pools and to account for losses from the ecosystem. Plant and microbial uptake of P and the rate of organic matter decomposition can be limited by the availability of other nutrients, particularly N, K, and S. Hence it is desirable that models of ecosystem P dynamics also consider these other elements (Frossard *et al.*, 1995).

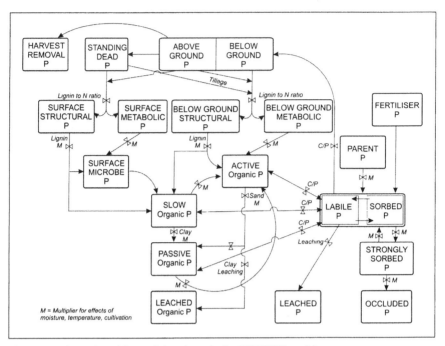

Figure 2-10 The pools and flows of phosphorus in the CENTURY model
Source: Frossard *et al.*, (1995)

2.3 Phosphorus transfers from terrestrial to aquatic ecosystems

Fundamentals of phosphorus transport and transformation

The transfer of P from terrestrial to aquatic environments in runoff can occur in soluble and particulate forms (Figure 2.11). Total dissolved phosphorus (TDP) consists of inorganic orthophosphate (DP-PO$_4$) and dissolved organic (DOP) compounds and complexes (see Box 2.A for phosphorus forms in water). Total particulate P (TPP) encompasses all primary and secondary mineral P forms plus organic P (POP), includes P sorbed by mineral and organic particles eroded during runoff. It constitutes the major proportion of P transported from cultivated land, *viz.* 75-90% (Sharpley *et al.*, 1995).

Runoff from grass or forest land carries little sediment, and is, therefore, generally dominated by (DP-PO$_4$). While DP-PO$_4$ is, for the most part, immediately available for biological uptake (Wetzel, 1983), TPP can provide a long-term P source for aquatic plant growth (Carignan & Kalff, 1980; Wildung *et al.*, 1974). In the past, most studies have measured only DP-PO$_4$ and total P (TP) transport in surface runoff.

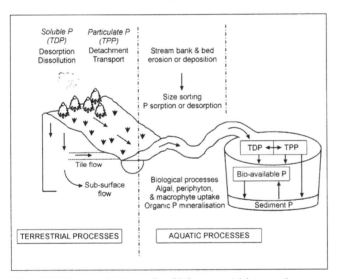

Figure 2-11 Processes in the transfer of P from terrestrial to aquatic ecosystems
Source: Adapted from Sharpley *et al.,* (1995)

The first step in the transfer of TDP in runoff is desorption, dissolution, extraction, and mineralisation of P from soil, plant, stream bed, or aquatic biota. Sources of TPP in streams include eroding surface soil, stream banks, and re-suspension of bed sediments. The primary source of sediment in watersheds with a permanent vegetative cover, such as forest or pasture, is from stream bank erosion. This sediment will have characteristics similar to the sub-soils or parent material of the area, which are often P deficient (Sharpley *et al.*, 1995; Withers *et al.*, 2001).

During detachment and transport of sediment, the finer-sized fractions of source material are eroded preferentially, thus, the sorptivity and P content of suspended sediment is greater than that of the source soil on a mass basis. Phosphorus exports in eroded material from arable land are usually above 0.3 g P/m^2.a while leaching accounts for about 0.01 g P/m^2.a (Hillbricht-Ilkowska *et al.*, 1995). These authors also reported P leaching rates as high as 0.09 g P/m^2.a.

Riverine export of phosphorus

Several approaches have been used to estimate the riverine flux of materials into lakes, estuaries or oceans (Van Bennekom & Salomons, 1981; Schlesinger & Melack, 1981; Meybeck, 1982; Wollast, 1983; Meybeck, 1988; Melack, 1995). Phosphorous is naturally present in very low amounts in rivers and has been estimated at 55 mg P/m^3 total dissolved phosphorous (TDP) by Stumm (1973) who attributed 20 mg/m^3 to DP-PO_4, based on solubility products of phosphorus-bearing minerals, and an equal amount for organic phosphorus. TDP includes the organic form (DOP). Garrels *et al.*, (1973) adopted a value of 50 mg P/m^3 without giving more details. Total particulate phosphorous (TPP) represents, *viz.* 95% of P naturally carried by rivers, of which 40% is in organic forms (POP) (Meybeck, 1982). Stumm (1973) estimated additional phosphorus due to man's activities to 15 mg P/m^3.

When Stumm's global average dissolved P concentration of 55 mg/m^3 was multiplied by a total annual riverine discharge of 33 000 10^3 Mm^3/a, a value of 1.82 Mt P/a was obtained. Lerman *et al.*, (1975) computed a total P export to the oceans of 20 Mt P/a as the product of the combined rates of mechanical and chemical denudation of continents and mean P content of crustal material.

Richey (1983) offered up-dated values for dissolved (1.5 to 4 Mt P/a) and particulate (17 Mt P/a) riverine export of P to the oceans but did not document clearly the basis for these numbers. Richey (1983) also tabulated estimates of fluvial fluxes for ten geopolitical regions based on total discharge for each region and a range of P concentrations in the rivers of each region, but he did not include a summary of data for each region which would allow evaluation of the estimates.

The first well documented and systematic analysis of the concentrations and transport of P by rivers of the world was done by Meybeck (1982). Although he tried to use annual mean values, not all were discharge-weighted, and few data were available for several important tropical rivers (i.e. Amazon, Zaire, Orinoco and Niger). Concentrations of DP-PO_4 and TDP were obtained for 19 and 12 rivers, respectively, judged to be uncontaminated by human activities. The rivers with DP-PO_4 values had an annual discharge of 10 200 10^3 Mm^3/a and an average DP-PO_4 concentration of 10 mg/m^3. Based on a DP-PO_4 to TDP ratio of 0.4, the average TDP concentration was 25 mg/m^3. Concentrations of DP-PO_4 and TDP were available for an additional 29 and 23 rivers, respectively, judged to be contaminated by human activities (see Table 2.2 and 2.3 for a sample of some rivers considered to be polluted and unpolluted).

Table 2-2 Phosphorus concentrations in some major unpolluted rivers

River	P-PO$_4$ mg/m^3	TDP mg/m^3	Basin area (Tm^2)	Flow (Gm^3/a)	Population density persons per Mm^2	% urban
Temperate						
Gloma	6.8	20	0.04	19	12	73
Colombia	-	15	0.67	249	10	74
Sweden	10	-	0.25	117	-	-
Finland	-	55	0.23	67	-	-
Volga	11		1.35	260	50	66
Tropical						
Sumatra-Borneo	7	-	0.15	213	-	-
Niger	13		1.21	191	20	19
Zaire	24	60	3.82	1253	12	37
Zambezi	10	-	1.34	224	15	31
Solimoes	15	25	3.00	2400	-	-
Negro	6	8	0.76	1400	-	-
Amazon	12	20	6.30	5500	1	70
Orange	9.1	-	1.02	12	20	56

- = Missing data
M = Mega (10^6); G = Giga (10^9); T = Tera (10^{12})
Source: Meybeck (1982); Meybeck *et al.,* 1989; Caraco (1995); GEMS (2002)

Table 2-3 Phosphorus concentrations in some major contaminated rivers

River	DP-PO$_4$ mg/m^3	TDP mg/m^3	Basin area (Tm^2)	Flow (Gm^3/a)	Population density persons per Mm^2	% urban
Europe						
Rhine (The Netherlands)	217	465	0.19	69	300	86
Meuse (Belgium)	235	487	0.04	10	250	96
Scheldt (Belgium)	1240	1860	-	-	250	96
Danube (Romania)	15	50	0.81	201	90	52
Thames (Britain)	2475	-	0.02	3	400	92
Trent (Britain)	1335	-	-	-	-	-
North America						
Potomac (USA)	-	109	-	-	-	-
Mississippi (USA)	-	302	3.22	515	30	74
Missouri (USA)	59	-	-	-	-	-
South America						
Magdalena (Columbia)	64	-	0.24	238	30	67
Parana (Argentina)	67	-	2.80	504	10	67
Asia						
Java (Indonesia)	23	-	-	-	-	-
Chao Phrya (Thailand)	580	-	0.11	28	-	-
Huang He (China)	11	-	0.75	105	200	21
Murray-Darling (Australia)	140	-	1.073	11	2	86

- = Missing data
M = Mega (10^6); G = Giga (10^9); T = Tera (10^{12})
Source: Meybeck (1982); Meybeck *et al.,* (1989); Caraco (1995); GEMS (2002)

Martin & Meybeck (1979) provided numerous determinations of TPP, but few of POP, and they calculated an average of 1.15 mg P/g of suspended sediment. Particulate organic phosphorous (POP) has been even less studied than particulate organic nitrogen (PON) or particulate organic carbon (POC). According to a few studies the PON/POP ratio is around 2.5. The composition of the organic particulate matter in rivers has, therefore, been estimated close to $C_{22}N_{2.5}P_1$ by weight (Meybeck, 1982).

Total fluvial transport of P to the oceans was calculated by Meybeck (1982) using Baumgartner & Reichel's (1975) value of 37 400 Gm^3/a for runoff from exoreic basins and an average concentrations for the different P fractions. Meybeck, (1982) reported transport of TDP by natural rivers to be 1.0 Mt P/a. Meybeck's (1982) estimate for the dissolved P flux from rivers polluted by human activity doubled the figure for the TDP flux from all rivers to 2 Mt P/a. To determine transport of TPP, Meybeck, (1982) combined an estimate of the flux of total suspended material (i.e. 17 500 Mt/a) with that for P content/g of sediment (1.15 mg P/g) to obtain a value of 20 Mt P/a. Although Meybeck (1982) did not consider the influence of human disturbance on particulate transport, the assumed estimate of the flux of total suspended material was higher than the 14 500 to 15 500 Mt/a determined by Milliman & Meade (1983) for suspended sediments and bed load in disturbed and undisturbed rivers.

Other estimates of total P transport to the oceans range from 24 to 39 Mt P/a (Froelich *et al.,* 1982; GESAMP, 1987). Howarth *et al.,* (1995) assessed the estimates of Meybeck (1982) and GESAMP (1987) and suggested 22 Mt P/a as the best estimate for total riverine flux to the oceans. Howarth *et al.,* (1995) suspected that Meybeck's (1982) estimate for P flux from '*natural*' rivers may have been high and, if so, that the influence of human activity may have been under-estimated in the GESAMP (1987) analysis and instead suggested a P input into oceans of 8 to 9 Mt P/a prior to agricultural development and population increases. The error in Meybeck's (1982) analysis is attributed to a high estimate for the total suspended sediment load in relatively unpolluted rivers of 17 500 Mt/a, and possibly a higher value of 37 400 Gm^3/a for runoff. Howarth *et al.,* (1995) concluded that human activity has increased the P flux to the oceans by some 2.5 or 3-fold, from 8 or 9 Mt P/a to 22 Mt P/a.

However, all these estimates remain uncertain for several reasons (Meybeck, 1988). The lack of measurements of particulate P in a sufficiently large number of rivers with consistent techniques and with adequate sampling of the non-uniform distribution of particulates within a river is the largest sources of error (Melack, 1995). Very few details on computation methods and on basic data are given. Specific forms of P, particularly the organic and particulate ones, are generally not detailed (Meybeck, 1982).

Furthermore, impounding rivers reduces transport of P by increasing sedimentation of particulate P within reservoirs, but may also augment transport of P below dams by increasing erosion of channels, floodplains and deltas deprived of sediment-laden waters. The net result of impounds on P inputs to the world's oceans has not been quantified, but estimates for several large rivers and many reservoirs indicate the likelihood of a net decrease in P transport from land to oceans (Petts, 1984; Meade, 1988).

Atmospheric deposition

The global P cycle is unique among the cycles of the major biogeochemical elements in having no significant gaseous component (Schlesinger, 1997).

Wet deposition consists of the dissolved components in precipitation originating from two sources: i) components derived from cloud processes during nucleation and ii) washout or scavenging of aerosol particles and dissolution of gases in raindrops as they fall. Phosphorus content in precipitation is not commonly reported in literature, although wet deposition may contribute relatively large annual inputs of P into forest ecosystems, especially in regions affected by air pollution. The average P content of rains has been estimated with a set of unpolluted stations as around 5 mg P/m^3 for DP-PO4 and 10 mg P /m^3 for TDP (see Table 2.4) (Meybeck, 1982; Duce *et al.*, 1991).

Dry deposition is the gravitational sedimentation of particles during periods without precipitation. These particles include: aerosols, sea salts, particulate material, and adsorbed/reacted gases captured by vegetation. Impaction is the capture of particles moving horizontally in the air stream by the vegetation canopy. This filtering effect may accumulate large quantities of nutrients in some environments. Dry deposition values are also usually very low, ranging from 20 to 50 mg P/m^2.a (Graham & Duce, 1979, White, 1979; Duce, 1983 and Duce *et al.*, 1991).

Table 2-4 Phosphorus contents in atmospheric precipitation

Station	DP-PO$_4$ (mg/m^3)	TDP (mg/m^3)
Thonon (France)	9	18.5
HBEF (New Hampshire)	2.5	2.8
Rio Negro	1.3	9.2
Manaus (Amazon)	3	11
Ghana	22	-
Lake Maggiore (Italy)	8	14
Hausen (Switzerland)	51	87
Dubendorf (Switzerland)	1.9	14
World	5	10

Source: Meybeck (1982)

Key global fluxes and stocks of phosphorus

Previous evaluations of the global P cycle have identified the key fluxes and reservoirs, and have provided some estimates of their relative magnitudes (Stumm, 1973; Lerman *et al.*, 1975; Pierrou, 1976, Meybeck, 1982; Richey, 1983; Degens *et al.*, 1991; Melillo *et al.*, 2003). The use of phosphorus fertilisers in modern agriculture and the eutrophication of fresh waters from run-off and effluent discharge are the most visible results of human intervention in the P cycle.

Meybeck (1982, 1988) and Meybeck & Helmer (1989) estimates of global P fluxes and stocks are probably the most quoted for '*natural*', relatively unperturbed rivers and terrestrial ecosystems.

The knowledge of primary reservoirs and fluxes of global P-cycling are important in this dissertation as reference and input to Chapter 3, 4 and 5. Table 2.5 provides a summary of the major global reservoirs and fluxes of the phosphorus cycle. The data is primarily obtained from Richey, (1983) with some modifications and updates based on literature and independent calculations presented in this dissertation. The global P-cycle following from the data in Table 2.5 is shown in Figure 2.12.

Table 2-5 The major reservoirs and fluxes of the global phosphorus cycle[7]

Reservoirs	Value Mt P	Reference
Atmosphere		
Particulates over land	0.025	(1)
Particulates over oceans	0.003	(1)
Land		
Biota	2 600	(2)
Soil	96 000-160 000	(3)
Mineable rock	38 000	(4)
Fresh-water (dissolved)	90	(5)
Ocean		
Biota	50-120	(6)
Dissolved (inorganic)	80 000	(7)
Detritus (particulates)	250	(8)
Sediments	840 000 000	(9)

Fluxes	Value Mt P/a	
Atmosphere (land) to atmosphere (ocean)	1.0	(1)
Atmosphere (ocean) to atmosphere (land)	0.3	(1)
Atmosphere to land	3.2	(1)
Atmosphere to ocean	1.3	(1)
Land to atmosphere	4.2	(1)
Ocean to atmosphere	0.3	(1)
Marine dissolved to biota	600-1 000	(10)
Marine detritus to sediment	2-13	(11)
Terrestrial biota to soils	200	(12)
Mineable rock to soil	14	(13)
Fresh-water (dissolved) to oceans	2	(13)
Fresh-water (particulate) to oceans	20	(13)
Ocean to anthroposphere	0.2	(14)

(1) The atmospheric reservoir and fluxes are directly from Graham & Duce (1979), Duce (1983) and Duce *et al.*, (1991) who summarised extensive measurements of atmospheric P concentrations and deposition rates in marine and continental regions. The redox potential of most soils is too high to allow for the production of phosphine gas (PH_3) (Bartlett, 1986), except under specialised, local conditions (Devai *et al.*, 1988). The global flux of P in phosphine is probably <0.04 Mt/a (Gassmann & Glindemann, 1993).

(2) Estimates of the terrestrial biota were calculated from estimates of C mass of $0.45 \ 10^6$ Mt C (Bolin, 1979) $0.83 \ 10^6$ Mt C (Whittaker & Likens, 1975), and C/P atomic ratios of 500 (Stumm, 1973) to 822 (Deevey, 1970), though Pierrou (1976) used dry-weight biomass conversions. These yielded a most-likely value of 2 600 Mt P, with a range of 1 400 to 4 300 Mt P.

(3) Inorganic P in the soil has been computed from a total land area of 130 000 Gm^2 and a soil depth of 0.6 m (Lerman *et al.*, 1975) to 1.0 m (Pierrou, 1976), with a P content of 0.10 to 0.12 % (Taylor, 1964) and a soil density of 1 000 kg/m^3. Total soil P, including about 10% organic P (Bohn, 1976), is thus 90 000 to 160 000 Mt P, depending on soil depth.

(4) The amount of P in mineable rock, as P_2O_5, has been defined as that minimum amount which is economically recoverable (see section 2.4). The estimates of reservoir size and consumption rate are from Fantel *et al.*, (1985); Stowasser (1986); Steen (1998) and FAO (2001). In year 2000 the global annual rate of P-fertiliser consumption was estimated at 14 Mt P/a.

(5) Pierrou (1976) calculated the P in fresh-waters from a total volume of $720 \ 10^6 \ Mm^3$ and a mean concentration of 0.12 g P/m^3. The concentration value is probably uncertain by a factor of 2 (Richey, 1983).

(6) Phosphorus in the marine biota is calculated generally from applying the Redfield ratio[8] of C/P = 106 and carbon estimates of 2 000 Mt C (Williams, 1987) to 5 000 Mt C (Bolin, 1979).

---

[7] Some values possess considerable uncertainty. Examination of the underlying assumptions and sources of error for each term is crucial as given in the original reference.

(7) The previous estimates of dissolved inorganic P in the oceans have used mean concentrations of 0.08 to 0.10 g P/m^3 and mean depths of 3 000 to 3 500 m (Richey, 1983). GEOSECS data (Takahashi *et al.,* 1981) suggest a concentration of 0.062 g P/m^3.

(8) The inventory of marine detrital P can be calculated as the amount of particulate carbon (30 000 Mt C; Mopper & Degens, 1978) times a detrital C/P atomic ratio of 120 (Broecker, 1974), or 250 Mt P.

(9) Stumm (1973) and Lerman *et al.,* (1975) calculated the phosphorus in sediments from a geochemical mass balance.

(10) Estimates of the photosynthetic uptake by marine biota have been obtained by applying the Redfield ratio to productivity data, which range from 25 000 Mt C/a to 40 000 Mt C/a (De Vooys, 1979), or 600 to 1000 Mt P/a. The release of P by decomposition is assumed to be equal to photosynthetic uptake.

(11) Emery (1968) assumed that the P content of sediments is 0.092% and that 1 cm of solid sediment forms every 6000 years, for a rate of 13 Mt P/a. Assumptions of steady state, as calculated by the others (Froelich *et al.,* 1982; Kelderman, 1984; Mach *et al.,* 1987; Ruttenberg, 1990; Schlesinger, 1997), indicate that this figure might be high, and that a value of 2 Mt/a is more appropriate.

(12) The uptake and release of P by terrestrial biota has been estimated from productivity estimates of 30 000 Mt C/a (Bolin, 1979) to 50 000 Mt C/a (Whittaker & Likens, 1975) and the C/P atomic ratios of 500 to 822. A mean estimate of 200 Mt P/a results.

(13) The river run-off of P to the oceans includes both natural and human-influenced leaching and particulate erosion products, less that which is retained or consumed within the river. The most recent assessment is by Howarth *et al.,* (1995) and the suggested 22 Mt P/a is used as the best estimate for total riverine flux to the oceans (refer to main text of this section).

(14) Howarth *et al.,* (1995) estimate that 0.2 Mt P/a is harvested in fish catch and other seafood.

The concentration of DP-PO$_4$ in the surface oceans is low, but the large volume of the deep sea accounts for a substantial pool of P. Although the total dissolved reservoir in the oceans is about 80 000 Mt P, only about 50 Mt P are contained in marine biota. This is due in large part to much of the P being below the euphotic zone (Richey, 1983; Melillo *et al.,* 2003). The mean residence time for reactive P in the oceans, relative to the input in rivers or the loss of sediments, is about 25 000 years (Ruttenberg, 1993). Thus, each atom of P that enters the sea may complete 50 cycles between the surface and the deep ocean before it is lost to sediments (Schlesinger, 1997).

The uptake and release of phosphate by terrestrial plants is about 200 Mt P/a, while the equivalent marine flux is 1 000 Mt P/a. Given the lesser marine biomass, the turnover rate in the oceans is much greater than on land (Bolin & Cook, 1983; Schlesinger, 1997).

---

[8] Nutrient ratios are commonly used to infer nutrient limitation. One approach is to compare the molar ratios of dissolved inorganic nutrients to those in the biomass of phytoplankton, defined as the Redfield ratio = C:N:Si:P = 106:16:16:1 (Redfield, 1958).

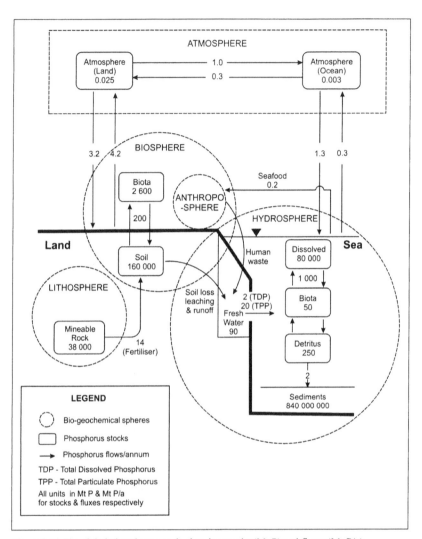

Figure 2-12 The global phosphorus cycle showing stocks (Mt P) and flows (Mt P/a).
Source: Original conception by Richey (1983), but with modified data and layout

The amount finally sedimenting (2 Mt P/a) in the oceans is only a small fraction of the production-mineralisation flux – roughly equivalent to the delivery of TDP to the oceans by rivers (Howarth *et al.*, 1995). On a time scale of hundredth of millions of years (10^8 years), these sediments are uplifted only as a result of tectonic movement of the Earth's crust and subject to rock weathering, completing the global cycle. Most of the P in rivers is derived from the weathering of sedimentary rocks, and it represents P that has made at least one complete journey thought the global cycle (Griffith *et al.*, 1977).

2.4 Global phosphorus reserves and applications

Reserves

Phosphorus belongs to the class of cycles called lithospheric cycles in which the principal and most important reservoir (*source*) is the lithosphere. Modern P fertilisers are almost entirely derived from geological deposits of apatite-containing phosphate rock, with small deposits from recent guano being exploited locally (Cook *et al.,* 1990). Estimates of the world reserves of phosphate rock - apatites and phosphorites – have been attempted by geologists and mineralogists but discoveries of new deposits are being frequently announced so that the estimates have to be amended from time to time. Despite the fact that forecasting the expected life of existing P deposits is an inexact science and depends on many factors, global P reserves are regarded as finite, with present recoverable reserves calculated to last for about 100 years (Barber, 1989; Runge-Metzger, 1995; Steen, 1998). Potassium (K) is assumed to last for about 300 years (IFA & UNEP, 1998; 2002). Production of nitrogen (N) fertilisers requires energy, as does the reduction of nitrogen in sewage treatment plants. Oil and gas, the most important energy resources for production of nitrogen fertilisers, have been calculated to last for about 40 and 60 years, respectively.

Estimating reserves is not easy for a number of reasons. Estimates are often based on different criteria; a country or company may consider its estimated reserves confidential and commercially sensitive; possible future changes in technology and costs of production are very difficult to forecast; and there is no certainty about future rates of consumption. It is therefore difficult to say just how long the world's phosphate supply might last (Steen, 1998; Johnston & Steen, 2000).

Many reserve[9] or resource[10] estimates are subjective as they depend on standards and criteria assumed by the data provider in determining the circumstances that might render a deposit economically useful. Hence, it is to be noted that discrepancies exist. For example, the United States Geological Survey (USGS) has defined *reserves* as those exploitable at a cost below $ 35 per metric tonne and the *reserve base* as deposits that can be processed at a cost below $ 100 per metric tonne (Fantel *et al.,* 1985; Stowasser, 1986).

In 1986 global phosphate rock *reserves* were estimated to be about 15 000 Mt i.e. at a cost of less than US$ 35 per metric tonne (see Table 2.6). The global *reserve base* was estimated at about 38 000 Mt (at costs ranging from less than US$ 18 per metric tonne to US$ 100 per metric tonne). Table 2.6 provides a regional breakdown of rock phosphate *reserves* and *reserve base* as estimated in the mid 1980's.

---

[9] A *reserve* is usually defined as a mineral deposit of established extension that is – or could be - profitably mined under prevailing costs, market prices and technology.
[10] A *resource* is considered to be a deposit of less well defined size which is not now economically exploitable but which could potentially become so, if there was a sufficiently favourable change in costs, prices or technology.

Table 2-6 World phosphate rock reserves and reserve base

Region and major producing countries	Reserves[a]	Reserve base[b]
	(Mt)	
North and Central America		
United States	1 543	5 951
Canada		44
Mexico		12
Total	1 543	6 127
South America		
Brazil	44	386
Colombia		110
Peru		154
Total	44	650
Europe		
Former U.S.S.R	1 433	1 433
Other		154
Total	1 433	1 587
Africa		
Algeria		276
Egypt		871
Morocco	7 604	22 040
Western Sahara	937	937
South Africa	2 865	2 865
Other	264	231
Total	11 670	27 220
Asia		
China	231	231
Jordan	132	562
Syria		198
Other	55	485
Total	418	1 476
Oceania		
Australia		551
Nauru	11	11
Total	11	562
World total	15 119	37 594

a) Cost less than US$ 35 per metric tonne. Cost includes capital, operating expenses, etc. and a 15% rate of return on investment. Costs and resources are as of January 1983.
b) Cost less than US$ 100 per metric tonne.
Source: Adapted from Stowasser (1986); Steen (1998)

Although the world has potentially vast *inferred* and *hypothetical* resources (15 000 Mt and 38 000 Mt of phosphate rock respectively[11]), comprehensive economic production thresholds for these resources have not been calculated. The amount of phosphate rock reserves and resources in individual countries are rather uncertain as this type of data is often privileged information and documentation concerning many deposits is simply not available. Furthermore, there is generally a lack of information concerning the extent of exploration and criteria used to determine the economics of production (Sheldon, 1987; Steen, 1998). Other deposits probably would likely be discovered. Deep phosphate rock deposits might also hold promise if economically acceptable means for recovering them without excessive surface disturbance can be developed (IFA & UNEP, 1998). Another alternative is to '*mine*' sea water for its phosphorus and other minerals (Sheldon, 1987).

---

[11] Sheldon (1987) estimates 112 000 Mt as the actual and inferred reserve base.

Estimates of current P reserves that can be exploited vary from 250 years to as little as 100 years. If known potential reserves are taken into account, the forecast may be as long as 600 to 1000 years (Johnston & Steen, 2000). The traditional approach to forecasting phosphate fertiliser demand has been to use a linear model. However, most forecasters often restrict their assumptions to the next five to ten years and this provides little help for calculating the approximate lifetime of phosphate reserves. The most conservative and simplistic approach for assessing the lifetime of phosphate reserves is to calculate how long the reserves will last based on present consumption. This provides a basis for simple comparisons but does not say very much about anticipations of future development. The reason for these approaches is the difficulty in predicting demand that depends largely on the development of market economy and political circumstances (Steen, 1998; UNIDO, 1998).

The current fertiliser industry forecast on the development of global phosphate consumption suggests an annual increase of approximately 3% until around 2010 (see next section for more detail on fertiliser production and consumption). Some fertiliser manufacturers have extended their forecast up to 2015, suggesting an increase in phosphate fertiliser consumption of 2.8% per year before it begins to level off. The view that crop yields might increase by some 2-2.5% per year matches the long-term historical cereal yield trends. With this view, it is assumed that agricultural production will continue to keep pace with, or slightly exceed, the global population growth and hence the world per person crop production would remain stable or slightly increase.

This would lead to a first estimate of a 2.5% annual growth in phosphate consumption over the long term. This would, in turn imply an additional consumption of 10 Mt of P_2O_5 between the years up to 2010, an additional 13 Mt between 2011-2020, and close to 20 Mt between 2021-2030. In total, this would amount to an annual consumption of around 100 Mt of P_2O_5 in 2050. This is more than twice the current consumption in global agriculture and would equal an average supply of some 7.0 g P_2O_5/m^2.a which certainly would be regarded as an oversupply of phosphate (Steen, 1998; Johnston & Steen, 2000).

A workable assumption is that Western agriculture, i.e. in the developed world, would not need to supply more phosphate than that removed by the harvested crop. In approximately 50% of the remaining agriculture it would be feasible to improve the soil phosphate status, and in the other 50% slightly more than replacement would be a relevant policy (IFA & UNEP, 1998). Furthermore, it could be assumed that the crop uptake efficiency of phosphate fertilisers could improve in the future due to improved farming technologies, nutrient management and plant breeding.

Using these assumptions, the phosphate fertiliser consumption in developed agriculture could stabilise at its present level of around 2.0 to 2.5 g P_2O_5/m^2.a on average with a possible slow increase starting after 2010. Further, recycling of nutrients will improve in the developed world, somewhat reducing the need to add mineral phosphates. In those developing countries where it is necessary to improve soil fertility, it could be necessary to supply some 30-50% more phosphate than crop requirements for a period of perhaps 30-50 years. After this, it would be necessary to maintain the phosphate status of the soil as in countries where this is done currently. In other countries, where the soil phosphate status should be maintained, it may be necessary to add 10-30% more phosphate than is removed by the harvested crop. This scenario gives a second more realistic estimate for annual P_2O_5 consumption in 2050 of around 70 Mt (Steen, 1998). The development of world agricultural phosphate consumption until 2050, according to the above scenarios, is summarised in Figure 2.13a.

By applying the phosphate consumption growth rates estimated in the phosphate use in agriculture, world fertiliser use would reach about 60-70 Mt P_2O_5/a by 2050. It is concluded that global phosphate resources extend, for all intents and purposes, well into the future, but that depletion of current economically exploitable reserves can be estimated at somewhere from 60 to 130 years (Steen, 1998). In essence, using the median reserves estimates and under reasonable predictions, it appears that phosphate reserves would last for at least 100 years (see Figure 2.13b).

There might appear therefore that there is no supply crisis of P at present, but there are a number of issues of concern besides P being a limited resource. Phosphate rock processing yields enormous quantities of waste (rucksack) [12] that are potentially recoverable, but which are virtually un-recovered at present (Ayres & Ayres, 1996). This result in unnecessary destruction of land [13], both for the disposal of phosphate process waste and for the mining of other materials (gypsum and fluorites) that could easily be extracted from these wastes. A further problem of concern is that mining waste like phospho-gypsum (five tonnes per tonne of phosphorus pentoxide produced) contains trace quantities of cadmium, a very toxic metal, which is not removed by the fertiliser manufacturing process and consequently ends up in agricultural soils where it can enter the human food chain. In addition phospho-gypsum contain traces of radioactive radium, uranium and thorium, trace metals arsenic, lead, mercury, nickel, vanadium chromium, copper and zinc (Mortvedt & Beaton, 1995; Ayres & Ayres, 1996; IFA & UNEP, 1998; Steen, 1998).

---

[12] Phosphoric acid (H_3PO_4) is the end product of phosphate rock processing. For example the largest producer of crude phosphate rock in the USA in 1992 mined about 155 Mt of rock. This was beneficiated by washing and floatation to yield 47 Mt of marketable rock. This material contained about 14 Mt of phosphorus pentoxide (P_2O_5). The phosphate rock was reduced by the so-called 'wet process' – essentially digestion by sulphuric acid to fertiliser grade phosphoric acid with 11 Mt of phosphoric acid content. It is worth noting that approximately 0.002 m^3 of water is used in the process for each kg of phosphate rock processed (Ayres & Ayres, 1996).

[13] In mining activities, the land surface and subsurface is disturbed by such activities as the extraction of ore, the deposition of overburden, the disposal of beneficiation wastes and the subsidence of the surface.

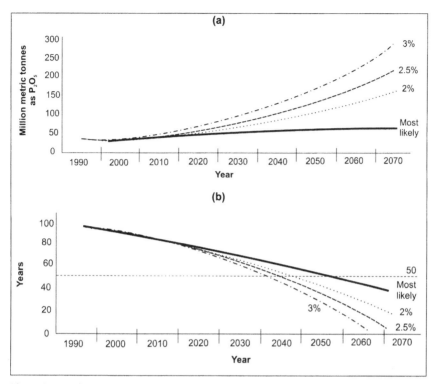

Figure 2-13 Relationship between various scenarios for P_2O_5 consumption (a) and the lifetime of reserves (b) assuming 2%, 2.5%, 3% and the most likely increase in P_2O_5.
Source: Modified from Steen (1997); Johnston & Steen (2000)

Of immediate concern is the mobilisation of P from particulate (stable form) to soluble (unstable forms) and subsequent diversion, as it were, from the terrestrial into the aquatic environment. This is partly due to excessive fertilisation of croplands and manure production with the net result, in macroscopic terms, being a transfer of P from concentrated point (land) *sources* to dispersed (water and land) *sinks* (Beck *et al.*, 1994; IFA & UNEP, 1998; Isherwood, 2000).

In year 2000, the global production of phosphate was around some 40 Mt/a of phosphorus pentoxide[14] (P_2O_5), derived from roughly 140 Mt of marketable rock concentrate. Around 80% of phosphates produced by the world's industry were used in fertilisers, with a further 5% being used to supplement animal feeds, 12% in synthetic detergent manufacture and about 3% of the total consumption used in diverse applications such as metal surface treatment, corrosion inhibition, flame retardant, water treatment and ceramic production (Harben and Kuzvart, 1996 CEEP, 1997; 1998; Steen, 1998; FAO, 2001). The three major producing countries USA, China and Morocco, produced approximately two thirds of global phosphate production. Morocco alone accounted for one third of the international trade in phosphate rock (see Table 2.7).

---

[14] Phosphorus pentoxide dissolved in water is phosphoric acid, the active ingredient in most phosphate fertilisers (e.g. 'super-phosphates'). It is not used, generally, in pure form. 1.0 kg as P is equivalent to 2.2919 kg P_2O_5. This conversion factor is used throughout this dissertation.

Table 2-7 Major phosphate rock producing countries in year 2000

Country	% of total
United States	30
Morocco	17
China	14
Former Soviet Union	8
Israel and Jordan	7
Tunisia	6
Brazil	3
South Africa	2
Other	13

Year 2000 phosphate rock production was approximately 140 Mt/a
Source: Modified from Johnston & Steen (2000)

Today there is a range of phosphatic fertilisers available to farmers (see Table 2.8). Some contain only phosphorus, others contain two or more nutrients. Manufacturers often produce a variety of such fertilisers which contain nitrogen, phosphorus and potassium in various proportions (usually called blends or compound fertilisers). The proportions are adjusted to meet the needs of a specific crop and to allow for the level of plant available nutrients in the soil. The average world prices of di-ammonium phosphate (DAP) have declined since the mid 1990's with current prices averaging approximately US$ 0.20/kg as illustrated in Figure 2.14.

Table 2-8 Some characteristics of common soluble phosphate fertilisers

Fertiliser and acronym	Primary P compounds	Nominal grade P_2O_5 content (%)
Mono-ammonium phosphate (MAP)	$NH_4H_2PO_4$	52
Di-ammonium phosphate (DAP)	$(NH_4)_2HPO_4$	46
Triple super-phosphate (TSP)	$Ca(H_2PO_4)_2.H_2O$	46
Single super-phosphate (SSP)	$Ca(H_2PO_4)_2.H_2O$	18-20
Nitric phosphate (NP)	$CaHPO_4, NH_4H_2PO_4\ Ca(NO_3)_2$	20
Urea-ammonium phosphate (UAP)	$CO(NH_2)_2, (NH_4)_2HPO_4, NH_4H_2PO_4$	28
Ammonium polyphosphate (APP)	$(NH_4)_3HP_2O_7.H_2O$	34

Source: Hedley *et al.*, (1995); Johnston & Steen (2000)

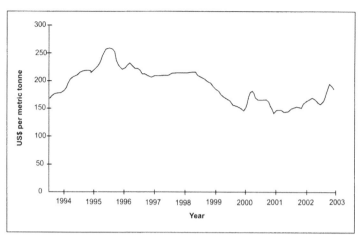

Figure 2-14 Average global prices of di-ammonium phosphate (DAP) from 1994 to 2003
Data source: IFDC (2003)

About 12% of the global phosphoric acid produced is used to manufacture sodium tripolyphosphate, or STTP ($Na_5P_3O_{10}$), a soap[15] and detergent[16] builder. Production of these chemicals has been declining since at least 1980. This appears to be due to the introduction of a competing type of detergent builder, namely zeolites together with polycarboxylic acid. These have virtually taken over the market in Germany, Italy, Switzerland and the Netherlands, because of concerns about the eutrophication of lakes, rivers and streams.

Phosphate soaps, detergents and dishwashers[17] have also lost some market share in the UK from 67% in 1991 to 54% in 1993. Declining use of phosphate also occurred in France, Belgium, Sweden and Denmark. However the so-called 'compact detergents', introduced in 1988, may have actually reversed the trend (Ayres & Ayres, 1996). Whereas traditional detergents contained 20-25% STTP, the compact detergents increase this to 50% (see Table 2.9).

---

[15] Soaps are alkali salts of long-chain fatty acids. They emulsify soils, microbes and liquids and detach them from surfaces, and act as a surface action agent (surfactant) to reduce the surface tension of water, making it wetter.
[16] Soaps are alkali salts of long-chain fatty acids. They emulsify soils, microbes and liquids and detach them from surfaces, and act as a surface action agent (surfactant) to reduce the surface tension of water, making it wetter.
[17] Dishwashers are similar to detergents and differ in the use of non-ionic surfactants for low foam formation. They also contain fillers, anti-cracking agents and chlorine generating agents.

Table 2-9 Composition of detergents

Components	Standard detergents		Compact detergents	
	With phosphates (%)	Without phosphates (%)	With phosphates (%)	Without phosphates (%)
Sodium tripolyphosphate STTP	20 to 25	0	50	0
Zeolite	0	25	0	20 to 30
Polycaboxylates	0	4	0	5
Organic phosphonates	0 to 0.2	0.4	0	0.2
Sodium silicate	6	4	5	4
Sodium carbonate	5	15	4	15 to 20
Surface active agent	12	15	14	15
Sodium perborate	14[a]	18	10[b]	13
Activator	0 to 2	2.5	3	5
Sodium sulphate	21 to 24	9	4	5
Enzymes	0.3	0.5	0.8	0.8
Anti-redeposition agent	0.2	0.2	0.3	0.3
Perfume	0.2	0.2	0.2	0.2
Water	10	5	8	5
Total	100	100	100	100

    a)   Tetrahydrate perborate is used in standard detergents.
    b)   Monohydrate perborate is used in compact powders. Its power is much higher than
        Tetrahydrate perborate.
Source: Ayres & Ayres (1996)

Linkages between fertiliser production and food security

The increase in life expectancy, reduced child mortality and improved farming methods, which have led to increased food production, have resulted in a rapid and exponential world population growth over the last 150 years, from 1 billion to just over 6 billion in year 2000. World population is currently growing by approximately 1.5% per annum, 80 to 85 million/a, or 250 000 people/day (UNDP, 2001, 2003). This trend will not continue indefinitely. The latest United Nations World Population Projections to 2050 suggest that a slowing down of population growth may be already occurring with a median projection of 9.4 billion by 2050 (UNDP, 2001, 2003). The population growth is expected to be concentrated in the developing regions of the world, mainly Africa and Asia, while in the developed countries growth will be very slow.

The combined effects of rising population and wealth will inevitably increase the demand for higher dietary standards and higher-grade foodstuffs. Consequently, the portion of meat and possibly also of dairy products in the diet will increase. But, even if a higher dietary standard could be afforded it is not likely that there will be a total change to Western-like diets because many communities are likely to retain their traditional cooking to a great extent (Steen, 1998). As demand for food increases, this may result in bringing into agricultural use more land, but certainly will bring a requirement for increased yields, thus increasing fertiliser demand. Hence, agricultural phosphate use may increase faster than world population.

The modern fertiliser industry is little more than 150 years old and was first envisaged by the German chemist J. von Liebig. He set down his principles of plant nutrition and production in "*Chemistry in its application to agriculture and physiology*" (Liebig, 1840) where he stressed the value of mineral elements derived from the soil in plant nutrition and the necessity of replacing those elements to maintain soil fertility (Johnston & Steen, 2000).

Most of the increase in food production in the second half of the last century has come from existing land through yield increases made possible by the seed-, water- and fertiliser-based 'Green Revolution'. The International Food Policy Research Institute (IFPRI) recently estimated that the world's farmers will have to produce some 36% more grain in 2020, compared with 1997 (IFPRI, 1997, 2001). However, the expansion of the cereal area is unlikely to exceed 5%, since almost two-thirds of this growth is expected to occur in the poor regions of sub-Saharan Africa. Inevitably, most of the higher production must come from higher yields-per-unit area, which will require a correspondingly larger quantity of plant nutrients. Generally the 'nutrient budget' is negative. Nutrients in soils are mined, leading to soil degradation and erosion (Stoorvogel *et al.,* 1993). In southern Mali, for example, about 60% of farmers' income is, in effect, derived from stripping the soil of its nutrients (IFA & UNEP, 2002).

Discussions about mineral fertilisers and sustainable development must be based on a clear awareness of the differences between developed and developing countries. Developed countries are mature markets, where food security is assured and overproduction (over-application of synthetic fertilisers) not unheard of. Major soil fertility problems have been solved and environmental concerns have become as important politically as food supply, if not more so (Conway & Pretty 1991; Bumb & Baanante, 1996; NRC 1989; Gruhn *et al.,* 2000). Developing countries on the other hand continue to face food insecurity and serious problems of soil degradation exacerbated by nutrient mining. As a result, in many developing countries, fertiliser supply is as politically sensitive as food supply itself (Stoorvogel & Smaling 1990; Loftas, 1995; Tandon 1998; FAO, 1999b; Smil, 1991; 1999, 2001; IFA & UNEP, 2002,)

Agricultural conditions are also very different. Most developed countries lie in temperate zones where soils retain organic matter for long periods, and many developing countries are in the tropics where soil organic matter breaks down as much as four times faster (ITC & FAO, 1995; Smil, 1999; 2001). In addition many developing countries face harsh climatic conditions, population pressure, land constraints, and the decline of traditional soil management practices (Eicher & Baker; 1982; Gruhn *et al.,* 2000).

Restoring, maintaining and preferably increasing soil productivity are essential for sustainable food production. A proper and a balanced nutrient supply is important for maintaining soil productivity and reducing soil degradation. Fertilisers can have a positive effect on the environment in several ways (Laegreid *et al.,* 1999; Gruhn *et al.,* 2000).

Hedley *et al.,* (1995) outlined the key points of global P fertility management. Soil P deficiency has been identified as a major constraint preventing upland soils being used to produce food for the growing populations of developing countries in tropical and subtropical regions (IRRI, 1980; Grant, 1981; FAO, 1987a; Stangel & Von Uexkull, 1990; Smaling, 1993; FAO, 1999b). Overcoming this deficiency will involve strategies to (a) maximise yields on current arable land and (b) bring additional areas of adverse P deficient soils into production (Stangel & Von Uexkull, 1990; Von Uexkull & Mutert, 1995) and stop urban sprawl over agricultural land.

Many small scale farmers in developing countries cannot afford to pay for expensive manufactured fertilisers and technology developed for their efficient use (FAO, 1986a; 1986b; 2000a). Fertilisers must either be supplied as aid or low cost alternatives need to be developed and promoted (Hedley *et al.,* 1995; Runge-Metzger, 1995; Sanchez *et al.,* 1997). The increased urbanisation of populations creates greater dependence upon mineral fertilisers as a greater proportion of wastes from food distribution and consumption is generated remote from farms.

To design appropriate strategies to address these issues requires quantitative information on soil P status, crop responsiveness to P amendments, and agricultural practices that alter soil P status (Walker & Syers, 1976; Hedley *et al.,* 1995).

Global fertiliser use soared from less than 14 Mt/a in 1950 to 136 Mt/a in 2000 reaching peak consumption in 1988 of 146 Mt/a. Consumption is now stable or declining in the industrialized countries but demand is still rising in the developing world and by 2020 it is estimated to reach 250 nutrient Mt/a (IFA & UNEP, 2002). The major driving force is increasing food production, driven in turn by increasing human population and the growing demand for livestock products, particularly in developing countries (Alexandratos, 1988, 1995; Pepper *et al.,* 1992; Alcamo, 1994). There is no substitute for phosphorus in agricultural uses; although, its use in detergents has been reduced by the substitution of other compounds. Figure 2.15 shows the breakdown of global fertiliser consumption in terms of the macro-nutrients N[18] (Nitrogenous fertiliser as N), P (Phosphate based fertiliser as P_2O_5) and K (Potash fertiliser as K_2O) as from 1961. It is apparent also in Figure 2.15 that the ratio of N/P as fertiliser applied has increased substantially over the past 40 years. This has largely been dictated by technological advances in production of nitrogenous fertiliser.

Table 2.10 indicates that wheat is the largest user of fertilisers world-wide and that cereals use approximately 55% of the world's fertiliser. The next largest users are oilseeds and pasture. The estimates are believed to be reliable and should be used to reflect the general magnitude of usage by a crop rather than an exact measurement (IFA & UNEP, 1998).

---

[18] The content of nitrogen, phosphorus pentoxide (P_2O_5) and potassium oxide (K_2O) in fertiliser forms the main basis of its commercial value. It may also contain other macro-nutrients such as calcium, sulphur, and magnesium, as well as micro-nutrients like boron, iron, manganese and zinc, which all affect its value.

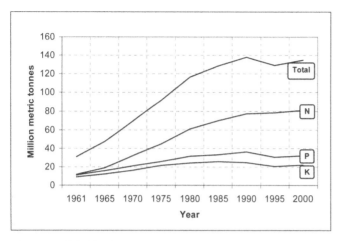

Figure 2-15 Global fertiliser consumption from 1961 to 2000 in Mt/a
Data source: FAO (2001)

Table 2-10 Estimated world fertiliser usage by crop grouping

Crop group	% of world usage
Wheat	20
Maize	14
Rice	13
Barley	4
All other cereals [a]	4
Oilseeds [b]	12
Roots and tubers	6
Fruits and vegetables	5
Sugar	4
Fibres	4
Other crops [c]	3
Pasture [d]	11
Total	100

a) Oats, milo, millet, rye, triticale and teff
b) Primarily potatoes
c) Includes cocoa, coffee, tea, tobacco and pulses
d) Includes grassland, hay, fodder, silage, etc.
Source: IFA & UNEP (1998)

The increasing use of mineral fertilisers in developing countries started in the 1960s and by 2000, accounted for about 60% of the world total, compared with 12% in 1960 (see Figure 2.16). This trend is continuing. As populations increase in developing countries, agriculture production and the development of fertiliser use is a high priority – and rising on both national and international political agendas.

Fertiliser use is uneven: in some places it is overused, in others too little is applied, and in still others they are optimally used. These differences occur within and across countries and even seemingly 'optimal' total applications may hide imbalances between individual nutrients (IFDC, 1992; Park, 2001).

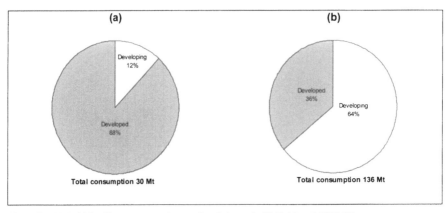

Figure 2-16 World fertiliser consumption, regional shares in 1960 (a) and 2000 (b)
Data source: IFA & UNEP (2002)

P remains in short supply over large parts of the globe, where economic and political constraints mean that naturally low P availabilities are only insufficiently supplemented by imported fertilisers. Figure 2.17 indicates that during the five year period of 1985 to 1989 a large part of the continents applied less than 0.5 g P/m^2.a on its cropped arable land. Globally fertiliser is also applied more liberally than necessary for plant growth, usually to ensure that crops are not underfed. Although some of the excess builds up in farmland in the short run, it is estimated that 20-50% of the phosphorus applied to the soil is lost through, and much more lost through soil erosion, surface runoff and leaching thereby polluting waterways and degrading ecosystems (Runge-Mertzger, 1995).

Figure 2-17 Use of mineral phosphate fertiliser (5 year average) 1985 to 1989 in g/m^2.a as P
Source: Adapted from Runge-Mertzger (1995)

2.5 Impacts of open-ended phosphorus flows

Increased fertiliser use

In the mine-to-crop continuum, it is the application of fertilisers which represents the biggest potential environmental effect. The challenge is to retain the benefits of fertilisers while minimising adverse consequences (Maene, 2000). Insufficient and excess fertiliser application as well as unbalanced use can have unwanted environmental impacts. Proper use is made more challenging by the diversity of site specific factors which influence local requirements (IFA & UNEP, 2002).

Apart from crop exports, large P losses occur with soil erosion. Erosion losses are particularly severe in the tropical highlands, but even moderate annual P losses from erosion can carry large total economic costs (Milliman & Meade, 1983; FAO, 1986; Stocking, 1986). For Zimbabwe, Stocking (1986) estimates a maximum P loss from commercial farms of 0.8 g/m^2.a which amounts to an economic loss of US$ 690 million/a (average value of P-loss through soil erosion recorded was 0.054 g/m^2.a). In the export-oriented regions, substantial losses are reported from New Zealand (Ward *et al.*, 1990).

In Europe, the integration of P from organic sources (plant, animal and human manures) into plant nutrition management, although desirable from an ecological point of view, is still financially not attractive to farmers as costs of collection, storage, treatment and timely distribution are high (see Box 1.B, Chapter 1).

Between 1950 and 2000 it is estimated that about 800 Mt of fertiliser as P were applied to the Earth's surface, primarily on croplands (see Figure 2.15) (Carpenter *et al.*, 1998; FAO, 2001; IFA & UNEP, 2002). During the same time period, roughly 300 Mt of P were removed from croplands in the form of harvested crops (FAO, 2001). This is calculated from the cumulative growth of arable land (see Appendix A) and the average P removal from cropland through the harvested portion of different crops (approximately 1-3 g/m^2) (see Appendix B).

Some of this produce was fed to livestock and a portion of the manure from these animals was reapplied to croplands, returning some of the harvested P (about 50 Mt) to the soil (refer to Box 1.B in Chapter 1). In Europe, intensive animal production on farms with little land produces manure in excess of the nutrient requirements of crop and pasture lands (Sibbesen & Runge-Metzger, 1995). The excess P from the animal manure, if it is not directly dumped into waste waters, accumulates in the surface soils which lead to increased leaching and erosion losses of P to aquatic environments. The excess P has either remained in soils or exported to surface waters by erosion or leaching. The majority of applied P remains on croplands, with only 20 to 50% leaving by export to surface or ground waters (150 Mt). It is likely, therefore, that about 400 Mt of P has accumulated in the worlds croplands (Gumbo *et al.*, 2002a).

The standing stock of P in the upper 0.10 m of soil in the worlds croplands[19] is roughly 1 300 Mt (Carpenter *et al.*, 1998). That means between 1950 and 2000 a net

---

[19] Global average P content of soil is 0.10% per kg (1 g/kg as P), and a soil density of 1 000 kg/m^3 (White, 1979; Johnston and Steen, 2000).

addition of about 400 Mt occurred which translates to an increase of the P content of agricultural soils of about 30% (Schlesinger, 1997; Carpenter *et al.,* 1998). P-based detergent manufacturing has also increased since 1950 and an estimated total of 80 Mt (refer to previous section on P uses) has been used by households and industry (Pretorius, 1983; Heynike & Wiechers, 1984; Ayres & Ayres, 1996). The estimated P fraction which passes through the household system in the form of food and detergent products has amounted to about 330 Mt. Assuming that only a small fraction of the P in food ingested by human beings is absorbed and assimilated in human tissue and bones, then almost 100% of the P reaching the households has been disposed of either via the sewage or solid waste fluxes (Gumbo *et al.*, 2002a). Figure 2.20 illustrates that close to a half of the mined P (500 Mt) since 1950 has found its way into the aquatic environment (oceans and fresh water lakes) or buried in sanitary landfills (*sinks*).

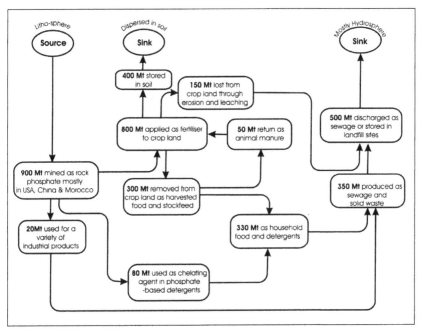

Figure 2-18 The source to sink of P with respect to the global economy and resource use since 1950 (Mt as P_2O_5)
Source: Gumbo *et al.*, (2002a)

The P status of the world's major rivers (refer to section 2.3) reflects land use in their drainage basins (Caraco, 1995), but the intricate relationships between land use patterns, surface and subsurface transport of P, and aquatic P loads are only now being explored in detail (Hillbricht-Ilkowska *et al.,* 1995; Sharpley *et al.,* 1995; Tiessen, 1995).

Many agricultural wastes are utilised as fertiliser because of their proximity to the farm. Much improvement, however, can be made in the way that municipal and industrial wastes are utilised. Most wastes have inherent nutrient value and if processed correctly can be used to increase crop yields. Sewage sludge is a major P source but unfortunately frequently is contaminated with heavy metals from industrial sewage (Kofoed *et al.,* 1986; Ekvall, 1995; Van der Ryn, 1995). Strict regulations are required to ensure its safe use (Kirkham, 1983; Hall & Williams, 1983). Large volume industrial wastes important for their P content and generally free from heavy metal contamination are wastes from the sugar, palm oil, rubber (Hedley *et al.,* 1990) and milk (Gregg & Currie, 1992) processing industries. These can be important P sources in tropical and subtropical P-deficient soils.

Once arrested in a well-aerated soil, P-bearing materials are known to be rarely leached into underlying ground waters (Tunney *et al.,* 1997). On this basis, land systems would appear to be the preferred destination of the P fluxes passing through a city, notwithstanding the fact that this is still a distortion of '*natural*' state of affairs (given the mined origins of this particular flux). This obviously has the potential benefit of substituting other forms of agricultural P inputs. It is also eminently sensible to convert the urban flux of soluble P back into particulate form during wastewater treatment, rather than attempting to manipulate the deposits in river and lake sediments (Beck *et al.,* 1994).

2.6 Un-balanced world trade

Soil P-mining

Increases in productivity or exports of commodities require external nutrient inputs if they are not to cause a decline in fertility. External P inputs have become available on a large scale with the mining of phosphate deposits and the wide-spread availability of the commodity 'phosphate'. This has decoupled patterns of supply, consumption and waste production from natural nutrient cycles, and has made them dependent on economics (Tiessen, 1995; Beaton *et al.,* 1995).

Poverty associated with low investments in soil conservation is regarded as one of the major reasons for the widespread phenomenon of soil mining. Under such circumstances, myopic behaviour is rational from the individual point of view, but unfortunately results in the socially undesired consequence that natural resources will be exploited (Larson & Bromley, 1990; Cropper, 1988; Perrings, 1989).

Inter-continental trade of P-fertilisers, raw materials and intermediates plays a major role in the movement and redistribution of P. Also important, but often overlooked, is the transfer of P in agricultural products that are exported and imported from one region to another and used as food or for industrial purposes. This movement of P has important consequences for soil fertility and productivity in both the exporting and importing regions (Beaton *et al.,* 1995). The P contained in agricultural commodities exported from a region, country or farms are completely lost from the local P cycle. These losses may be partially offset by P imported in food used by man and animals, and by fertiliser. The net gain or loss of P from the agricultural system is influenced by the disposal practices used for human and animal wastes, and the amount of P present in non-food products such as rubber, tobacco and fibre.

In a global perspective P is not lost from the system as discussed in section 2.3, it is simply transferred from one region or medium to another. The purpose of this section is to examine the movement of P in fertilisers and various agricultural commodities throughout the world, and to identify areas of net loss and gain through regional P-budgets. Year 2000 was used as a base year. The methodology used was based on similar calculations conducted by Beaton *et al.,* (1995).

Regional phosphorus transfers

Regions identified and used included: Africa, Asia, Central America and Caribbean, China, Europe, North America, Oceania, and South America (see Appendix A for the country listing and regions). These eight geographical areas and the P-budgets calculated were based on the online FAO statistics for production, trade and commerce, food balances, fertilisers and land use (FAO, 2001). The selection of the regions was therefore influenced by the data structures in the FAO online database FAOSTAT.

Vegetal products or commodities were divided into 12 categories:
- Cereals - (excluding beer, wheat, barley, maize, rice, rye, oats, millet, sorghum and other cereals)
- Starchy roots - (potatoes, cassava, sweet potatoes, yams and other roots)
- Sugar-crops and sweeteners- (sugar cane, sugar beet and other sweeteners)
- Pulses – (beans, peas, soyabeans and other pulses)
- Oil-crops and vegetable oils – (tree-nuts, ground nuts, sunflower seed, rape and mustard seed, cotton seed, linseed, palm kernels, sesame seed, oilseed cake, olive oil and other oils)
- Vegetables – (tomatoes, cabbages, onions and other vegetables)
- Fruits - (bananas, apples, grapes, raisins, pears, peaches, pineapple, canned pineapple, dates, oranges, lemons, and other citrus, excluding wine)
- Stimulants - (coffee and tea, cocoa products, cocoa paste, chocolate, cocoa beans, cocoa powder)
- Spices – (peppers, pimentos, cloves and other spices)
- Beverages - (alcoholic, fermented, beer, wine, alcohol non-food)
- Fibre crops (cotton lint, jute, sisal, flax fibre, and silk)
- Tobacco – (tobacco leaf and other tobacco products)

Similarly animal products were grouped into 6 general categories. These included:
- Meat – (beef, pork, mutton, goat meat, poultry meat, fresh or frozen, dried and prepared meat meal)
- Offal – (liver, heart, stomach, and other meat products)
- Fats – (animal fats, fish oil, fish liver oil)
- Milk (fresh, dried, and concentrated, excludes butter)
- Eggs - (fresh, preserved, liquid, and dried)
- Seafood – (fish, fresh water, demersal, pelagic, marine, crustaceans, cephalopods, molluscs, and other aquatic products including plants).

The P content of each vegetal and animal product group was calculated using average P concentrations within each group (Tables B.1 and B.2 in Appendix B). Net P balances for vegetal products, animal products and fertiliser material were calculated as the difference of total P in imports minus total P in exports. A negative net P balance indicates a loss of P from the region (i.e. export of P exceeds import of P) while a positive balance indicates a gain of P. Appendix A (Tables A.3 to A.11) provides details of the regional trade in vegetal and animal, food and non-food products for the year 2000.

Trade in P commodities includes rock phosphate, phosphoric acid and manufactured fertilisers. Regional import and export figures of P-based fertiliser for year 2000 are shown in Table 2.11 together with production and consumption figures. The FAO (2001) data for P is quoted as P_2O_5 and hence to convert to elemental P the figures were divided by 2.2919 (see footnote 14 this Chapter). Figure 2.19 shows that in year 2000 Asia and China accounted for more than 50% of the global P-fertiliser consumption whilst Africa, Oceania, Central America and the Caribbean combined accounted for less than 10% of the total consumption.

Table 2-11 Regional P-fertiliser production, consumption, import and export for the year 2000

Region	All units in Mt as P/a				
	Production	Consumption	Imports	Exports	P-Net [a]
Africa	1.11	0.41	0.19	0.89	-0.70
Asia	3.01	3.97	1.39	0.38	1.00
C. America & Caribbean	0.19	0.22	0.17	0.11	0.06
China	2.92	3.76	0.97	0.13	0.84
Europe	2.46	1.81	1.20	1.79	-0.59
North America	3.27	1.94	0.28	1.92	-1.65
Oceania	0.42	0.69	0.36	0.06	0.30
South America	0.69	1.40	0.85	0.04	0.81
Total	**14.07**	**14.19**	**5.40**	**5.33**	**0.07**

a)   P-Net = P-import - P-export).
Note that ideally, Production + Import = Consumption + Export but because of no records of change of stock in each region and inaccuracies in data, the P-fertiliser trade balance does not add up.
Source: Data derived from FAO (2001)

Figure 2.20 shows that on a unit area of arable land[20] basis, China has the highest application rate of P as P_2O_5 and Africa has the least application rate. The world average fertiliser application rate in year 2000 was 1.0 g/m^2.a as P calculated on the basis of arable land.

---

[20] Arable land is defined by FAO (2001) as land under temporary crops (double-cropped areas are counted only once), temporary meadows for mowing or pasture, land under market and kitchen gardens and land temporarily fallow (less than five years). The abandoned land resulting from shifting cultivation is not included in this category. Data for arable land by region is shown in Appendix A. The data is not meant to indicate the amount of land that is potentially cultivable.

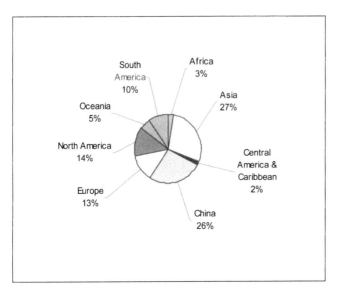

Figure 2-19 Comparison of regional P-fertiliser consumption in 2000

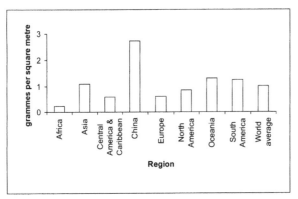

Figure 2-20 Regional P-fertiliser application as P_2O_5 per unit of arable[21] land in 2000

Phosphorus transferred in world vegetal products[22] is shown in Figure 2.21. Four major crop commodities accounted for more than 90% of the P traded in the world in year 2000. Cereals dominated, accounting for about 50% of global P movement, followed by oil-crops and vegetable oils (38.8%), stimulants (2.8%), and fibre crops (2.4%). The export and import of P in fruit, vegetables, tobacco, sugar crops and sweeteners, spices, beverages and other minor crops (2.2%) may however have been

---

[21] Arable areas generally receive more nutrients and are more prone to P losses by surface runoff and wind erosion than permanent grassland areas.

[22] The vegetal commodities evaluated did not include any lumber products and forage crops. The P contained in lumber commodities was considered a minor contributor to global P transfer. For example, the P concentration of wood ranges from about 0.02 to 0.1 mg/g and most wood has a density from about 330 to 450 kg/m³ (Beaton *et al.*, 1995). Assuming a P concentration of 0.05 mg P/g, a density of 400 kg/m³ and a year 2000 world industrial round-wood production of 3 400 Mt or 8 500 Mm³ (FAO, 2001), industrial wood accounts for about 170 000 Mg/a of P. This represents about 2% of the average total global P transferred in year 2000.

important in certain regions, but individually represented less than 1% of P traded on a global scale. Table 2.12 shows the relative importance of the various crop groups in each region during the year 2000. Only those commodity groups representing at least 3% of the import or export totals, as measured by the average total P transferred[23] ('000' Mg) in each region, were included.

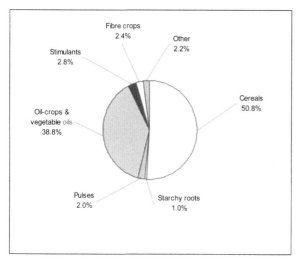

Figure 2-21 Total P in world crop commodity trade (% of total average P-transferred) in 2000

World trade balance

Collectively, global P in all imported crop groups amounted to about 1.97 Mt/a in year 2000. Total global P in exports for the same time periods were very similar at 1.91 Mt/a. Thus, there is a close agreement between the calculated amounts of total P transferred in imports and exports, although for oil-crops up to 20% difference was observed.

Seafood, milk products, and meat are responsible for more than 95% of the global P exchange in animal commodities (see Figure 2.22). Total P transported on a global basis in animal product imports in year 2000 was 0.204 Mt/a. P in exports for the same time period was 0.201 Mt/a. Europe, North America, Oceania and South America were the regions where exports of P in livestock commodities exceeded imports.

---

[23] Average total P transferred is half the sum of exports and imports.

Table 2-12 P imports and exports of the major crop products by region for year 2000

Region	Crop group	Imports 000 Mg/a	% of total	Crop group	Exports 000 Mg/a	% of total
Africa	Cereals	151.5	79.2	Oil-crops	14.6	31.5
	Oil-crops	30.8	16.1	Stimulants	13.1	28.3
	Other	9.1	4.7	Cereals	8.0	17.3
	Total	**191.4**	**100.0**	Fibre crops	7.9	17.0
				Other	2.8	5.9
				Total	**46.4**	**100.0**
Asia	Cereals	398.9	52.3	Oil-crops	148.0	44.4
	Oil-crops	308.2	40.4	Cereals	132.6	39.8
	Fibre crops	25.5	3.3	Fibre crops	16.5	5.0
	Other	29.9	4.0	Other	35.8	10.8
	Total	**762.5**	**100.0**	**Total**	**332.9**	**100.0**
C. America & Caribbean	Cereals	60.2	52.3	Stimulants	4.2	26.5
	Oil-crops	41.1	40.4	Cereals	3.5	22.1
	Fibre crops	3.6	3.3	Oil-crops	3.1	19.6
	Other	2.2	4.0	Vegetables	1.7	10.7
	Total	**107.1**	**100.0**	Fruits	1.4	9.0
				Other	2.2	12.1
				Total	**16.1**	**100.0**
China	Oil-crops	116.7	73.3	Cereals	47.6	72.2
	Cereals	34.9	21.9	Oil-crops	9.9	15.0
	Other	7.7	4.8	Pulses	2.4	3.7
	Total	**159.3**	**100.0**	Other	6.0	9.1
				Total	**65.9**	**100.0**
Europe	Oil-crops	251.9	45.7	Cereals	286.1	61.0
	Cereals	210.4	38.2	Oil-crops	133.9	28.6
	Stimulants	29.3	5.3	Other	48.6	10.4
	Other	59.5	10.8	**Total**	**468.6**	**100.0**
	Total	**551.1**	**100.0**			
North America	Cereals	31.3	36.6	Cereals	374.2	59.8
	Oil-crops	31.1	36.3	Oil-crops	224.8	35.9
	Stimulants	11.4	13.3	Other	26.7	4.3
	Fruits	3.9	4.6	**Total**	**625.7**	**100.0**
	Other	7.8	9.2			
	Total	**85.5**	**100.0**			
Oceania	Cereals	4.3	42.3	Cereals	65.2	64.5
	Oil-crops	3.7	36.7	Oil-crops	18.0	17.8
	Fibre crops	1.0	9.5	Pulses	11.1	11.0
	Stimulants	0.6	5.7	Fibre crops	5.1	5.1
	Other	0.6	5.8	Other	1.7	1.6
	Total	**10.2**	**100.0**	**Total**	**101.1**	**100.0**
South America	Cereals	79.0	72.8	Oil-crops	152.4	59.4
	Oil-crops	21.4	19.7	Cereals	87.2	34.0
	Fibre crops	3.4	3.1	Stimulants	8.6	3.3
	Other	4.7	4.4	Other	8.3	3.3
	Total	**108.5**	**100.0**	**Total**	**256.5**	**100.0**

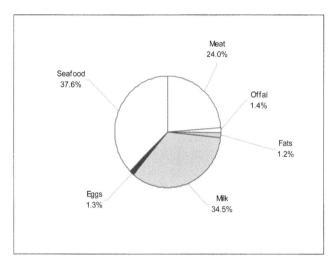

Figure 2-22 Total P in world animal commodity trade (% of total average P-transferred) in 2000

Imports and exports of P in fertiliser materials in year 2000 are shown in Table 2.11. North America Africa and Europe were the largest exporters of fertiliser P while Asia, China, and South America were the largest importers.

Regional net P balances (P-imports - P-exports) for all commodities for the year 2000 are shown in Figure 2.23. North America, Oceania, and South America exported more P in crop commodities than other regions of the world (Figure 2.25a). Conversely, Africa, Asia, Central America and the Caribbean, China and Europe were net importers of vegetal commodities. Trade in animal products is orders of magnitude smaller that of vegetal commodities with Asia being the largest importer of animal products (Figure 2.25b). The negative balance in fertiliser trade shows more P was exported in fertiliser materials from Africa, North America and Europe than was imported (Figure 2.25c). Total imports and exports of P in fertiliser materials amounted to about 5.40 Mt/a as P total 5.33 Mt/a as P, respectively.

The global net P balance for vegetal, animal, and fertiliser commodities is combined in Figure 2.25d. Trade in inorganic P is dominant. In 2000, processed fertiliser P, accounted for 71% of world P trade. In comparison, P in vegetal commodities accounted for 26% and livestock commodities for only 3% of world P trade, respectively.

Asia, Central America and the Caribbean, China, Oceania and South America were the primary *sinks* for global P transfer in 2000, whilst Africa, North America and Europe were the dominant P *sources* (see Figure 2.29). Africa (predominantly countries north of the Sahara) and North America were large exporters of P. The USA was responsible for P exports from the latter region. Phosphorus in fertiliser exports was dominant in both of these regions, accounting for about 95% of total P exports in Africa and about 75% of P exports in North America.

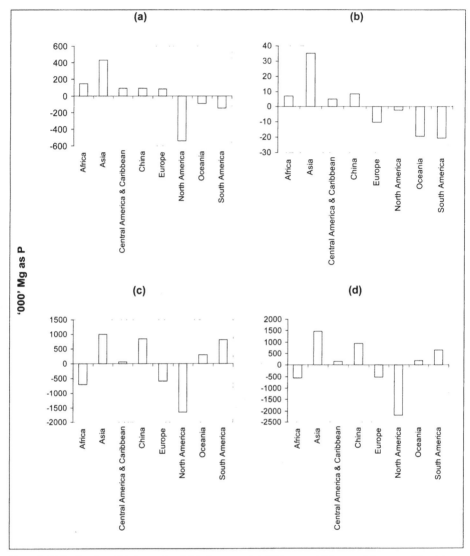

Figure 2-23 Regional net P balances (total P in imports – total P in exports) for vegetal commodities (a) animal commodities (b) fertiliser commodities (c) and combined of vegetal, animal and fertiliser commodities in year 2000 in 10^3 Mg/a.

Phosphorus in exports of agricultural commodities is important in that it represents a loss of soil P and a potential decline in soil productivity. This source of P withdrawal (Table 2.12) can be compared with the fertiliser P consumption for each region (Table 2.11). On regional basis fertiliser consumption in year 2000 greatly exceeded vegetal export of P, thus off-setting the potential negative impact of removal in agricultural exports. However, these estimates do not include the P required for crop production or P lost through waste disposal or other non-agricultural commodities. In addition, many soils throughout the world are deficient in P and require fertilisation. Caution should be exercised when interpreting gross regional data for local situations. In some areas, the Canadian prairies for example, crop removal exceeds P fertilisation for adequate crop production (Beaton *et al.,* 1995).

The regional and global P transfer balance would not be complete without making comparisons on a per person equivalent. From the global P balance illustrated in Figure 2.23 the regional population estimates for year 2000 (see Appendix A) can be factored in to produce a global P balance per person in each region (see Figure 2.24). In year 2000 each African, European and Northern American had a net export of 0.70 kg, 0.70 kg and 6.90 kg as P, respectively. Whilst on the other hand an Asian, Chinese, Central American and Caribbean, Oceanian and South American gained 0.60 kg, 0.90 kg, 0.70 kg, 6.20 kg and 1.90 kg as P, respectively.

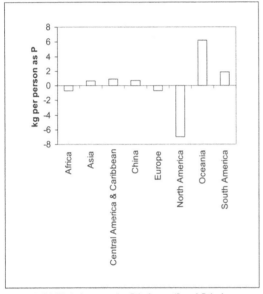

Figure 2-24 Global per capita P balance ([total P in imports – total P in exports]/population) of vegetal, animal and fertiliser commodities in year 2000

Limitations to the balance

The budgets calculated in this dissertation provide an indication of net gains and losses of P in vegetal, animal and fertiliser products on a regional and global basis. The difference between imports and exports in year 2000 on a global basis was about 138 000 Mg/a (or 0.138 Mt/a) of P, or about 2% of the total global P trade for the commodities included in the analysis. This discrepancy may partially be explained by rounding errors in the handling of the data set and partially by incomplete trade data for some commodities. In addition the discrepancy could be due to differences in fiscal years for trade statistics among countries and to the selection of assumed P concentrations in crop and livestock commodities. With the collapse of the Soviet Union, the FAO database classifies some of the former Soviet block countries as European and some as Asian. Similarly statistics for North America are divided into two, namely the developing and developed regions. Therefore the definition of geo-political boundaries is not exact and consistent and this could have been a source of error in the budget calculations.

The approach did not account for P associated with trade of canned or packaged foods, or forages as data on these products was not readily available and therefore not included in the calculation of net balance. Other sources of P may also have been inadvertently neglected.

In addition to incomplete P data, the net P balance presented in this section has other limitations. Only large geographic regions were considered and P movement within regions could be important locally. Net loss of P from a region need not indicate a loss in soil fertility and productivity. Africa, for example was a large net exporter of P due mainly to the large movement of rock phosphate to other regions. Stock changes need to be included in a complete P balance. Overall P balances in Africa are estimated to be negative because of the low application rates of mineral fertilisers (Breman, 1990; Stoorvogel & Smaling, 1990; Van der Pol, 1992), although the methodologies of such studies can be questioned (Runge-Metzger, 1995). If these negative trends are not reversed, agricultural productivity of the continent will decline further.

Another important source of P loss not accounted for is waste disposal. Effects of internal movement and disposal of P within a region may outweigh the long-term effect of P loss from a region through commodity trade. In rural areas, human waste and refuse may find its way back to the soil, but in urban areas much of the P will be lost in sewage discharged to water bodies or as sludge, placed in land-fills.

Much of the P in animal feed will remain in the agricultural system and enhance soil fertility when animal wastes are applied to soil, but industrial feed lots often uncouple this P cycle and cause direct P discharges as detailed in the case of the Netherlands in Chapter 1.

Despite the possible inaccuracies and problems associated with a global P budget, the net balances calculated should reflect the majority of inter-continental transfer of P. Inclusion of P found in other trade commodities would further refine the trends described, but would not likely change the pattern of global P transfers presented here.

2.7 Conclusions

Given the escalating population growth, land degradation and increasing demands for food, achieving sustainable agriculture and viable agricultural systems is critical to the issue of food security and poverty alleviation in most, if not all, developing countries. Efficient use of P-based fertiliser, in particular, is fundamental to the sustained productivity and viability of agricultural systems worldwide. At present, nutrient use efficiency is low in most countries. In developed countries considerable progress has been made but more can be done. Influencing the use of mineral fertilisers is difficult in view of the very large number of farmers that are involved. Furthermore, the manufacturer is often separated from the end user by a distribution system that has to handle millions of tonnes of material.

Most evidence rejects the hypothesis that ecosystems naturally occur in a 'steady state' with respect to biogeochemical cycling. The complexity of chemical cycling makes ecosystem processes quite chaotic, so some elements rise in availability and some fall in an unpredictable fashion. Many of the facets of 20th and 21st century human society have radical effects upon biogeochemical cycling. These effects may be subtle as well as blatant. Transportation of crops around the worlds represents wholesale depletion or supplementation of the mineral resources of a region. Cultivation of crops, which are rapidly harvested, accelerates loss of phosphorus from the soils necessitating fertilisation. The P-cycle is being uncoupled through such actions.

Estimates of world phosphate reserves and availability of exploitable deposits vary greatly and assessments of how long it will take until these reserves are exhausted also vary considerably. Modest calculations indicate that the current economic reserves would last at least 100 years at consumption rates of about 14 Mt P/a as P. Whilst there is no supply "crisis" for the phosphate industry's main raw material (phosphate rock), it is clear that reserves are finite. Furthermore, it is commonly recognised that the high quality reserves are being depleted expeditiously and that the prevailing management of phosphate, a finite non-renewable source, is not fully in accord with the principles of sustainability.

The mining of rock phosphate and its agricultural, industrial and domestic uses have increased during the last few decades from less than 30 Mt/a in 1961 to about 140 Mt/a in year 2000. Other activities of modern societies such as clearing of forests, extensive cultivation and urban waste disposal and drainage systems have enhanced the transport of P from terrestrial to aquatic environments to an estimated 22 Mt P/a. As a consequence of these activities, concentrations of P in rivers and supply of P to lakes and estuaries have increased almost three-fold from an estimated 8 Mt P/a of pre-industrial and intensive agriculture era. The influence of human activity on the riverine P flux is still poorly known despite the efforts of several research based on a series of assumptions and estimates of land erosion, fertiliser use, sewage discharge, production of animal waste, and waste production in food processing. The processes of P export and transport are very complex, and detailed data and knowledge of these processes are seldom available without lengthy and expensive investigations. What is clear however is that the natural P-cycle is now greatly modified by human activities and is no longer at steady state.

Most phosphorus utilised by man is still stored in the Earth's surface (dispersed arable lands, in waste dumping sites, as pollution of underground waters and lakes, as increase of organic matter and in the surface water). This evolution is now a major environmental problem for it has induced a widespread eutrophication of surface waters (rivers, lakes, and estuaries) resulting in overproduction of organic matter, oxygen consumption, metal release from sediments and other consequences

In general, global P management results in pronounced imbalances in certain parts of the world. On one hand, phosphorus is a constraining factor for increased agricultural production in those regions which are simultaneously most seriously affected by chronic or temporary food shortages. On the other hand, demand for mineral P fertiliser has stabilised in those countries where the stock of soil P has already been raised substantially and where the P-throughput is the highest. Here, huge food surpluses are produced which cause significant distortions in the world agricultural market. Total P transported on a global basis in crop and animal products (imports and exports) is in excess of 2 Mt P/a.

Urbanisation as well as population increases will likely raise concentrations of P from land to aquatic systems. Urbanisation is associated with increased point source loading and is also correlated with higher fertiliser use. The contribution of P from sewage into the aquatic environment globally is small (estimated at less than 0.5 Mt P/a) as compared to P from agricultural lands. But at local scale P in sewage is significant in terms of pollution of finite freshwater sources (see 'Problem statement in Chapter 1 and more detail in the following Chapter). Therefore at a global scale (macro P-flux analysis) P transfers and transformations might not reveal the precise impact of man on its *natural* cycle, but at smaller scales (micro P-flux analysis) e.g. river basins, the distortion of the natural P-cycle have had dramatic consequences.

3 Description of study area

3.1 Phosphorus reserves and use in Zimbabwe

Socio-economic profile and water availability in Zimbabwe

UNDP (2003) classifies Zimbabwe (Figure 3.1) as a country with a medium human development index. The Human development Index (HDI[1]) measures the overall achievements in a country in three basic dimensions of human development: longevity, knowledge and decent standard of living. Table 3.1 provides some important human development indicators for Zimbabwe. Zimbabwe is a semi-arid country with an average annual rainfall of 736 mm/a spread over a land area of 391 000 Mm^2. In relation to available renewable fresh water resources the country falls within what is termed a 'water stress' category (Gleick, 2000; Hirji *et al.*, 2002; FAO, 2003). According to FAO (2003), annual internal renewable fresh water totals are approximately 15 000 Mm^3/a of which 30% is currently utilised. Of this amount, the annual ground water potential is approximately 1 000 to 2 000 Mm^3/a. Although a small resource, this forms a disproportionately important source of water for poor rural communities in the drier lowveld.

Figure 3-1 Map of Zimbabwe showing the major urban centres

Table 3-1 Human development indicators for Zimbabwe

Indicator	Value
Estimated population in 2000 (million)	12.4
Urban population (% of total)	35
Population under 15 (% of total)	45
Life expectancy at birth (years)	43
Adults living with HIV/AIDS (% age 15-49)	25
GDP per person (US$)/a	2 870
Adult literacy (% age 15 and above)	33
Gross enrolment[2] (%)	65
Population using adequate sanitation facilities (%)	80
Population using improved water sources (%)	85

Source: PASS (1995); WHO (1997); UNAIDS (2000); UNDP (2001, 2003)

Information on renewable water availability and use by sector is shown in Table 3.2 and Figure 3.2. Countries with renewable freshwater availability of less than 1700

---

[1] The HDI ranking, ranks 162 United Nations member countries. Other 29 member countries are not ranked as well as for two non-members, Switzerland and Hong Kong.
[2] The number of students enrolled (tertiary, secondary and primary levels), regardless of age, as a percentage of the population of official school age.

m³/p.a are said to experience water stress and those with less than 1000 m³/p.a, water scarcity. This water scarcity benchmark has been accepted as a general indicator of water scarcity by the World Bank and other analysts (Gleick, 2000; IMWI, 2000; Engelman & LeRoy, 1993). It is taken as the approximate minimum for an adequate quality of life in a moderately developed country.   It assumed to satisfy the requirements of agriculture, industry, domestic use and energy production.

Table 3-2 Water availability in Zimbabwe

Water availability	Value
Total renewable freshwater available[3] (Mm³/a)	15 000
Water per person in 2000 (m³/p.a)	1 210
Estimated population in 2025[4] (million)	20
Water per person in 2025 (m³/p.a)	750

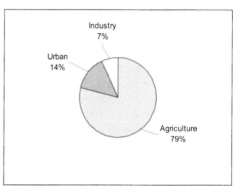

Figure 3-2 Water use by sector in year 2000 as a percentage of total freshwater withdrawals

Water requirements for food production are a constraint to the continued growth of agriculture in Zimbabwe. Plant nutrients are also a limiting factor as 80% of the soils (sandy and granitic in origin) are inherently infertile, highly weathered and leached. The soils are commonly coarse-grained and they do not easily retain water or nutrients.   There is a general tendency towards acidic soil conditions (Grant, 1981; Nyamapfene, 1991), particularly in the higher rainfall areas. Under these conditions, Al and Fe are present in forms that can fix phosphate, and so they effectively make the phosphate unavailable to plants. Thus large amounts of fertiliser have to be added to satisfy the Al and Fe needs and to provide nutrients for the crops.   These soils require regular applications of fertiliser to ensure adequate yields of crops that are harvested annually.   Also because of the relatively short growing periods nutrients need to be applied in a readily available water-soluble form.

Locally the country is recognised as falling into three zones: the Lowveld (below 900m), the Middleveld (900 m – 1 200 m) and the Highveld (above 1200 m).   These zones represent about 36%, 40% and 24% of the total land area respectively. They are demarcated not only on the basis of elevation, but also on the basis of associated variations in the physical environment; namely natural vegetation, temperatures, soils, water supply and animal life.

Phosphorus deposits

Zimbabwe is self-sufficient in terms of rock phosphate, with a current consumption of about 40 000 Mg/a as triple or single super-phosphate, which drives its agro-based economy.   The deposits occur within igneous rocks in carbonatite complexes of late

---

[3] This is renewable fresh water (including ground water that) is generated within the geo-political boundaries of Zimbabwe each year and excludes water that flows in from neighbouring countries.
[4] A population growth rate of 3% per annum is used to estimate the population in 2025. This has been adjusted to account for the current prevalence of HIV/AIDS.

Palaeozoic to Mesozoic age, and meta-carbonatites believed to be of Achaean age and in addition cave accumulations of bat guano. The locations of these are indicated in Figure 3.3. No phosphate deposits of sedimentary origin are known in Zimbabwe (Barber, 1989; Fernandes, 1989). All the deposits contain concentrations of apatite, but only the Dorowa complex is exploited for phosphate.

The Dorowa carbonitite complex is situated approximately 150 km south-east of Harare on an alkali ring in the Save catchment, in Buhera District. Exploration for deposits began as early as 1937. The chemical composition of apatite from Dorowa exhibits presence of fluorine, but very little chlorine (Table 3.3). It is estimated that fluoro-apatite comprises approximately 53% of the concentrate, hydroxy-apatite 22%, and the remainder is probably carbonate-apatite. The total amount of concentrates produced since the open cast mining operations began is in excess of 5 million tonnes. Production figures from 1964 to 1987 indicated in Figure 3.4 suggest that in recent years over 1 million tonnes of ore have been mined annually and between 130 000 and 150 000 tonnes of concentrates have been produced at an average grade of 35.05% P_2O_5.

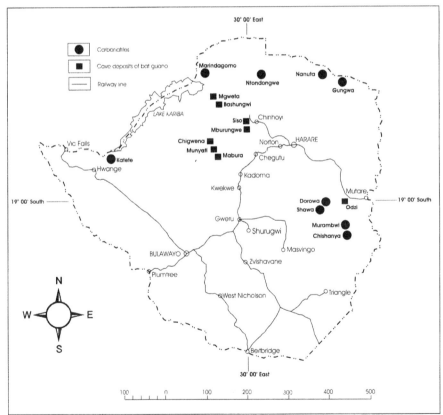

Figure 3-3 Localities of the known carbonatite structures and cave deposits of bat guano in Zimbabwe
Source: Barber (1989)

Super-phosphates are produced at Zimphos, Msasa in Harare, by reacting ground phosphate rock with sulphuric acid or phosphoric acid. The production of super-phosphates started in 1928 using raw bones as a source of phosphate (Fernandes, 1989).

Table 3-3 Chemical composition of Dorowa apatite concentrates

Element	%
CaO	46.3
P_2O_5	35.5
Cl	0.01
F	1.68
OH	0.63
CO_2	4.1

Source: Fernades (1989)

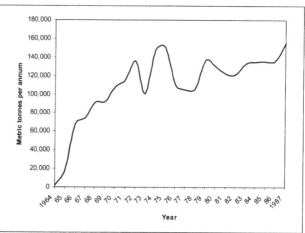

Figure 3-4 Annual production of phosphate concentrate by Dorowa Minerals Limited from 1964 to 1987

Single super-phosphate is produced by reacting phosphate rock containing less than 1% CO_2 with sulphuric acid in the ratio of 0.64:1. The product contains 18% P_2O_5. Triple super-phosphate containing 42% P_2O_5 is made by reacting phosphate rock with 50% phosphoric acid. The super-phosphates are sold both as straight fertilisers and as compound fertilisers. There are two manufacturers of compound fertiliser in Zimbabwe, Zimbabwe Fertiliser Company and Windmill Private Limited. Both companies have similar granulation and bagging facilities and they formulate 13 compounds to supply the specific needs of crops grown by Zimbabwean farmers (Table 3.4).

Table 3-4 Composition of compound fertilisers produced in Zimbabwe

Compound	N	P_2O_5	K_2O	S	B	Zn
A	2	17	15	10.0	0.1	-
B	4	17	15	9.0	0.1	-
C	6	17	15	7.5	0.1	-
D	8	14	7	6.5	-	-
J	15	5	20	3.5	0.1	-
L	5	18	10	8.0	0.25	-
M	10	10	10	6.5	-	-
P	10	18	0	6.5	-	-
S	7	21	7	9.0	0.04	-
T	25	5	5	5.0	-	-
V	4	17	15	8.0	0.1	-
X	20	10	5	3.0	-	-
Z	8	14	7	6.5	-	0.8

Source: Fernandes (1989)

Phosphates and agriculture in Zimbabwe

Agriculture has played an important role in the economy of Zimbabwe both before and after Independence and the use of fertiliser is well established (Barber, 1989). Agricultural production in Zimbabwe comprises of a number of commodity crops and animal products. Crop production includes food crops like maize and industrial crops such as cotton and tobacco. Zimbabwe like other African countries endeavours to be self-sufficient in terms of agriculture and ensure an adequate supply of food and fibre for the indigenous population (FAO, 1986a; 1986b; Sanchez *et al.*, 1997).

It is recognised that the route to self-sufficiency in agriculture is by the application of appropriate fertilisers and pesticides to ensure good yields. Yet the use of fertiliser in Africa falls well below the norm for optimum yields of agricultural produce (see Figure 2.17 and 2.18) (Sanchez, 1976; Eicher & Baker, 1982; FAO, 1986a; 1986b; Rockstrom, 1997; Smaling, 1997). High import prices contribute to the low level of fertiliser use in Sub-Saharan Africa. High fertiliser prices arise from small procurement orders (tenders for less than 5 000 Mg are common), weak bargaining power, coupled with high transportation costs due to poor infrastructure and international marketing costs (Isherwood 1996; Gruhn *et al.*, 2000). Special mixes tailored for African needs, and other micro-nutrient additions, such as sulphur or boron, may add an additional cost to the price (Coster, 1991). The constraints listed above are well known and it is against this background that alternative sources of plant nutrients e.g. organic manures, un-processed rock phosphates, animal and human manure should be examined and evaluated.

The breakdown of major land use categories in Zimbabwe is given in Table 3.5 and Figure 3.5. The country is divided into five agro-ecological regions (Figure 3.6) based on the interaction of climate, soils, slope and secondary terrain features. Natural regions I up to III are suitable for intensive crop and livestock production. The optimal, commercial food production area is concentrated on the soil-rich, rain-fed plateau, whilst the lowveld, the site of highly populated communal lands is an arid soil-poor environment unsuitable for extensive agricultural production.

Table 3-5 Major land use categories in Zimbabwe

Land use	Area (Mm^2)	As % of total
Agriculture	107500	27.5
Pasture	172000	44.0
Forests/woodlands	82000	21.0
Urban	2000	0.5
Water bodies/rock	3800	1.0
All other land	23800	6.0
Total	391000	100

Source data: FAO (2001); Whitlow (1988)

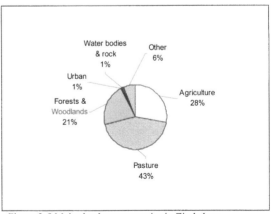

Figure 3-5 Major land use categories in Zimbabwe

The farming systems in Zimbabwe are Large Scale Commercial Framing, Small Scale Commercial Farming, Communal Farming (peasant agriculture), Resettlement Farming and Urban and Peri-urban Agriculture. Generally, most of the farming land is found in the natural regions IV and V (over 60%).

Region	Area (Mm²)	Rainfall (mm/a)	Recommended farming activities
I	7 000	>1000	Specialised & diversified farming
II	58 600	750-1000	Intensive crop farming
III	72 900	650-750	Semi-intensive mixed farming
IV	147 800	450-650	Semi-extensive livestock farming
V	104 400	<450	Extensive livestock farming (ranching)

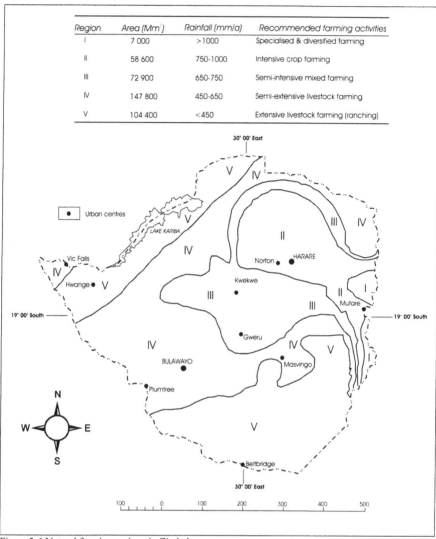

Figure 3-6 Natural farming regions in Zimbabwe
Source: Adapted from CSO (2000)

Maize is the most important food crop in Zimbabwe and in 1996, farmers produced some 2.6 Mt. The 1998 season produced only 1.42 Mt, year 2000, 2.1 Mt and in 2001 1.5 Mt this decline in production is attributed to drought and huge reduction in cropping area due to the land reform process respectively. For the year 2000 the average yield from the 14 200 Mm² under cultivation was 0.148 kg/m². Crop production for year 2000 is given in Table 3.6 and the head count of livestock in Table 3.7 (FAO, 2001). Detailed food balance for the year 2000 is given in Table A.13, Appendix A.

Table 3-6 Crop production of major commercial crops (2000)

Commodity	Production (Mt/a)
Sugar crops	4.23
Cereals	2.53
Oil crops	0.49
Cotton lint	0.33
Tobacco leaf	0.23
Fruits	0.23
Starchy roots	0.21
Vegetables	0.15

Source: CSO (2000); FAO (2001)

Table 3-7 Livestock head count in Zimbabwe (2000)

Livestock	Number (000's)
Cattle	5 560
Goat	2 950
Sheep	630
Pigs	450
Horses	26
Donkeys & mules	109
Chickens	17 500

Source: CSO (2000); FAO (2001)

From the classification system proposed by Nyamapfene (1991), the major soil types in the Harare environs are Kaolinitic (Figure 3.7). They are moderately to strongly leached soils. Clay fractions are mainly inert together with appreciable amounts of free sesquioxides of iron and aluminum. The soils are either classified as fersiallitic (mixed clay) or paraferrallitic (inert clay) or orthoferrallitic (very inert clay) soils.

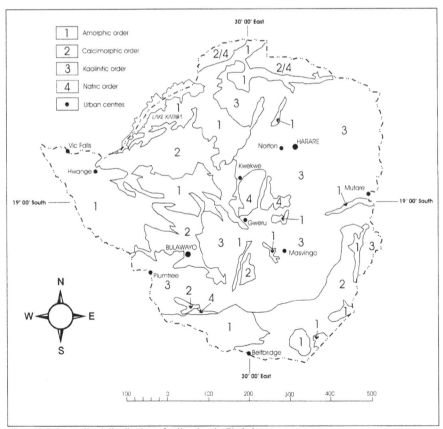

Figure 3-7 Generalised distribution of soil orders in Zimbabwe
Source: Adapted from Nyamapfene (1991)

The important consideration is that all these formations are relatively rich in ferro-magnesium minerals and therefore give rise to clayey soils that are red, reddish brown to yellowish red in the well-drained areas. Most of the agriculturally important red soils of Zimbabwe are derived from these formations (Nyamapfene, 1991).

Zimbabwe is experiencing a number of environmental problems, which are mainly a result of human activities. The single biggest problem is that of land degradation, which is emanating from excessive concentrations of human and livestock populations in ecologically marginal, dry and fragile soils mainly in communal, small scale commercial and resettlement areas (Whitlow, 1988). Soil erosion, which is also a result of deforestation and poor farming practices, is severe in the communal and resettlement areas (Figure 3.8). Soil loss due to land degradation, gold panning, riverbank cultivation and deforestation to has resulted in river and dam siltation. This has resulted in low soil fertility in most communal areas due to continued cultivation with less or no fallow periods to allow the land to regenerate its fertility.

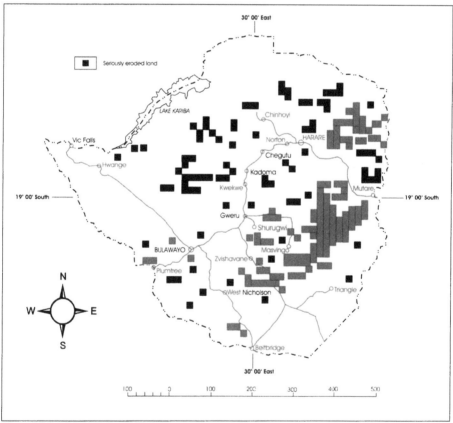

Figure 3-8 Land degradation in Zimbabwe
Source: Adapted from Whitlow (1988)

The most detailed research on nutrient losses in Zimbabwe has been that conducted by Stocking (1986). Using the so-called Soil Loss Estimation Model for Southern Africa (SLEMSA) (Elwell, 1975; 1977; Elwell & Stocking, 1982) mean rates of soil erosion were predicted under the five categories of farming and land use in Zimbabwe (see Chapter 4). Using regression models the loss of soil was shown to be significantly correlated to nutrient losses. Table 3.8 shows the results of calculated nutrient losses for four levels of soil erosion and for two soil groups in Zimbabwe (Stocking, 1986).

Table 3-8 Nitrogen, phosphorus and organic carbon losses for two soil groups in Zimbabwe

Soil erosion g/m^2	Losses in g/m^2					
	N		P (as P_2O_5)		C	
	Soil 1	Soil 2	Soil 1	Soil 2	Soil 1	Soil 2
300	0.63	0.29	0.05	0.03	4.62	3.21
1 500	3.15	1.46	0.23	0.22	23.10	16.05
5 000	10.50	4.85	0.78	0.88	77.00	53.50
7 500	15.75	7.28	1.16	1.40	115.50	80.30

  a)   Soil 1 is all Zimbabwe other than, Soil 2, granite sandveld.
  b)   Approximately 60% of Zimbabwe's soils are granite derived and would fall into Soil group 2
       (see Chapter 4 for details on the soil groupings).
Source: Stocking (1986)

The demand for fertilisers in Zimbabwe is expected to remain stagnant or decline in the near future. The maximum consumption rate of 180 000 metric tonnes (as N, P & K) was achieved in the 1981-82 growing season after independence due to a rapid rise in the demand for fertiliser in the communal sector. Droughts and the land reform process are likely to lower the demand. Exports of fertilisers have been very small amounting to approximately 2 000 metric tonnes/a as N, P & K (approximately 500 metric tonnes/a as P). Figure 3.9 shows the fertiliser use pattern in Zimbabwe from 1961 to 2000 and Figure 3.10 shows the import and export of P-based fertiliser since 1961.

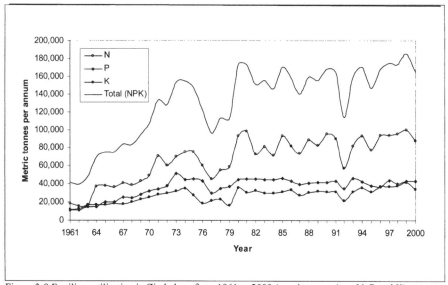

Figure 3-9 Fertiliser utilisation in Zimbabwe from 1961 to 2000 (metric tonnes/a as N, P and K)
Data source: FAO (2001); CSO (2000)

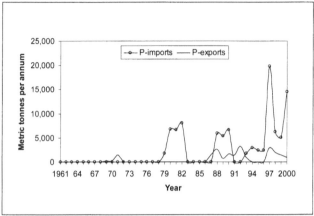

Figure 3-10 P-based fertiliser import and export variation in Zimbabwe from 1961 to 2000 (metric tonnes/a as P_2O_5)
Data source: FAO (2001)

To illustrate the level and intensity of fertiliser use in Zimbabwe some comparisons are in Table 3.9 with the Netherlands, a country described in Chapter 1 as having livestock manure problems due to intensive methods of stock farming. The two countries have comparable human populations with the Netherlands having about 116 million head count of livestock (Table 1.3) compared to Zimbabwe's 27 million in addition (Table 3.7).

Table 3-9 Population, land and fertiliser use in the Netherlands and Zimbabwe[5]

	Netherlands	Zimbabwe
Population-Estimates in year 2000 (000)		
Total	15 864	12 627
Rural	1 667	8 168
Urban	14 197	4 459
Land use in year 2000 (Mm^2) [a]		
Total area	4 153	39 076
Permanent crops	34	130
Permanent pasture	1 012	17 200
Non arable and non permanent	2 444	35 335
Land area	3 388	38 685
Agricultural area	1 956	20 550
Arable and permanent crops	944	3 350
Arable land	910	3 220
Fertiliser use, production, import and export in year 2000 (Mg as P_2O_5/a)		
P-fertiliser consumption	54 000	43 400
P-fertiliser production	188 000	30 000
P-fertiliser imports	130 000	14 500
P-fertiliser exports	180 000	1 100
Total fertiliser consumption (as N, P and K)	418 000	165 300
Total fertiliser production (as N, P and K)	1 325 000	101 400
Total fertiliser imports (as N, P and K)	601 700	70 500
Total fertiliser exports (as N, P and K)	1 307 000	6 600

a)   Definitions of land use classes are given in Appendix A.
Source: FAO (2001)

[5] The use of manure as fertiliser is not included in the fertiliser production and use statistics.

The land area of Zimbabwe is about ten times that of the Netherlands but in relation to P-fertiliser usage the Netherlands has an application rate of about 28 g/m^2 as P_2O_5 on its agricultural land as compared to 2 g/m^2 on Zimbabwe's arable land. Such comparisons are useful in making predictions of sustainable development and consumption patterns (ecological footprints as described in Chapter 1).

A number of ambitious studies on quantifying nutrient balances at national or continental scale in sub-Saharan Africa have been conducted (Smaling *et al.*, 1993; Stoorvogel *et al.*, 1993; World Bank, 1996; World Bank & FAO, 1996) (refer to Chapter 2). Major assumptions have been made regarding processes for both spatial and temporal system boundaries. The limitation of most of the methodologies used is that, calculating nutrient balances (transfer functions and regression equations) quickly becomes an exercise involving *'black boxes'* nested within other *'black boxes'* – something too easily forgotten when a single number is generated and published at the end.

In this dissertation despite the methodological limitations, an attempt has been made to calculate P-balances for year 2000 with respect to the Zimbabwean territory. The data used for the calculations has been presented in this Chapter and Chapter 1 and 2. Detailed data on import and export of vegetal, animal and fertiliser products and food balance for the year 2000 is presented in Table A.12 and A.13 in Appendix A. Figure 3.11 depicts the estimated P-fluxes and stocks for Zimbabwe in year 2000. Not all possible P-fluxes are shown but only the major ones are represented whose data was available.

The relative magnitudes of P-fluxes and stocks shown in Figure 3.11 are important for setting up monitoring concepts, early recognition of P-resource demand, environmental impacts and sustainable resource use. The MFSA picture is also important in evaluating the effect of technical or management measures in mitigating environmental impacts or decisions on sustainable resource use.

While P-balances for the Zimbabwean territory shown in Figure 3.11 are indispensable for elaborating the P-flows and stocks in different systems, compiling accurate scenario requires methodologically complex tools. The difficulties of missing and inaccurate data suggest that the P-balances in Figure 3.11 should be treated with caution and informed scepticism. For example national studies have made aggregate generalisations about the extent of P-losses thus misrepresenting soil fertility dynamics and risking feeding into an unrealistic 'crisis narrative'. The calculated P-balance is essentially a historical snap-shot of year 2000 and the calculated values provided are meant to form a basis for future research and can be improved once more accurate data becomes available. The values presented have varying degrees of uncertainty i.e. whilst others are within a $\pm 10\%$ standard deviation, for others it could be more.

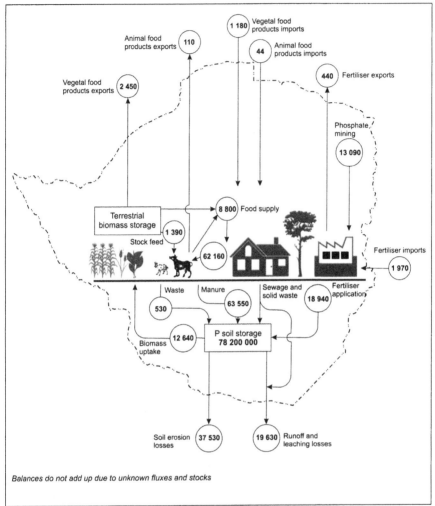

Figure 3-11 Estimated P-flows and stocks with reference to Zimbabwean territory in year 2000 (in Mg or Mg/a as P)

a) Total P in the soil storage has been computed from a total land area of Zimbabwe of 391 000 Mm^2 (Table 3.5) and a soil depth of 0.2 m (plough layer, see Chapter 4) with a P content of 0.10% per kg (1 g/kg as P) of soil, which is the global average (refer to Chapter 2) and a soil density of 1 000 kg/m^3. Total soil P, is thus $1.0 \times (391\ 000 \times 10^6 \times 0.2) \times 1\ 000 = 78\ 200\ 000$ Mg.

b) Total P uptake and storage in the terrestrial and aquatic biota has not been estimated.

c) The river run-off of P to the Indian Ocean via the Limpopo and Zambezi rivers consists of both natural and human-influenced leaching, surface runoff, sewage flows and particulate erosion products, less that which is retained or consumed within the rivers. The particulate soil erosion flux is calculated by assuming a moderate erosion rate of 1 500 g/m^2 (Table 3.8) and a corresponding P loss rate of 0.22 g/m^2 as P_2O_5. Using the land surface area of Zimbabwe the erosion P loss is = $(0.22/2.2919) \times 391\ 000 \times 10^6 = 37\ 530$ Mg/a as P. It is like that this value is an overestimate because the effect of impounding reservoirs and riverine deposition and uptake by aquatic biota are not taken into account (refer to Chapter 2, section 2.3).

d) Dissolved P losses (leaching, surface runoff and sewage discharges) can be estimated from the average P-concentration of the Zambezi river water (Table 2.2) of 10 g/m^3 as ortho-P and the average annual river flow contribution of Zimbabwean land mass to the total runoff of the Zambezi and Limpopo Rivers of 4 500 Mm³/a (Kabell, 1984) i.e. P runoff losses = $(10.0/2.2919) \times 4\ 500 \times 10^6 = 19\ 630$ Mg/a as P.

e) P import and export of food, fibre and fertiliser products is summarised in Table A.12 in Appendix A.

f) The domestic food supply, stock feed and waste from food processing P-fluxes are calculated using the figures in food balance sheet for Zimbabwe in year 2000 (Table A. 13, Appendix A) and the corresponding P-content values for each food or fibre item (Table A.12). The food, stock feed and waste P-fluxes calculated are 8 800 Mg/a; 1 390 Mg/a; and 530 Mg/a as P respectively.

g) Manure derived from livestock is calculated using Table 3.7 and assuming that the excretion rate of P of the livestock in Zimbabwe is similar to that of livestock in the Netherlands (see Table 1.5, Box 1.B in Chapter 1; Gerriste & Vriesma, 1984). Cattle, horses, donkeys and mules are considered as one group of animals, whilst goats, sheep are considered as veal calves. The total manure production calculated in this way is 63 550 Mg/a as P. This figure is likely to be an overestimate as P is used extensively in the Netherlands as stock-feed and hence the livestock P-excretion rates are likely to be higher than those of animals, which feed on pasture.

h) Since the P stored in the terrestrial biota has not been estimated it can be assumed that the balance of P reflected as animal excretion in manure is taken up from natural grazing areas where there is no synthetic P-fertilisation i.e. P-uptake of animals consists of harvested stock feed from P-fertilised farm areas (1 390 Mg/a indicated above) and from natural grazing areas (63 550 − 1 390 = 62 160 Mg/a). The figures indicate that the bulk of animals rely on natural grazing rather than stock feed derived from crop and other products.

3.2 Water and P-fluxes in the Lake Chivero basin

Introduction

The Lake Chivero catchment (2 250 Mm^2) lies within the Upper Manyame River Basin (UMRB), covering the City of Harare and its satellite towns, Chitungwiza, Ruwa, Norton and Epworth. Increased population growth (Figure 3.12) within the catchment has imposed a severe strain on the assimilative and regenerative capacity of the Lake system since it started impounding water in 1952. Lake Chivero in combination with 3 other reservoirs in the UMRB as depicted in Figure 3.13 form an important resource network which sustains the population, drives the industrial hub of Zimbabwe and also supports the ecosystem which directly benefits a large part of the population.

A configuration of sub-river basins in the Upper Manyame River Basin is shown in Figure 3.14. Settlement densities are among the highest in the country, with the Harare-Chitungwiza urban areas accounting for almost 50% of the national urban population and 22% of the total country population (Table 3.8). Harare urban has a settlement density of about 3 300 people/Mm^2. Densities in the rural parts of the catchment are a lower average of about 30 people/ Mm^2.

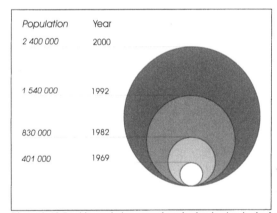

Population	Year
2 400 000	2000
1 540 000	1992
830 000	1982
401 000	1969

Table 3-10 Urban space

Urban area	Area Mm^2	
	1992	2000
Harare City	447	638
Chitungwiza	43	43
Epworth	11	15
Ruwa	34	40
Total	535	736
Population density (No. per Mm^2)	2 900	3 300

Source: CSO (1992); UNDP (2001; 2003); JICA (1996).

Figure 3-12 Rapid population growth and urbanisation in the Lake Chivero sub-basin

Physical characteristics

The Department of Meteorological Services carries out an intensive monitoring of meteorological and hydrological parameters with a network of stations throughout the country. Meteorological data of Harare City monitored at Belvedere Station during the last 40 years are summarised in Figure 3.15. The climate in the study area is very seasonal with distinct three wet and dry seasons. Spring is a hot and dry season from September to November and sporadic strong rainfall is unlikely occur. Average daily temperature is approximately 22 °C ±6 °C. Summer is classified as the rainy season with hot and wet conditions from December to April. Average daily temperature is about 20 °C ±6 °C. The remaining period of the year is a cold and dry season. Average temperature is approximately 16 °C ±6 °C. Mean total annual rainfall is approximately 786 mm/a with annual fluctuations ranging from 440 mm/a to 1 220 mm/a. During the summer season approximately 80% of the total annual rainfall is observed.

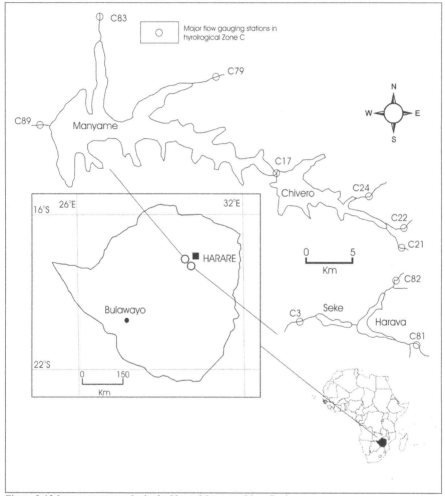

Figure 5-15 Important reservoirs in the Upper Manyame River Basin

The former Hydrological Branch of the Department of Water Development (now Zimbabwe National Water Authority) also maintains a network of river and groundwater flow monitoring stations. Flow rate of rivers fluctuates very seasonally. Large volume of river flow is normally observed during summer season (December to April), while the minimum flow occurs in winter (May to August) wherein compensation water from the upstream dams is released and dry weather sewage effluent is discharged into the rivers. Mean annual runoff of the major rivers in the basin is shown in Figure 3.16. Manyame River (including its major tributary, Nyatsime River) occupies as large as 60% of the total gauged flow into Lake Chivero, while Mukuvisi and Marimba Rivers contribute to the rest of the flow.

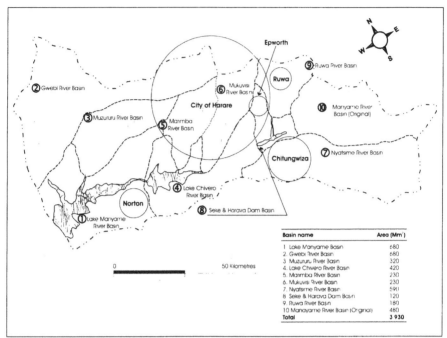

Figure 3-14 The Upper Manyame River Basin and its sub-basins

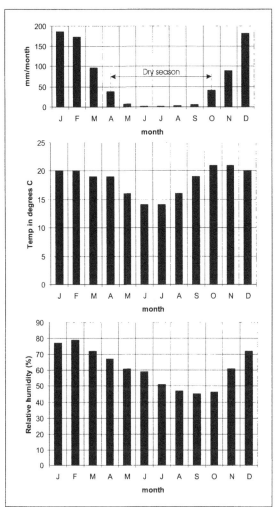

Figure 3-15 Meteorological characteristics of study area: 40 year monthly average of rainfall (mm/month), temperature (°C) and relative humidity (%) as measured at Belvedere in Harare

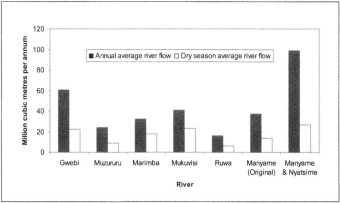

Figure 3-16 Average and dry season river flow in the Upper Manyame River Basin (Mm³/a)

A water balance for Lake Chivero based on average river flow data from 1990 to 2000 in shown in Table 3.9 and illustrated in Figure 3.17. The discrepancy in the balance is attributed to errors of estimation and measurement, particularly the influence of groundwater flow (JICA, 1996, Nhapi *et al.*, 2002)

Table 3-11 Water balance for Lake Chivero in year 2000 based on 10-year data

Item	Gauging station	Flow Mm³/a
Inflow		
Manyame River[6]	C21	153.4
Marimba River	C24	32.8
Mukuvisi River	C22	40.7
Precipitation over the lake	Belvedere 786 mm/a	20.8
Direct area runoff	-	13.2
Total inflow		**260.9**
Outflow		
Lake spillway	C17	90.0
Raw water abstraction	City of Harare	95.2
Evaporation	DWD, onsite evaporation pan	42.7
Total outflow		**227.7**
Inflow - outflow		33
Discrepancy as a % of inflow		13

Data source: JICA (1996); Nhapi *et al.*, (2002)

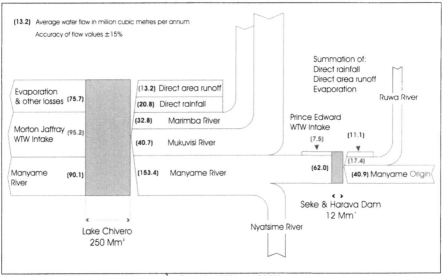

Figure 3-17 Water flow balance in Mm³/a for Lake Chivero (2000)

Natural Environment and ecology

The area falls into Natural Region II in the Agro-Ecological classification of Zimbabwe and is considered as the intensive farming region of the country. As a result of intensive agriculture and urban development the natural environment has been modified throughout most of the area. Most of the area is underlain by granitic geology which has given rise to light textured sandy soils, while in some areas where

---

[6] Manyame river flow with reference to Lake Chivero, includes it tributaries emanating from, Manyame Original, Ruwa and Nyatsime sub-basins

basic rocks, extending from the Mazowe valley up to the northern edge of Harare, resulted in heavier textured more clayey soils.

Land use in the Lake Chivero catchment is broken down into cultivated land (cropped and fallow) grazing land, forest land, and developed area (residential, industrial and commercial) as shown in Table 3.12 and depicted in Figure 3.18. About two thirds of the catchment is vegetated and two thirds of the remaining third is cultivated. There is a reasonably well defined sectoral structure to the city of Harare in terms of the distribution of high-income areas of low-density housing (typically 5 000 m^2 plots per family house) and low-income areas (500 m^2) of high density housing, where population densities as high as 9 000 persons/Mm^2 can be found. In general, the population densities rise from north-east to south-west.

Table 3.13 presents the number of major livestock in the area surveyed by the Department of Veterinary Services of Ministry of Agriculture in 1997. Besides listed livestock, a large number of poultry is also raised in the area.

Table 3-12 Land use in the Lake Chivero catchment area

Land use category	Area Mm^2	Percentage
Grazing land	1 444	67.6
Cultivated land	420	23.0
Developed area	535	7.8
Reservoirs	32	1.5
Other	10	0.1
Total	2 250	100.0

Source: JICA (1996)

Table 3-13 Number of major livestock in the area

Livestock	Number
Cattle	135 000
Goat	12 000
Sheep	6 200
Pigs	21 000
Horses	2 200

Source: JICA (1996)

The area is generally gently undulating featureless plateau with altitude ranging from 1 300 to 1 500 m above sea level. The area is generally underlain by Archaean age rocks forming a part of the Zimbabwe Basement Complex. The upper part of the area is underlain by rocks of the Older Gneiss Complex containing relatively small inclusion of schistose rocks being comprised of meta-sediments and meta-volcanics of Bulawayan Age, while relatively small part of the upper extremity is underlain by granite. Harare City including its industrial area lies on the outcrop and sub-outcrop of these rocks.

In broad vegetation terms the area falls into the miombo belt that occurs all over the central African plateau between 800 m and 1 800 m above sea level and where annual rainfall is in the range 500 mm to 1 800 mm (Wild & Barbosa, 1967). A feature of the miombo woodlands is the bright coloured red, purple and green foliage of early spring. Most of the woody species flower before the rains.

Before development took place most of the large mammals found in Zimbabwe also occurred within the Manyame catchment but agricultural and urban development have made the area unsuitable for the larger wild mammals except where fenced game parks have been established.

The eutrophication of the river system and Lake Chivero in particular has led to large amounts of algae and floating aquatic plants, particularly water hyacinth (*Eichornia crassipes*), water lettuce (*Salvinia molesta*), and water fern (*Azolla filiculoides*). The floating aquatic plants interfere with commercial fishing and recreational use of Lake Chivero in particular (Marshall, 1997; Magadza, 1997). Methods of combating the water hyacinth have included spraying with chemicals and manual removal, biological control using a weevil, *Neochetina eichhoniae* has previously shown promising results.

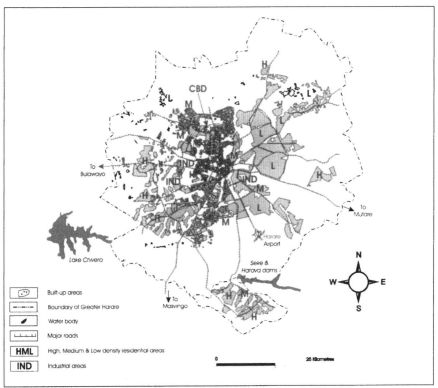

Figure 3-18 Land use map of Harare and Chitungwiza

Rooted macrophytes such as *Lagarosiphon major* are also very common in the lakes and these aquatic systems also have a diverse population of aquatic fauna. About 28 species of fish are known from the Upper Manyame, (Marshall, 1982; Magadza, 1997) and Lakes Chivero and Manyame both support commercial fisheries as the eutrophic nature of the impoundments leads to high fish productivity. A feature of the larger reservoirs is the very marked thermal stratification that occurs as the surface water warms during the summer causing two discrete layers to form. The lower layer remains colder but accumulates nutrients from the decay of sinking organic matter and becomes anaerobic. At the beginning of the cool season, stratification breaks down as the surface layer cools, the layers mix and trapped nutrients from the lower layer are released resulting in algal blooms. De-oxygenated water from the lower coming to the surface may also cause fish deaths as speculated at Lake Chivero in March-April 1996 (Moyo, 1997).

Phosphorus has been studied extensively in aquatic systems because it is considered important in the control of eutrophication (Vollenweider, 1981; Walker, 1983; Smil,

1990; Codd *et al.*, 1992; Grobbelaar & House, 1995). After analysing the total P, total N and epilimnetic chlorophyll in 493 lakes, McCauley *et al.*, (1989) concluded that the relationship between nutrients and chlorophyll is sigmoid. Their analysis also showed that total N accounted for a significant proportion of the variability in chlorophyll content, especially at high summer total P concentrations. These characteristics have been used to place lakes into different trophic categories (see Table 3.14).

Phosphorus is usually limiting productivity in freshwater ecosystems (Likens, 1972), whereas N generally limits primary production in marine systems. Phosphorus has been shown to contribute to the eutrophication of many freshwaters (Lean, 1973) and its control is most probably the best strategy for lake management and limiting eutrophication (Toerien *et al.*, 1975; Toerien, 1977).

Table 3-14 General ranges of phytoplankton production, total P, total nitrogen and chlorophyll a of lakes of different trophy

Trophic state	Primary Productivity mg C/m^2.day	Total P	Total N mg/m^3	Chlorophyll *a*
Ultra-oligotrophic	<50	<5	<250	<0.5
Oligotrophic	20-100	5-10	250-700	0.3-3
Mesotrophic	100-300	10-30	500-1000	2-15
Eutrophic	>300	10-50	500-2500	10-500
Hyper-eutrophic	>1000	30-5000	500-15000	>100

Source: Grobbelaar & House (1995)

Water supply and pollution control

Raw water is being drawn from four major lakes formed by dams in Seke, Harava, Chivero and Manyame reservoirs for water supply to the Harare metropolitan area Figure 3.13). Important morphometric data for the water supply reservoirs is given in Table 3.15. Water from Lake Chivero which is more polluted is usually blended with the less polluted water from Lake Manyame before treatment. Average daily raw water abstractions from the two Lakes since 1982 are shown in Figure 3.19. During drought years abstractions on Lake Mnayame is increased because of treatment difficulties of Lake Chivero water caused by reduced dilution. Lake Chivero receives the bulk of the wastewater from the Harare metropolis upstream.

The direct use of river water is minimal due to limited flow available during dry season. Lakes and reservoirs are also utilised for recreation and commercial fishery purposes. The water supply service for the satellite areas of the City is provided by means of bulk water supply. The combined present water demand is about 400 000 m^3/day.

Table 3-15 Main water supply sources for the Harare metropolis

Impoundment[7]	Year commissioned	Capacity (in Mm^3)	Maximum surface area (in Mm^2)
Harava	1972	9.3	2.15
Seke	1929	3.6	1.10
Chivero	1952	250	26.30
Manyame	1976	490	81.00

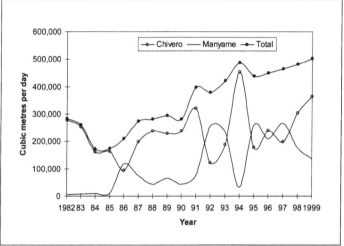

Figure 3-19 Raw water abstractions by City of Harare from Chivero and Manyame reservoirs since 1982

Two existing water treatment works (WTW), Prince Edward and Morton Jaffray, employ conventional water treatment systems which include sludge blanket clarifiers and rapid sand filters. The design capacity of the Morton Jaffray WTW and Prince Edward is 614 000 m^3/day and 90 000 m^3/day respectively. However Prince Edward WTW is operated intermittently to supplement peak demand since the safe yield of Harava and Seke reservoirs is limited to 23 000 m^3/day. The quality of water supplied to the City of Harare and adjoining towns is to a large extent reliant on the quality of the raw water abstracted for treatment. The deterioration of raw water quality has affected the operation of treatment plants. The Morton Jaffray waterworks, for instance, requires high chemical dosage (activated carbon and a polyelectrolyte in addition to aluminium sulphate, lime and chlorine), which is beyond its full capacity of chemical handling and dosing equipment.

The urban centres within the basin have a high sewerage connection ratio compared to most cities in developing countries. About 95% of the population is served by water-borne public sewerage system. Some stands in low-density areas have on-site treatment facilities, i.e. septic tanks. In application of septic tanks, the minimum stand size is principally regulated at more than 4 000 m^2. However, this restriction is loosened to 2 000 m^2 when soil test results are favourable to locate septic tanks. In Epworth the majority of houses are using Ventilated Improved Pit (VIP) latrines

---

[7] Note that since 1978 high quality effluent from Biological Nutrient Removal (BNR) modified activated sludge wastewater treatment plants has been discharged into Lake Chivero as an indirect wastewater reuse strategy to augment water supplies. At present about 125 000 m^3/day of effluent is discharged (see Table 3.16).

sponsored by the national government. Presence of unacceptable 'pit latrines' is quite limited.

Due to overload (increased population densities) and lack of maintenance of infrastructure specific sections experience sewer blockages on a frequent basis. The overflow of sewage from manhole to streets is due to deposit of sand and sludge in sewers and manholes associated with solid waste management and habitual behaviour of residents, aside from the limited capacity of sewer lines. The industrial wastewater is also a major cause of over-loading of the sewerage system and treatment works and adversely affecting their treatment efficiency. Since the river systems leading to Harare's main water supply sources pass through densely populated and heavily industrialised areas the quality of urban storm runoff at some instances is worse than sewage effluent. This complicates the water quality management efforts, because wastewater treatment plant management query the validity of strict effluent quality control when nothing obvious is done to solve diffuse source pollution.

In the Lake Chivero river basin, there are seven sewage treatment works employing different treatment systems, namely Marlborough (serving a population of 15 000), Donnybrook (134 000), Ruwa (20 000), Zengeza (465 000), Firle (946 000), Crowborough (670 000) and Hatcliffe (34 000) respectively. Norton (80 000) lies within the upper Manyame river basin downstream of Lake Chivero. The details of the sewage treatment works are provided in Table 3.16. A schematic drawing of the rivers, lakes and water uses in the area is shown in Figure 3.20.

Table 3-16 Existing sewage treatment works in the Greater Harare area and satellite towns

Works	Year first built	Type [a]	Design capacity (m^3/day)	Present influent flow (m^3/day)	River discharge [c] (m^3/day)	Irrigation use (m^3/day)
Firle	1960	TF & BNR	144 000	152 000	83 500	68 500
Crowborough	1957	TF & BNR	54 000	91 300	19 800	48 600
Marlborough	1952	WSP	2 000	2 000	-	1 000
Donnybrook		WSP	5 500	5 500	-	3 200
Hatcliffe[b]		EA	2 500	2 500	2 000	-
Zengeza		BNR	20 000	20 000	20 000	-
Ruwa		WSP	5 300	5 300	-	3 300
Norton		WSP	4 000	5 000	-	3 000
Total	-	-	237 300	281 600	125 300	127 600

a) Type of treatment system refers to: TF=Trickling Filters; BNR=Activated Sludge Biological Nutrient Removal; WSP=Waste Stabilisation Ponds; EA=Extended Aeration.
b) Hatcliffe STW discharges its effluent into the Mazowe catchment which adjoins the UMRB to the north of Harare.
c) Although at some instances records and design criteria indicate discharge to farms usually operational problems result in increased river discharges either of partially treated or untreated wastewater.

Source: JICA (1996); Nhapi *et al.*, (2002)

The first step in controlling the causes of eutrophication was adopted in Zimbabwe in the 1970's through the introduction of legislation which limits the P-concentration in effluents, derived from point sources such as treated domestic and industrial wastewater, discharging into sensitive catchments to 0.5 g P/m^3 as total phosphorus (Water Act, 1998; SI274/2000, 2000). However the control of P-loads derived from urban storm runoff, and control measures focusing on clean and cleaner production, waste minimisation and local recycling need to be investigated so as to justify the massive financial investments that have and are still being made in terms of installation of Biological Nutrient Removal (BNR) sewage treatment works.

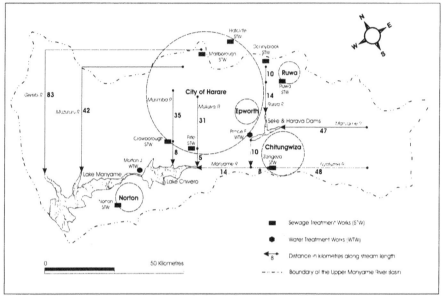

Figure 3-20 Water and sewage treatment works in relation to rivers and lakes

Increasingly, municipal wastewater treatment plants are required to remove phosphorus from their effluent, where treated wastewater is used to augment potable water supplies, at significant extra cost as with the case of Lake Chivero (JICA, 1996). Although Zimbabwean legislation, Government Notice 687 of 1977 (later revised in Statutory Instrument 274 of 2000), calls for the removal of nitrogen and phosphorus from effluents discharged to surface waters, P concentrations in wastewater are generally considered too low to facilitate direct recovery by biological techniques (CEEP, 1998). Moreover the P-fraction is enmeshed in contaminated sludge since the sewerage system mixes of industrial and municipal waste waters.

The competent authorities for water quality examination and monitoring are the Zimbabwe National Water Authority (ZINWA, formerly Ministry of Rural Resources and Water Development) through the Pollution Control Unit (PCU) at central government level and the City of Harare at local government level (ZINWA, 1998). Table 3.17 provides a summary of the water quality monitoring programmes conducted the two agencies.

Although water quality is periodically monitored by these agencies, their frequency or interval, sampling points and water quality parameters being monitored are not consistent with each other. Inter-agency co-ordination is necessary to achieve an efficient monitoring and smooth implementation of administration procedures.

Table 3-17 Water quality monitoring programme in the Lake Chivero basin

Agencies	PCU within ZINWA	City of Harare, Department of Works
Objective	• Monitoring of water pollution in rivers • Policy making for pollution control • Assess the efficiency of water pollution control methods • Development of the natural water data base	• Monitoring of pollution in public water bodies (rivers and lakes) • Monitoring of effluent discharged from the Sewage Treatment Works • Monitoring of effluent discharged from the factories
Location and number of surveillance stations	• 12 self recording water level surveillance stations • 20 river sampling points • 15 boreholes operated by the PCU	• 48 points on the rivers • 3 points on the reservoirs • 5 Sewage Treatment Works • Factories
Frequency or interval	Once a month	Once to 12 times a year depending on stations
Water quality parameters	pH, OA, Total Alkalinity, Cl⁻, SO_4^{2-}, Mg^{2+}, Ca^{2+}, NH_4-N, NO_3-N, PO_4-P, Fe, Mn, Pb, Cr, K, Na	pH, OA, DO, EC, Hardness, Total Alkalinity, Cl⁻, SO_4^{2-}, Mg^{2+}, Ca^{2+}, NH_4-N, NO_3-N, PO_4-P, Fe, Mn, Pb, Cr, K, Na, Albuminoid

Figure 3.21 shows the location of major permanent water quality sampling points in the Lake Chivero river basin. Most of the water quality data is derived from grab samples that represent instantaneous values of pollutant concentrations. In most cases a single instantaneous value cannot be used to represent average conditions; furthermore, it does not describe the variability of the pollutant concentration. In addition, much of the data is collected during dry weather conditions and is not useful for assessing storm water pollutant discharges.

Several studies have been conducted to establish and characterise pollution loads in the Lake Chivero basin particularly the nutrients Nitrogen and Phosphorus (Thornton, 1980; Thornton & Nduku, 1982; Marshall, 1982; JICA, 1996; Moyo 1997; Hranova et al., 2002; Nhapi et al., 2002 see also Box 1.J in Chapter 1). The JICA report (1996) is one of the most comprehensive and holistic studies conducted to date. Pollutant mass flow calculations were presented together with several options and scenarios for mitigating the pollution load. Pollution loads considered include point and non-point sources of pollution, domestic, industrial, agricultural and natural sources. The pollution loads generated were used in conjunction with a number of equations and mathematical formulations (Streeter-Phelps model for self purification capacity of rivers and the Vollenweider model for eutrophication) to predict the impact of discharges on the four major lakes in the UMRB. The reconstructed P-balance for Lake Chivero based on JICA (1996) with adjustments (to reflect the year 2000 situation) derived from the other studies is presented in Figure 3.22.

It is clear that P is accumulating in the Lake sediments in particular and that the Lake water quality (average 0.6 ±3 g/m^3 TP; Nhapi, 2004) might not improve even when drastic measures are taken to reduce further P-inputs because of high internal loading. Limited studies have been conducted on determine the P enmeshed in sediments of the Lake and hence the P storage indicated is only of the P in the water column and is calculated by multiplying the P-concentration of lake water by its full supply capacity.

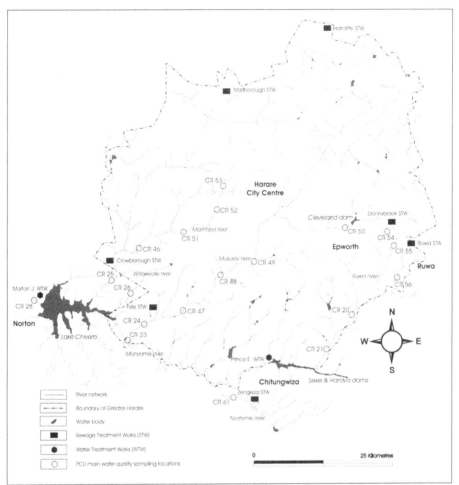

Figure 3-21 The Harare metropolis drainage network and location of major water quality sampling points

The P-balance for Lake Chivero system indicates that in year 2000 the Lake accumulated some 411 Mg as P of phosphorus (ΔP). Overall the system is not sustainable as P-inflows into the lake far exceed the P-outflows. The total P-inflows amount to some 676 Mg/a as P or approximately 2 Mg/day as P which is about a third of the total P-based fertiliser imports in Zimbabwe (refer to Figure 3.11) and represents about 4% of the total national P-fertiliser consumption. The bio-available P in the water column was calculated from the full supply capacity of the lake and the average P-concentration to give a value of 150 Mg. The P stored in the sediments is unknown.

The P-balance for Lake Chivero for year 2000 has a number of limitations as the P-balance diagram for the Zimbabwean territory (Figure 3.11). In calculating the P-fluxes and stocks most of the processes have been treated as '*black boxes*'. It also represents a snap-shot in one year and for example does not represent the river low and high flows. However it is essential for making management decisions and focusing of further research.

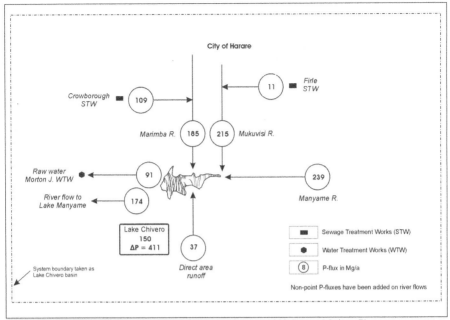

Figure 3-22 The P-balance diagram for Lake Chivero for year 2000 (Mg or Mg/a as P)
Source: Adapted from JICA (1996); Nhapi *et al.*, (2002); Nhapi (2004)

The 189 Mm^2 Marimba River catchment area a sub-catchment of the Lake Chivero basin is of particular interest in this dissertation, since it contains the micro-study catchment described in section 3.4. The Marimba River and its tributaries (including Avondale stream) drain part of the Harare city centre (Figure 3.21). The suburbs and their areas falling in this sub-catchment, the population statistics and densities for years 1992 and 2000 are shown in Table 3.18. Crowborough sewage treatment plant and some rural farmland form part of this Marimba sub-catchment (see Chapter 4). The river and its tributaries drain south westwards to the Lake Chivero. The main gauging station on the river is C24 located just before it discharges into the Lake. The longest stretch of the river meanders for a total distance of about 50 km to the lake. Flow data for the catchment area are available for the last 47 years.

The southern most tributary, the Avondale stream originates from the University of Zimbabwe grounds at an altitude of 1 490 m above sea level to drain to Lake Chivero at an altitude of 1 380 m. Most of the elevated terrain are found in the low density residential areas in northern part of the in the Marimba River catchment area. The land slope in the catchment area is generally flat and ranges from 0.3% to 0.98%. Hence, the flow rate in the Marimba river system can be expected to be initially slow becoming rather sluggish as the river meanders towards the Lake.

Table 3-18 Population and density of suburbs in the Marimba River sub-catchment

Suburb	Density class [a]	Area (Mm²)	Population		Population density (persons per Mm²)
			1992 [b]	2000 [c]	
Alexandra Park	L	1.56	4549	6475	4138
Avondale	M	5.00	16569	23583	4717
Belgravia	L	3.99	3030	4313	1080
Belvedere	M	6.76	16471	23443	3467
Crowborough	H	2.15	-	43200	20112
Dzivarasekwa	H	28.95	58599	83405	2881
Kambuzuma	H	0.82	34413	48980	59717
Kuwadzana	H	3.11	76100	108314	34843
Mabelreign	M	13.70	32558	46340	3382
Marimba Park	M	32.55	7667	10913	335
Marlborough	L	2.86	3879	5521	1930
Milton Park	L	3.78	3824	5443	1438
Mount Pleasant	L	13.22	1324	1884	143
Mufakose	H	8.34	70148	99842	11978
Ridgeview	L	1.97	2275	3238	1645
Rugare	H	1.02	10800	15372	15114
Southerton	H	7.54	14766	21017	2789
Tynwald	H	12.75	10647	15154	1189
Warren Park	H	11.16	68200	97070	8694
Westwood	M	1.12	3275	4661	4171
Total or average	**H**	**162**	**441086**	**670168**	**4128**

a) Residential area density class: L = Low density, M = Medium density, H = High density
b) Results of the 1992 Census. Crowborough suburb was established after 1992.
c) A annual population growth rate of 4.5% was assumed. This rate is regarded as normal in most urban centres in Zimbabwe.
Data source: CSO (1992); JICA (1996)

3.3 Urban agriculture in Harare

Extent and policy guidelines

The history and expansion of urban agriculture in Harare is illustrated in Table 3.19. The problem of illegal cultivation of public land started as early as the 1950's and has gained momentum ever since (Mbiba, 1995; ENDA, 1996; Bowyer-Bower *et al.*, 1996). During the 1999 to 2000 rainy season the area under cultivation had increased tremendously to an estimated 70% of all open spaces in the city (see Figure 3.23). This was attributed to general population increase, continued high rate of rural-urban migration, a growing number of dependants per household, burgeoning unemployment, and inflation resulting in a lowering of income in real terms (PASS, 1995). According to quantitative data (Matshalaga, 1997), about 40% of the residents of Tafara, a high density suburb in Harare, engage in urban agriculture as a survival strategy. Some families are even able to produce enough to meet their annual requirements of maize-meal and enough to sell.

In socio-spatial terms there are two main sources of urban food production, namely, the private space of urban gardens (on-plot i.e. within the designated residential stand), and the public open spaces in and around the city proper (off-plot). Local Government reactions to urban food production (off-plot cultivation) have generally been proscriptive, partly because it spoils the modern image that many administrators want for the city and partly because urban planning and management is not structured to incorporate such activities. Hostility and repression are, therefore, still far too frequent in Harare, despite the growing but reluctant toleration of urban agriculture which has followed recognition of increased pressures on the poor (ENDA, 1995; PASS, 1995; Mbiba, 1995; 1998).

Table 3-19 Extent of cultivated public land in Harare (1955 to 1994)

Year [a]	Area of public land ('000' m²)	As % of total open space
1955	2 700	1.0
1965	10 700	4.0
1972	14 000	5.5
1978	37 000	14.0
1980	47 600	18.5
1990	48 200	19.0
1994	92 900	36.0

a) Year of aerial photograph interpretation
Source: Bowyer-Bower *et al.*, (1996)

The City has formulated policies embedded in the City of Harare Combination Master Plan (HCMP) allowing urban agriculture (HCMP, 1992). The Master Plan tends to promote planned and organised urban agriculture. The enforcement of legislation, either through the destruction of crops or levying of fines, has been mollified to some extent by the opportunities for residents to form co-operatives and, as such, to apply for permission to use designated land for the cultivation of crops. The procedures for this are lengthy and cumbersome, and relatively few groups have, in fact, applied (ENDA, 1995).

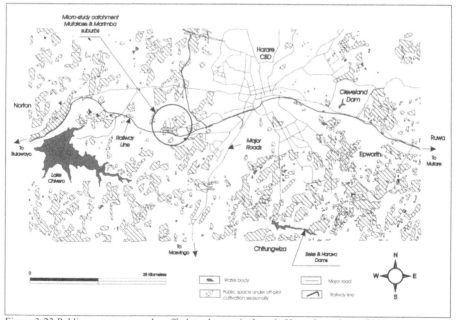

Figure 3-23 Public open spaces under off-plot urban agriculture in Harare's environs (2000)

Sites of widespread cultivation include road-side verges, vacant residential stands, vacant industrial sites, and parklands. By far the majority of the yet undeveloped open space in Harare is vlei land (57% of off-plot cultivated area; see Table 3.20) i.e. seasonally waterlogged drainage ways that are a natural feature of the environment (Thompson, 1972). Vlei land is often vacant because of its high shrink-swell capacity with drying and wetting, and its tendency to become waterlogged in the wet season, rendering generally unsuitable and or expensive to develop for building. In Harare two main types of clays occur: clay vleis developed on basic rock, which are mostly found in the northern half of Harare (largely the Gwebi, Umwindisi and the upper headwaters of the Marimba and Mukuvisi areas), and the sandy vleis occurring on granites, mostly on the southern half of Harare (major potions of the Mukuvisi) (Bowyer-Bower et al, 1996).

Table 3-20 Land type used for cultivation in Harare, 1994

Type of Land	Area cultivated (Mm^2)	% of cultivated
Railway reserve	8.0	9.2
Moderate slope	6.7	7.7
Steep slope	1.8	2.1
Roadside	8.3	9.6
Vlei	49.8	57.4
Other land	12.2	14.0
Total	86.8	100.0

Despite its widespread practice in Zimbabwe in general, growing crops on vlei land is forbidden by current national environmental protection legislation (Natural Resources Act, which has been superseded by the Environmental Management Act of 2002) which forbids cultivation within 30 m of any water course or wetland. The cultivation of public land is also forbidden by the City of Harare's Protection of Lands By-Laws of 1973, which forbid all cultivation on the municipal land that is done without the written approval of the municipal authorities, and allow for the destruction of crops grown without such approval (Bowyer-Bower, 1996).

Characteristics of on-plot cultivation

From a study survey conducted in 1995 (Smith & Tevera, 1997), more than 60% of Harare's residents were involved in house gardens or on-plot cultivation, 83% of whom cultivated food crops, while a further 23% kept livestock, of which 80% were chickens. Garden plots were generally much smaller that the off-plot sites (open spaces mostly cultivated illegally). The size ranged from about 14 to 25 m^2 in high-density suburbs and cultivation relied on piped municipal water supplies. Apart from labour and seed inputs almost all cultivators use some form of fertiliser in their garden plots (ENDA, 1995).

Organic fertiliser is more widely used than chemical fertiliser and a few who can afford use pesticides. Although maize is also grown in many gardens, the most popular crops produced are leafy vegetables (especially the variety known as rape) and tomatoes. There are also different patterns evident in terms of the use that is made of garden produce (Smith & Tevera, 1997). Table 3.21 illustrates that most of the crops are self-consumed.

Table 3-21 Use of crops grown in the house garden (on-plot)

Crop	% of households who self-consume 80% or more of crop	% of households who sell 60% or more of crop
Rape	71	20
Maize	80	27
Tomatoes	58	58
Spinach	70	35
Sweet potatoes	78	14
Cabbage	49	60
Beans	89	25

Source: Adapted from Smith & Tevera (1997)

In terms of urban management this form of agriculture is only problematic if the residents engage in livestock production, such as poultry. With such activities nuisances generated are in the form of noise, smell and possible health hazards. On-plot crop cultivation is generally not viewed as offensive by the local authorities (Mbiba, 1995; 1998).

Characteristics of off-plot cultivation

The difference between on-plot and off-plot is in the perception of ownership of land. Here land is perceived as to be '*public land*'. Anybody can utilise it without anybody else claiming individual title ownership of the land. Depending on circumstances, off-plot agriculture can be legal and illegal. It is legal if the agriculturists have a permit from the local authorities and vice versa if they do no have (Mbiba, 1995).

A large majority of the plots are located in the same suburb in which the cultivators reside. Cultivation depends almost entirely on rainfall. In terms of labour input, the majority of the families rely on their own contributions with some using hired labour, mostly from the neighbours in their local area and often in return for a share of the crop as well as for cash payment. In contrast to on-plot cultivation the majority of the off-plot cultivators use chemical fertiliser rather than organic fertiliser to increase yields. Off-plot urban agriculture is practised largely on poor soils which explain why cultivators invest in chemical fertilisers despite tenure insecurities (ENDA, 1995).

Crops cultivated in off-plot sites differ considerably from those grown in gardens, largely in terms of relative importance rather that the particular items grown (Smith & Tevera, 1997). Maize is overwhelmingly the most important crop and is mostly for self-consumption, although nearly a third of cultivators sell 60% or more of their produce (Table 3.22).

Table 3-22 Use of crops harvested from off-plot fields

Crop	80% or more self-consumed	60% sold	20% or more given away
Maize	65	30	30
Sweet potatoes	21	20	50
Groundnuts	-	31	60
Beans/pulses	33	44	33
Rape	20	30	-
Tomatoes	40	40	-

Source: Adapted from Smith & Tevera (1997)

Figure 3-24 Urban agriculture in Harare; prohibition sign in land reserved for a cemetery (a), use of mechanised soil tilling techniques (b), cultivation on steep slopes (c), stream bank cultivation (d), open space maize cultivation in a low density high income suburb (e), slash and burn techniques during land preparation (f)
Photographs: Gumbo (1998-2001)

3.4 Solid waste management in Harare

Golden Quarry landfill site was established on an abandoned gold mine. It lies between Bulawayo and Kirkman roads close to the National Sports stadium on the fringes of Warren Park North residential area, about 7 km west of the city centre. Before the closure of Pomona landfill site 12 km to the north of the city centre (see Figure 3.25), Golden Quarry received about 90% of the waste disposed of through landfill operations due to its proximity to sources of waste such as the central business district, industrial areas and most of high density suburbs (Madimutsa, 2000; Tevera, 1991).

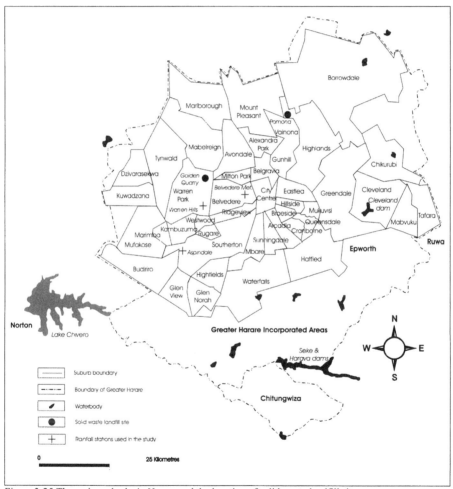

Figure 3-25 The major suburbs in Harare and the location of solid waste landfill sites

The waste disposal methods used at Golden Quarry are sanitary landfilling (controlled tipping) and open dumping for liquids. Sanitary landfill as practiced at the landfill site involves three key processes. First the waste is spread into thin layers, then it is compacted using bulldozers and landfill compactors and finally it is covered with at least 0.20 m of soil. Ideally the waste should be covered before the end of each day but because of landfill equipment shortages it is often left exposed for a couple of days.

Land filling at this site started in 1985 and was temporarily suspended in 1998, due to the continuous out break of fires. The landfill site drains into the Marimba River. The site covers an area of 0.4 Mm^2. There is no geological or engineering seal to prevent water resources pollution by leachate.

The average daily waste entering Golden Quarry is more than 700 000 kg/day (260 000 Mg/a) for solid waste and more than 100 m^3 of liquid waste (see Figure 3.26). The waste stream which is deposited at the Golden Quarry largely consist of solid materials such as metal products and scrap metal, glass, paper, plastic putrescent material. The waste stream also contains hazardous materials such as mercury or cadmium in old batteries, mercury in fluorescent lights, toxic chemicals, paints, used oil, asbestos waste, cleaning solvents, inks, dyes, and heavy metal contaminated sludge.

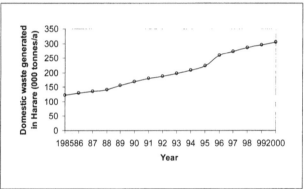

Figure 3-26 Quantities of domestic solid waste generated in Harare for the period 1985-2000 (thousand of tonnes/a)
Source data: Madimutsa (2000)

In Harare on average each resident discards about 0.5 kg/day of solid waste of which 60 to 80% is compostable or biodegradable waste. These figures are comparable to generation rates in other Third World Cities (refer to Chapter 1). About 50% of this is food scrap with a moisture content of about 50-70% (Table 3.23). The rest of the compostable material is made up of natural grass, tree leaves, weeds (yard waste) and crop residues (mostly maize stover and cobs). The latter predominate during the months of March to June i.e. during the harvesting period.

There is no comprehensive information on solid waste disposal systems in the future. Quantity of leachate emanating from these sites will depend on a number of factors. However, it can be assumed that it will increase in proportion to increase of population and waste generated in respective local authorities (Zata, 1996).

Table 3-23 Domestic solid waste characteristics for a typical high, medium and low density suburb in Harare

Category	Sunningdale	Mabelreign	Borrowdale
Paper and cardboard	11.7	13.5	23.1
Glass and Ceramics	0.8	2.3	6.5
Metals	0.8	1.1	4.0
Plastics	6.8	8.0	11.7
Leather and Rubber	0.3	0.0	0.4
Wood and bones	0.6	0.3	0.7
Organic compostable matter	75.9	73.6	47.7
Textiles	2.1	1.1	3.8
Miscellaneous	0.8	0.2	2.3
Total	100	100	100

Source: Madimutsa (2000)

3.5 The urban-shed Mufakose and Marimba suburbs

Choice of the micro-catchment

The micro study catchment falling within the Lake Chivero basin consists of Mufakose and Marimba Park suburbs, which lie within a distinct hydrological catchment area of 6.5 Mm^2. Mufakose and Marimba suburbs are situated in the western end of the city about 15 km from the city centre within the Marimba sub-catchment of the Lake Chivero basin. All the suburbs in the Marimba River sub-catchment are serviced by Crowborough sewage treatment works (Table 3.18)

The total population is believed to be between 95 000 and 105 000 inhabitants (projected from 1992 census, the median figure is thought to be closer to the actual value see Table 3.18) which translate into a population density of about 15 000 persons/Mm^2. There are an estimated 10 100 residential stands including flats and about 100 non-residential stands. The average occupancy per stand is estimated to be 9.9 people. The age structure is as follows: 0-14 years, 43%; 15-64, 54% and 65 years and over, 3 %. Most of the residents in Mufakose are in the low-income band with an estimated income per household of about US $50 per month (CSO, 1992; PASS, 1995).

Landuse

Much of the area is built up and almost all the remaining open spaces particularly near the edges are under cultivation yearly during the period November to March. Almost all of the properties are residential type with a few areas designated for commercial activities such as shops markets and other related business. The average plot sizes are 150-200 m^2 and average floor area of about 50-70 m^2 for the original housing units. However within the past 10 years a lot of backyard shacks or out-houses have mushroomed due to poor housing delivery in Harare.

In Harare it is estimated that about 70% of the population are lodgers i.e. they don't own either the house or property they reside on (HCMP, 1992; PASS, 1995, Matshalaga, 1997). The majority of the lodgers in the high-density areas are accommodated in these out-houses, which on average occupy about 25-40 m^2. On average therefore each person occupies between 7-10 m^2 of floor area and almost the same area of outdoor space. A large number of the out-houses are illegal i.e. they have not been built with City Council approval.   Inevitably this unplanned densification results in a strain on the infrastructure and also alters the local hydrology of the area (Gumbo, 2000b).   A summary of some of the characteristics of Mufakose and Marimba suburb are provided in Table 3.24.

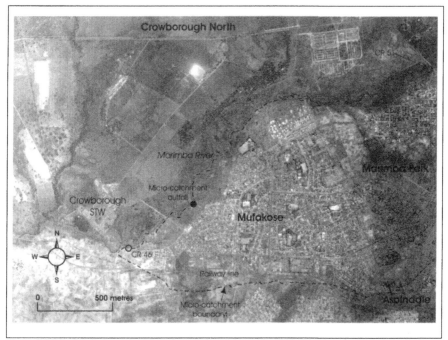

Figure 3-27 Aerial photograph of the micro-study catchment in 1997 (Mufakose and Marimba Park suburbs)

Water and sanitation

The area has its water supply coming from two main pipes with separate bulk meters. Readings from the bulk meters obtained from the City of Harare were inconsistent and not usable in this dissertation.   However water consumption figures used for billing purposes were found to be reliable and consistent[8]. The sewerage reticulation has two distinct out-fall sewers serving the area.   The sanitation system is entirely water borne. A separate system for sewage and storm-water exists.   However during rainy periods there is excessive infiltration of storm-water into the sanitary sewer system causing overloading and surcharging of the network.

The collection frequency of solid waste in the area is weekly.   However field observations indicated that there is wide spread and indiscriminate dumping of solid waste due to the irregular collection frequency by the city authorities.   In recent years solid waste collection has been exacerbated by fuel shortages, which have affected the whole country due to an economic down turn (see Figure 3.28).

---

[8] These readings are from consumer meters. Each consumer has an individual meter and a distinct account number. The recorded data does not take into account distribution losses and other forms of Unaccounted For Water (UFW) (see Chapter 4).

Table 3-24 Profile of Mufakose and Marimba suburbs in the mirco-study catchment in year 2000

Parameter	Value	Remarks
Population	100 000	Projected from 1992 census at an annual growth rate of 4.5% (CSO, 1992) (see Table 3.17). The urban annual population growth rate in Zimbabwe is between 3 and 6%. The higher values are attributed to high rural urban migration (PASS, 1995)
Average household income Mufakose	US$100	Income per month. Age structure: 0-14 years, 43%; 15-64, 54% and 65 years and over, 3 %.
Average household income Marimba Park	US$500	Income per month (PASS, 1995)
Number of households in Mufakose	9600	Semi-detached mostly, 150-200 m^2 in size (information from City of Harare Mufakose Housing Office)
Average floor area	50-70 m^2	City of Harare housing plans
Number of residential stands in Marimba	500	Plot size ranges from 1000-2000 m^2
Average floor area	100-150 m^2	City of Harare housing plans
Average occupancy per stand	9.9	For both suburbs
Average annual rainfall	820 mm/a	Long term average measured at Belvedere Meteorological Station within greater Harare
Average annual pan evaporation	1500 mm/a	Long term average measured at Belvedere
Total annual water consumption	2.1 Mm^3/a	As recorded from consumer meter readings (records kept by City of Harare Mufakose Housing Office)
Breakdown of water usage per household		The figures were established through a survey. From the year 2000 average water consumption rate of 2.1 Mm^3/a or 175 000 m^3/month the daily water use per person was calculated as approximately 60 l/p.d.
Bathing	25%	
Kitchen and other culinary functions	10%	For toilet flushing the cistern size varied from 9-12 litres. A significant volume of water is used for on-plot vegetable production (Hoko, 1999; Gumbo, 2000b). Some amount of grey water generated is also
Laundry	15%	
Toilet flushing (WC)	30%	
On-plot garden watering	20%	used occasionally for irrigating on-plot crops.
Size of area	6.5 Mm^2	System or hydrological boundary (calculated from topographical map). Two land use areas were defined i.e. impervious (built environment) and pervious (predominantly under cultivation). The two areas were calculated from a land use polygon map using ILWIS 2.2 (1998).
Total impervious area	2.1 Mm^2	About 32% of the total area from aerial photographs taken in 1997.
Impervious area consists of:		Off-street impervious area comprises of pavements and surfaced areas within plots. About 10-15% of the storm water flow ends up in the foul sewerage system
Streets (surfaced and un-surfaced)	15%	
Off-street impervious area	2%	
Roofs	15%	
Total pervious area	4.4 Mm^2	Comprises about 68% of the total area.
Pervious area consists of:		Open space is undeveloped or reserved for future development; it also consists of vleis, road reserves and stream-banks. About 60% of this area is under cultivation illegally or legally.
On-plot garden area	0.8 Mm^2	
Open public spaces	3.6 Mm^2	
Total area under cultivation	2.9 Mm^2	Both on-plot and off-plot, legal and illegal

A significant amount of solid waste is dumped into the foul sewer and storm drain systems. The storm-water in the area is drained by a combination of piped, lined and unlined open channels. The storm-water network is poorly maintained resulting in clogging and blocking of catch-pits and drains. The major problem affecting the performance of storm drains is un-collected solid waste material, which end up choking the gully pots and sewer pipes and channels (Figure 3.28). The storm-water is discharged untreated into the Marimba River, and ends up in Lake Chivero, the city's main potable water source.

Figure 3-28 Solid waste collection and disposal problems in the micro-catchment; un-collected solid waste accumulating on road verges (a), solid waste dumping in storm drains (b)
Photographs: Gumbo (2000)

A survey was carried out in 1998 and 1999 on domestic water usage in Mufakose (Hoko, 1999; Gumbo, 2000b). It included 43 households, which were monitored during this period. A questionnaire provided information on the number of occupants per stand, age groups, average water consumption, amount used for garden watering, toilet flushing and short comings of the system e.g. cuts in the water supply, sewer blockages and their frequency. The results of the survey indicated that a significant volume of water is used for on-plot vegetable production, as most residents grow vegetables as a coping strategy for food. Garden watering is significant during the dry winter season i.e. April to September (see Table 3.23).

The average monthly water use for the entire micro-study catchment in year 2000 was 175 000 m^3/month or 0.58 m^3/day per household (which translates to about 0.06 m^3/p.d or 60 l/p.d)[9]. According to the MoLG & SALA (1990b) and the HDS & BCHOD (1982) design manuals, the recommended average daily water consumption for a high-density residential suburb household is 850 l/day (maximum 1 300 l/day and minimum of 500 l/day). Therefore the year 2000 water consumption rate fell within the recommended design range, although more towards the minimum value of the range.

---

[9] In terms of the banded or stepped water tariff structure for the City of Harare in year 2000 this level of water use translated to an average water bill of about US$ 3.60/month which is about 3.6% of the estimated monthly household income in the same year (see Table 3.24).

3.6 Conclusions

This Chapter describes the study area through a systems analysis approach illustrated in Figure 1.7 in Chapter 1. The following important observations can be made:

- Zimbabwe like many developing countries has an agricultural based economy. It is perceived that for further development agricultural production would have to be increased and this can only be ensured through increased use of fertiliser. Over the past years the country has been self-sufficient in terms of P-based fertiliser (minimum imports). Preliminary analysis of the national P-balance indicates that P-mining is taking place (although in year 2000 the country was a net importer of P in terms of vegetal, animal and fertiliser commodities). The massive land degradation and loss of soil fertility is likely to be critical in the short-term and measures to mitigate soil loss and improve soil fertility are required.

- The Lake Chivero basin provides and interesting scenario in Zimbabwe where excessive P-fluxes induced by rapid urbanisation have resulted in severe pollution problems. The traditional waterborne sanitation system with its centralised end-of-the-pipe approach has offered limited solutions to the problem. About 4% of the annual national agricultural P-requirements can be met by the P-inflows into the Lake each year. Possible solutions require a more rigorous analysis focusing on the *sources* P and their *sinks*, and exploring whether decentralisation, recovery and reuse of P at household or neighbourhood could reduce the environmental impacts on the Lake ecosystem.

- The inherent uncertainty of Material Flow and Stock Accounting (MFSA) when applied to large areas with limited data is highlighted. The level of accuracy of the P-fluxes and stocks for each system described can only be improved through further research and debate.

- The micro-study area is introduced in preparation for Chapter 4. Urban agriculture is a major activity in Harare and within the micro-catchment. Urban agriculture offers an attractive destination for any P that could be recovered from the main waste P-fluxes, domestic sewage and organic solid waste.

4. Establishing fluxes and stocks in an urban-shed

4.1 Introduction

P-fluxes based on characterisation of input goods, processes, transformation, output fluxes and storage were established through measurement, field surveys and using literature values. The year 2000 was used as a base year for the analysis and it was assumed that the year represented an '*average year*'. Since the transport and transformation of P in a region is dependent on the water cycle, water flux balances were also established for the study area. The analysis distinguished all the rainbow colours of water (see Box 4.A: A rainbow of water; see also Chapter 1).

Box 4-A   A rainbow of water
Adapted from Savenije (1999) and Otterpohl (2001)

Water and sanitation has many colours. Falkenmark (1995) enriched the world of water resources with the "green water" concept: the water that vegetation uses through transpiration (in the absence of irrigation). Green water is here defined as the direct use of rainwater by plants after it has been stored in the soil's unsaturated zone. Unfortunately, green water is generally disregarded as a resource by engineers. Engineers are used to deal with "blue water". Blue water is the water that we can manage by engineering interventions and that we can allocate, re-allocate and measure by traditional monitoring. Blue water is the combination of surface water and "renewable" groundwater. The un-renewable or "fossil" groundwater is not part of the blue water and should be considered a mineral resource.

In addition there is "white water". White water is the part of the rainfall that feeds back directly to the atmosphere through evaporation from interception and bare soil. White water is sometimes considered as part of the green water, but that adds to confusion since green water is a productive use of water whereas the white water is non-productive. The white and green water together form the vertical component of the water cycle, as opposed to the blue water, which is horizontal (Falkenmark, 1995). In addition, the term white water can be used to describe the rainfall which is intercepted for human use, e.g. from roof catchments. With regard to household water use (used to clean and to nourish) water supplied can be considered and part of the blue water flux. "Black water" is waste water, the return flow of blue water used by humans. Black water consists of three components, namely; "Brown water" (faeces plus flush water), "Grey water" (waste water arising from the kitchen and bathroom i.e. not containing faecal material) and "Yellow water "(urine with or without flush water) (Otterpohl, 2001).

Finally, the last colour of the rainbow is the 'ultra-violet water', the invisible water, or the "virtual water". Virtual water is the amount of water required to produce a certain good. In agriculture, the concept of virtual water is used to express a product in the amount of water required for its production. The production of grains typically requires 2-3 m^3/kg, depending on the efficiency of the production process. Trading grains, implies the trade of virtual water (Allan, 1994; Savenije, 1999; 2000)

The micro study catchment area of 6.5 Mm^2 has an estimated population of about 100 000. There are 10 100 residential plots which translate into an average occupancy per stand of 9.9 people. In total, urban agriculture extends over an area of about 2.9 Mm^2 i.e. both on-plot and off-plot (see Chapter 3). P inflows into the "household" subsystem (mainly to do with the activities "to nourish and clean") have been established through mapping of monthly diet and detergent and soap usage of the inhabitants based on a national nutrition survey and a local solid waste study. The dynamics of urban agriculture had also been monitored for a period of two years documenting the amount of fertiliser and manure imported, the crop yields (as maize) and the quantity of phosphorus (both labile and non-labile) present in the soil after the harvest.

Equations describing the various processes and transformations have been developed from measured and collected data (Figure 1.12) and subsequently used in Stella Version 5, software by High Performance Systems (HPS) Inc (http://www.hps-inc.com). The use of Stella enables the imbibing of the systems thinking paradigm and simplifies the handling of stocks or reservoirs, flows (input and output), infrastructures and feedback loops (MFSA) of P and water fluxes within the micro study region. Stella is well documented in technical documentation available from HPS Inc (HPS, 1993). The software is also described by Hannon & Ruth (1994), and Soltzberg (1996). The software design philosophy is explained by Peterson (1994) and examples of its application in environmental systems are explained in Hannon & Ruth (1997) and Ford, (1999). One of Stella's most useful features is the ease of conducting a sensitivity analysis. This is a collection of simulations that reveals the importance of the model inputs (Ford, 1999) (see Appendix D).

For simplicity, the approach used in establishing the P-fluxes and stocks and setting up the Stella model was based on desegregation of the fluxes and stocks into four compartments, namely; the rainfall water balance, municipal water balance, household P-balance and agricultural P-balance.

4.2   The urban water flux and stock subsystem

Rainfall water balance

Water transport is one of the main ways of matter exchange between ecosystems. Hence, an understanding of the controls on the flux of water as the carrier and solvent of material in the landscape is essential. Erosion and nutrient cycling are natural processes which are linked to the hydrologic cycle. Soil generation and landscape development are in part products of weathering and the movement of sediment through the hydrologic cycle (Savoury, 1988). Water is a system. The annual water cycle from rainfall to runoff is a complex system where several processes (infiltration, surface runoff, recharge, seepage, re-infiltration and moisture recycling) are interconnected and interdependent with only one direction of flow: downstream (Debo & Reese, 1995; Savenije, 1995; 2002).

During the early stages of a rainfall event, part of the rain (R) is intercepted by foliage, branches, stems, ground surface and roof tops where it may be evaporated (I) back to the atmosphere (see Figure 4.1). As the precipitation continues, these surfaces reach their retention capacities and excess water drains to pervious soil surfaces or impervious surfaces.

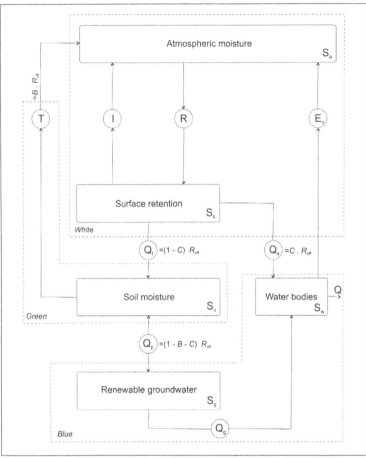

B – slope or relation between monthly effective rainfall and monthly transpiration
C – surface runoff coefficient dependent on land use and amount of effective rainfall
E_o – open water evaporation
I – interception
Q – discharge
Q_f – infiltration
Q_g – groundwater flow or seepage (±)
Q_p – percolation or capillary rise (±)
Q_s – surface run-on or run-off (±)
R – rainfall
R_{eff} – effective rainfall = R - I
S_a – storage in the atmosphere
S_g – storage in renewable groundwater
S_s – storage on the surface
S_u – storage in unsaturated soil
S_w – storage in water bodies
T - transpiration
Figure 4-1 The monthly rainfall water balance model for the micro-catchment (mm/month)

The capacity of the canopy to intercept water varies from 1 mm/day for grass to some 5 mm/day for coniferous forest (Savenije, 2004). However, several rainstorms or prolonged rainstorms during the day may cause more water to be evaporated from the leaf area per day (De Groen, 2002; Savenije, 2003).

Evaporation of intercepted precipitation may take place throughout a storm but is of primary importance after precipitation ceases. Generally, 20-35% of the annual precipitation in southern Africa is intercepted by the canopy (Pallet, 1997; Mare, 1998; Savenije, 2000; De Groen, 2002; Savenije, 2004). For an individual precipitation event, the amount of interception depends on the intensity and duration of the event (DMS, 1981; Schulze, 1987; Schulze et al., 1992). For the rainfall runoff process, interception is defined as that part of the rainfall that does not enter into any of the stocks contributing to transpiration or to the runoff process (S_w, S_g, and S_s) (Savenije, 1995, 1996; 1997).

Since there were no large surface water bodies in the micro-catchment, open water evaporation (E_o) was taken as equal to zero. It follows that the areal evaporation is the sum of intercepted water (I) and transpiration (T).

The monthly water balance for the micro-catchment is defined by:

$$\frac{dS}{dt} = R - (I + T) - Q \qquad \text{Eq. 4.1}$$

Where: S is the total sub-surface storage of the moisture in the catchment
           $S = S_u + S_g$ mm
           Q the average runoff from the catchment (including groundwater flow)
           $Q = Q_s + Q_g$ mm/month
           t average residence time of water in the soil, including ground water
                                                                          month

dS/dt, the change in the amount of water stored in the catchment, is difficult to measure. However, if the '*account period*' for which the water balance is established is taken sufficiently long, the effect of the storage term becomes less important, as rainfall (R) and Evaporation (E) accumulate while storage varies within a certain range (Viessman & Lewis, 1996; Savenije, 2003). When computing the storage equation for annual periods, the beginning of the balance period is preferably chosen at a time that the amount of water in store is expected not to vary much for each successive year. These annual periods, which do not necessarily coincide with the calendar years, are known as hydrologic - or water years. The storage equation is especially useful to study the effect of a change in the hydrologic cycle.

For a hydrological year, dS/dt may generally be neglected. On a monthly scale, the storage variation can be considered to consist mostly of groundwater (S_g). The storage variation of the intercepted water, the soil moisture and the surface water is generally small, unless there are major reservoirs in the basin (Savenije, 1997; 2003; 2004).

Estimating monthly rainfall

The long term average annual rainfall measured at Belvedere Meteorological Station within greater Harare is 820 mm/a (see Figure 3.13). Daily rainfall data for 1990 to 2000 from two rainfall gauging stations adjacent to the micro-study catchment was used in conjunction with Belvedere station data for the analysis. The two other stations are Aspindale and Warren Hills (see Figure 3.20). The daily rainfall records were aggregated into monthly totals and the variation of the monthly rainfall for the three stations was compared as depicted in Figure 4.2.

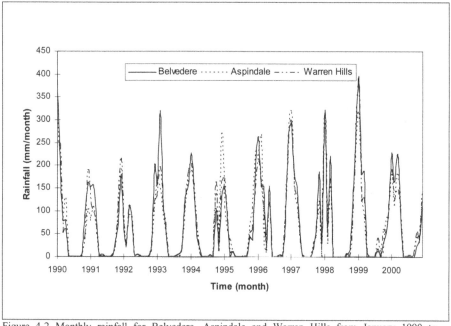

Figure 4-2 Monthly rainfall for Belvedere, Aspindale and Warren Hills from January 1990 to December 2000 (mm/month)

The eleven-year average monthly rainfall variation for Belvedere, Warren Hills and Aspindale station is depicted graphically by stock diagrams in Figure 4.3. The diagrams indicate the range of values (maximum and minimum monthly rainfall amounts recorded) about the average value for each station. Figure 4.3 shows that monthly rainfall varies greatly from year to year and thereby causing difficulties in selecting average monthly rainfall pattern. The calculated annual average for the eleven-year period for Belvedere, Warren Hills and Aspindale was 864 mm/a, 765 mm/a, and 735 mm/a respectively. The rainfall pattern recorded at the three stations is comparable in terms of magnitude and incidence, with Belvedere rainfall values generally exceeding those of Aspindale and Warren Hills. The year 2000 monthly rainfall was compared for the three stations (Figure 4.4a), with annual rainfall for Belvedere, Warren Hills and Aspindale being 971 mm/a, 834 mm/a, and 711 mm/a respectively.

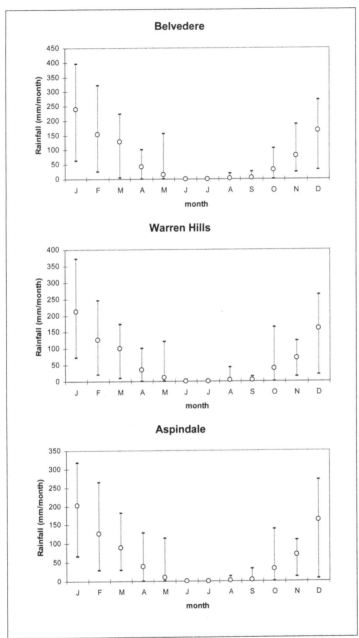

Figure 4-3 Average monthly rainfall and maximum and minimum values recorded at Belvedere, Aspindale and Warren Hills gauging stations (1990 to 2000) (mm/month)

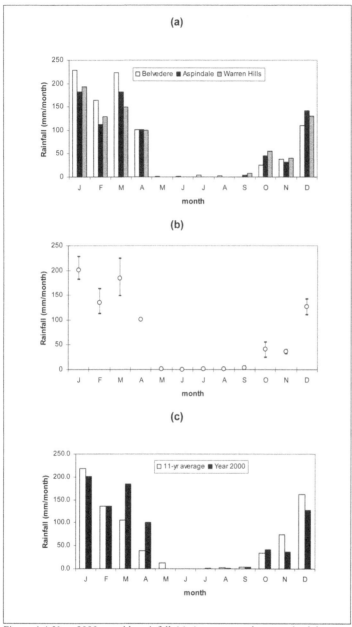

Figure 4-4 Year 2000 monthly rainfall (a) Average, maximum and minimum values (b) Year 2000 average rainfall compared to 11-year monthly average rainfall (c): for Belvedere, Aspindale and Warren Hills gauging stations (mm/month)

A stock diagram for the year 2000 monthly rainfall for the three stations (Figure 4.4b) indicates that the months of February and March experienced the largest differences between maximum and minimum rainfall recorded and the calculated average. Figure 4.4c compares the year 2000 rainfall with the eleven-year monthly average rainfall for the three stations. The eleven-year average rainfall for the three stations is 788 mm/a which is 5% less than the year 2000 annual rainfall of 834 mm/a. Hence by this comparison year 2000 was slightly above the normal year.

Overall rainfall estimation was limited by the condition that year 2000 rainfall represented an '*average year*' and that this input monthly rainfall (R) measured at the three stations was assumed to be evenly distributed over the area.

Estimation of interception

The white water component (I) consists of the direct evaporation from small stagnant pools, wet surface and foliage. Savenije (1997) showed that under the assumption that the soil moisture storage variation at a monthly or annual time step is small, the value for interception can be computed. Studies conducted in Zimbabwe by De Groen (2002) indicate that by using Markov[1] models monthly interception (I) is related to monthly precipitation (R) via an exponential function.

$$I = R\left(1 - \exp\left(\frac{-D}{\beta}\right)\right)$$ Eq. 4.2

Where: D is the daily interception threshold   mm/day
                β is the scale parameter of the exponential distribution   mm/day
                (see Box 4.B)

Equation 4.2 yields the median of the monthly interception for a certain monthly rainfall. For spatial scales of ~1 km the daily threshold (D) depends on local land cover conditions (vegetation cover, soil type, leaf area and climate) with typical values ranging between 3 and 5 mm/day. From the analysis of rainfall data from Harare, Masvingo and Bulawayo in Zimbabwe De Groen (2002) concludes that D is typically 3 mm/day and the results agree with other monthly interception models (Pitman, 1973; Clarke *et al.*, 1998).

The models described by De Groen (2002) were based on power relations between the monthly rainfall and the transition probabilities p_{01} and p_{11}. The rainfall water balance for this dissertation is based on these models. Table 4.1 summarises the calibrated parameters. The figures confirm that the relations between the monthly rainfall and the transition probabilities p_{01} and p_{11} are power functions. The p_{01} and p_{11} values adopted for the micro-catchment are those for Harare i.e. $q = 0.020$; $r = 0.55$; $u = 0.20$ and $v = 0.24$.

---

[1] Markov process' means that the probability of occurrence of an event in a certain time step depends on the state of the system in the previous time step. Markov rainfall models do not only exist at time steps of days, but also of hours (Hutchinson, 1990), months and years (Gregory *et al.*, 1993). Markov models have been proven applicable in many semi-arid and also in some more humid climates (De Groen, 2002).

Box 4-B   The scale parameter from Markov processes
after De Groen (2002)

For Zimbabwe a Markov process of two states (dry/rainy) is sufficient to describe the variability within a month. This means that the Markov process depends on two transition probabilities only: the probability of a rain-day after a dry day, p_{01}, and the probability of a rain-day after a rain-day, p_{11}. The transition probabilities p_{01} and p_{11} can be described by logistic or power functions of the monthly rainfall. For many locations in the world where two state Markov processes have been used, it has been shown that the following power functions between monthly rainfall (R) and the transition probabilities p_{01} and p_{11} are applicable.

$$p_{01} = q(R)^r \qquad\qquad\qquad (-) \qquad\qquad \text{Eq. 4.3}$$

$$p_{11} = u(R)^v \qquad\qquad\qquad (-) \qquad\qquad \text{Eq. 4.4}$$

where q, r, u and v are the only parameters that need to be calibrated at a few locations (~300 km) using daily rainfall series. The scale parameter β follows directly from the Markov process:

$$\beta = \frac{R}{n_r} = \frac{R(1 - p_{11} + p_{01})}{n_m p_{01}} \qquad\qquad \text{(mm/day)} \qquad\qquad \text{Eq. 4.5}$$

Where n_r is the number of rain-days in a month and n_m is the number of days in a month (days/month). For climates where occurrence of rain-days does not agree to a Markov process, another relationship between the monthly rainfall (R) and the number of rain-days n_r can serve to compute the scale parameter (β). The power functions are not only applicable in Zimbabwe, but also other locations in the world for which Markov rainfall models have been published.

Table 4-1 Calibrated model parameters for different rain stations

Location	$p_{01} = q\,R^r$		$p_{11} = u\,R^v$	
	q	r	u	v
Harare	0.020	0.55	0.20	0.24
Masvingo	0.030	0.43	0.20	0.24
Bulawayo	0.044	0.34	0.20	0.24
Peters Gate (SA)	0.094	0.33	0.034	0.40
Hyderabad (India)	0.092	0.38	0.024	0.53
Indianapolis (Indiana, USA)	0.129	0.30	0.045	0.42
Kansas (Missouri, USA)	0.129	0.27	0.061	0.30
Sheridan (Wyoming, USA)	0.216	0.22	0.084	0.30
Tallahassee (Florida, US)	0.127	0.29	0.017	0.55

Source: De Groen 2002

Effective rainfall is the part of rainfall that is not intercepted and subsequently feeds transpiration and runoff processes. At monthly time steps the expression for effective rainfall ($R_{eff} = R - I$) immediately follows from the equation for monthly interception Equation 4.2:

$$R_{eff} = R * \exp\left(\frac{-D}{\beta}\right) \qquad\qquad \text{Eq. 4.6}$$

Using the average monthly rainfall (R) recorded for the three stations (Belvedere, Aspindale and Warren Hills) from 1990 to year 2000, corresponding values of interception (I) and effective rainfall (R_{eff}) were calculated using Equation 4.2 and 4.5. The results from the model are depicted in Figure 4.5. Savenije (2000) estimates that a typical value of the residence time (t) for storage on the surface (S_s) is of the order of 0.8 days.

Estimation of overland flow

As rainfall reaches the surface it meets the first separation point (see Figure 4.1). At this point part of the rainwater returns directly to the atmosphere, which is called evaporation from interception (I). The remaining rainwater infiltrates into the soil until it reaches the capacity of infiltration. This is called infiltration Q_f. If there is enough rainfall to exceed the interception and the infiltration, then overland flow (also called surface runoff) Q_s is generated. Surface runoff or overland flow (Q_s) seldom occurs in non-perturbed forest soils because of the porous litter layer acting analogous to a sponge. The overland flow is a fast runoff process, which generally carries soil particles. Since the most fertile soil horizons occur at the surface of a soil, erosion of these horizons can result in a serious loss of the soil nutrient capital and reduction in stream water quality. A river that carries a considerable portion of overland flow has a brown muddy colour and carries debris (Pitman, 1973; Savenije, 2003).

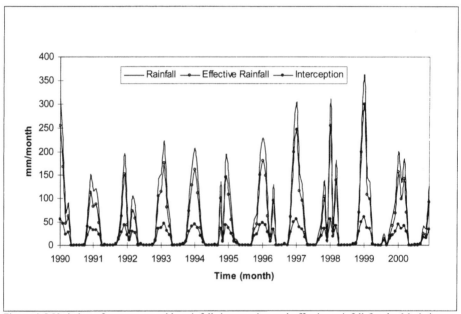

Figure 4-5 Variation of average monthly rainfall, interception and effective rainfall for the Marimba River catchment (1990-2000) (mm/month)

In general the relationship between overland flow or surface runoff (Q_s) and effective rainfall (R_{eff}) is given by the following expression:

$$Q_s = C * R_{eff}$$
Eq. 4.7

where C is a dimensionless effective runoff coefficient (-)

The runoff coefficient indicates the part of the effective precipitation that will become blue water (Q_s). From gauged data of Q_s and R, the effective runoff coefficient (C), can be calculated as follows:

$$C = \left(\frac{\sum Q_s}{\sum R_{eff}} \right)$$
Eq. 4.8

where $\sum R_{eff}$ and $\sum Q_s$ are the annual effective rainfall and annual runoff in mm/a.

The micro-study area did not have a flow gauging station and hence the value of C (the effective runoff coefficient) was determined indirectly by using the entire Marimba catchment in which the study area lies (see Figure 3.19). The Marimba river catchment is a sub-catchment of the Chivero basin covering about 190 Mm^2. The catchment can be classified as partially urbanised and the main stream meanders from urban to rural areas for about 50 km, from an altitude of 1 470 m above sea level before discharging into the lake at about 1 370 m (see Figure 3.19). Table 4.2 shows the breakdown of land use within the catchment. More than 50 % of the urban or municipal area was developed after 1980.

Table 4-2 Land use of the Marimba catchment derived from topographic series maps

Land use	Area (Mm^2)	% of total
Urban/ municipal	126	67
Rural	64	33
Total	190	100
Urban		
Industrial parks	5	3
Central Business District	3	2
Low density residential	49	26
High density residential	39	21
Open spaces (parks, cemeteries, gardens, golf courses)	30	15
Rural		
Cultivated farm land	29	15
Bush or forests and woodlands	35	18

Effective rainfall (R_{eff}) calculated (see Figure 4.5) from the average rainfall (R) for the three stations (Belvedere, Aspindale and Warren Hills) was used together with runoff data recorded (Q_{ob}) at Station C24 located at the mouth of Lake Chivero (see Figure 3.20) on the Marimba River. The observed runoff data was obtained from the Department of Water Development (DWD) for the entire period since the station was established. The average monthly river flow depth from 1953 to year 2000 is depicted in Figure 4.6. Using the corresponding data series of R_{eff} and Q_{ob} for the period 1990 to year 2000 and average value of $C = 0.26$ was obtained. Using this value of C surface runoff values were computed (Q_{co}) using Equation 4.7.

Chidavaenzi (2001) used the Soil Conservation Service[2] (SCS-SA) design flood estimation model for small catchment in southern Africa to estimate runoff from effective rainfall data (USDoA, 1986; Schulze *et al.*, 1992). Rainfall measured at the three stations was also used in conjunction with runoff recorded at Station C24 (1990 to 1999) to calibrate and validate the model (Hranova *et al.*, 2002). A lumped multiple linear regression model[3] was developed for a composite Marimba River catchment area (Chidavaenzi, 2001). The model was specific for the individual catchment area and required only readily available daily rainfall data as input to estimate runoff depth. A correlation coefficient of 85% was attained on average indicating a good fit between the observed and computed storm runoff depth[4]. An average value of $C = 0.24$ was obtained for R_{eff} and Q_{co} simulated using the SCS-SA approach.

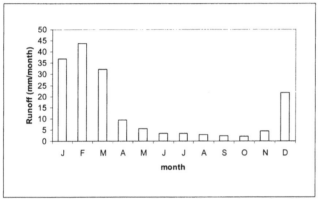

Figure 4-6 Average monthly runoff measured at gauging station C24 on the Marimba River (1953-2000) (mm/month)

Monthly observed flow (Q_{ob}) and simulated (Q_{co}) at point C24 for the two models (i.e. the SCS-SA model and the model described in this dissertation) were compared and illustrated graphically for the years 1990 to 1999 as shown in Figure 4.7. Overall there was a discernible trend and agreement between the observed flow depth and the simulated.

---

[2] The Soil Conservation Service (SCS) uses a curve number (CN) procedure for estimating runoff. The effects of land use and treatment and infiltration are embodied in the method (Viessman & Lewis, 1996). It does not directly estimate infiltration. The method was developed from studies of small agricultural watersheds. The major inputs are drainage area, average watershed slope, storm type and distribution, duration, watershed composite curve number and depth of rain and or its return period when designing storm drains (USDoA, 1986; Schulze *et al.*, 1992).

[3] Multiple linear regression is based on the assumption that there is a straight-line relationship between each independent variable and the dependent variable i.e. between monthly rainfall and monthly runoff.

[4] By plotting the monthly rainfall versus monthly runoff on a scatter diagram it was possible to determine whether there was correlation between rainfall and runoff during that particular month.

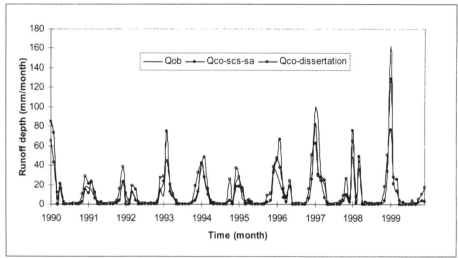

Figure 4-7 Comparison of observed (Q_{ob}) and computed (Q_{co}) runoff depth for Marimba River catchment (1990 to 1999) (mm/month)

Further analysis of the calculated R_{eff} and Q_{ob}, indicated that the memory of the catchment area system was about one month or less, meaning that the effect of rainfall on the river discharge was detected at C24 near Lake Chivero until one month after the end of the rains in April. Similarly the response time lag for the catchment area was approximately one month, meaning that the effect of the start of the rains was detected at C24 in November of each year, one month after the start of the rains in October (see Figure 4.8).

Limitations and sources of error in the estimation of C for Marimba River and its tributaries could have been due to urban return flows that include sewage effluent discharge i.e. although the rivers are naturally ephemeral, some sections of the river, flow throughout the year due to human interference in the natural water cycle (Figure 4.9). It was estimated that sewage effluent mostly emanating Crowborough sewage treatment works was about 2.5 mm/month (475 000 m³/month or 16 000 m³/day) which corresponds to the effluent during dry weather flow of about 19 800 m³/day (see Table 3.13).

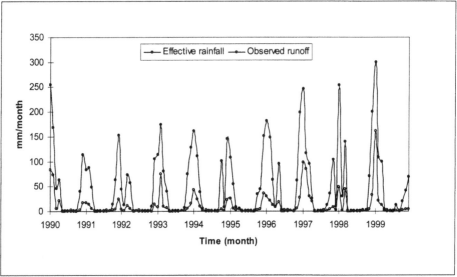

Figure 4-8 Comparison of monthly effective rainfall and observed runoff for the Marimba River catchment (1990-1999) (mm/month)

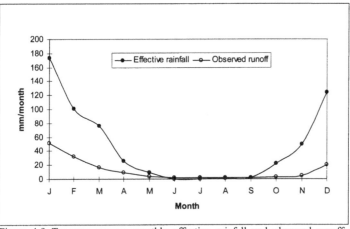

Figure 4-9 Ten-year average monthly effective rainfall and observed runoff for Marimba River catchment (1990-1999) (mm/month)

Finally the land use characteristics of the entire Marimba River basin might not have been fully hydrologically similar to the micro-study catchment. The land use characteristics of the micro-study catchment are given in Table 3.24 (Chapter 3).

Estimation of transpiration

Infiltration (Q_f) is the process by which water enters the soil. If the infiltration rate is greater than the precipitation intensity, all the water enters the soil profile and no overland flow occurs. Infiltration rates for surface layers of forest soils are generally in the range of 0.8 to 12.0 m/day, greatly exceeding most precipitation event intensities (Viessman & Lewis, 1996). From the water balance model (Figure 4.1) it follows that infiltration (Q_f) is the difference between effective rainfall and the component of effective rainfall which forms overland flow (Q_s), i.e.:

$$Q_f = (1 - C)R_{eff}$$
Eq. 4.9

Unsaturated soil water storage (S_u) is a function of infiltration capacity (Q_f) which is related to the surface area of soil particles (i.e., particle-size) and the amount of porosity occurring between these particles (i.e., soil structure). Soil pores occur across a wide range of diameters and are often categorised as macro-pores (>60 μm) and micro-pores (<60 μm). Water is present in macro-pores following precipitation events and is drained by the force of gravity. After water has freely drained, the soil is at field capacity and has a soil water potential generally between 0.01 to 0.03 MPa. Water in macro-pores is not available for plant use because it freely drains from the soil profile and is lost from the rooting zone. Water held in very small micro-pores (<0.2 μm) is held so tightly that plants are not able to extract if for use. The permanent wilting point is the soil water potential to which plants can effectively utilise water and corresponds to a soil water potential of approximately 1.5 MPa. Thus, the pores in the diameter range 0.2 to 60 μm are the primary storage pores for plant available water (i.e., water held between approximately 0.01 and 1.5 MPa). The distribution of pore sizes is primarily a function of the soil texture and structure (Savoury, 1988; Viessman & Lewis, 1996).

Mare (1998) in a study of the Mupfure catchment in Zimbabwe estimated that the average retention time (t) for S_u is approximately 5 months (Savenije, 1997; 2000). The monthly transpiration model used in this dissertation is based on daily transpiration model using the so-called spells approach (dividing the month into time steps of expected spell lengths) as suggested by De Groen (2002). The daily transpiration model relies on certain assumptions on the functioning of the soil water reservoir and these are listed in Box 4.C.

Equations relating soil moisture (S_u) (see Figure 4.10) and actual transpiration (T_{act}) have the form:

$$T_{act} = T_{pot} * \min\left(\frac{S}{S_b}, 1\right)$$
(mm/day)   Eq. 4.10

Where
S is the available soil moisture content, which is the actual soil moisture content minus the wilting point (mm)
S_b the available soil moisture content below which transpiration is soil moisture constrained (mm)

S_b is typically 50 to 80% of the maximum available moisture content in the root zone S_{max} (Shuttleworth, 1997; De Groen, 2002). The ratio depends on atmospheric demand and on vegetation. With any soil moisture content between S_b and S_{max}, the moisture evaporates at potential transpiration (T_{pot}).

Box 4-C   Assumptions for daily transpiration model
after De Groen (2002)

- Effective daily rainfall is the gross rainfall minus the daily interception as defined in Equation 4.5.
- Transpiration is equal to potential transpiration, unless the available soil water content is below a certain limit. This limit S_b is usually 0.5-0.8 of the maximum available soil moisture content S_{max}, depending on the soil and the vegetation. Below that limit the transpiration decreases proportionally to the available soil water content.
- Overland flow is not a limitation to soil moisture replenishment.
- Infiltration is assumed to benefit the soil water content homogeneously.
- Only when the soil water storage is saturated does rainfall recharge groundwater or contribute to overland flow.
- The soil moisture is not fed by groundwater in the rainy season.

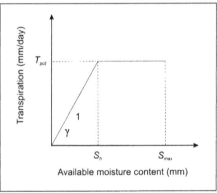

Figure 4-10 Typical form of relationship between soil moisture content and actual transpiration

The resulting relation between effective rainfall (R_{eff}) and transpiration (T) are almost linear for the lower range of effective rainfall. With increasing monthly rainfall the slope gradually diminishes until at a certain point maximum transpiration is reached and the slope is zero. The derivation of the monthly transpiration (T) expression is outlined in Box 4.D.

Box 4-D   Derivation of monthly transpiration model
after De Groen (2002)

The equation for monthly transpiration as a function of monthly rainfall and initial soil moisture condition has the following form:

$$T(R) = \text{Min}\big(A' + B * (R - I), T_{max}\big) \qquad\qquad \text{(mm/month)} \qquad \text{Eq. 4.11}$$

where

A' is the intercept with the $R = 0$ axis, which depends on the initial soil moisture S_{start} and the potential transpiration $T_{pot.d}$ (mm/month)

B slope of the linear part of the relation between monthly rainfall R and monthly transpiration T (-)

R monthly rainfall   (mm/month)

I monthly interception   (mm/month)

T_{max} monthly transpiration restricted by potential transpiration and the time it takes to progress from moisture constrained to unconstrained transpiration   (mm/month)

What remains is to derive slope B, intercept A and the maximum monthly transpiration T_{max} as a function of monthly rainfall R, maximum soil moisture content S_{max}, soil moisture content at which moisture constrained transpiration occurs S_b, initial soil moisture content at the start of the month S_{start} and potential transpiration $T_{pot.d}$

For notational convenience, the parameter γ (days) is transformed into a dimensionless parameter by dividing it with a time step of one month and with the number of days in a month n (days/month):

$$\gamma^\circ = \frac{\gamma}{n\Delta t} = \frac{S_b}{T_{pot} * \Delta t} \qquad\qquad \text{(-)} \qquad \text{Eq. 4.12}$$

where, as previously defined,

Δt is time in a month = 1   (month)

S_b available soil moisture content at the boundary between moisture constrained transpiration and potential transpiration   (mm)

T_{pot} monthly potential transpiration, which is an input parameter   (mm/month)

The monthly potential transpiration is the sum of the values of daily potential transpiration for all days in the month:

$$T_{pot} = \sum_{0}^{n} T_{pot} \qquad\qquad \text{(mm/month)} \qquad \text{Eq. 4.13}$$

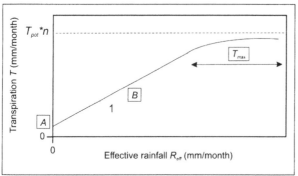

Figure 4-11 Monthly model for transpiration as described in Equation 4.11

As an illustration, with 30 days in a month the realistic values S_{max} = 120 mm, T_{pot} = 4 mm/day and S_b/S_{max} = 0.5, yield γ = 12 days. Hence in practical cases A is almost equal to $S_{start}/\Delta t$; all soil moisture is transpired if it does not rain for the whole month. For the purposes of this dissertation the value of the carry-over parameter A' is assumed to be zero. A value for $1/\gamma°$ of around 2 is quite realistic (De Groen, 2002).

The slope B is the rate (-) at which transpiration changes with effective rainfall. A change in effective rainfall affects the length of dry and wet spells and the average amount of rainfall on a rain-day. All these relate to effective monthly precipitation in a non-linear way, which is determined by the power relations between monthly rainfall and the conditional probabilities for rainfall occurrence p_{01} and p_{11}. The slope B of the linear part of the relationship between effective rainfall and monthly transpiration is given by Equation 4.14:

$$B = 1 - \gamma° + \gamma° \exp\left(-\frac{1}{\gamma°}\right) \qquad \text{(-)} \qquad \text{Eq. 4.14}$$

The relationship described by Equation 4.14 is depicted in Figure 4.12. The slope B is independent of the amount of soil moisture at the start of the month S_{start}. As indicated previously a realistic value of $1/\gamma°$ is in the order of 2, or $\gamma/(n\Delta t) \approx 0.5$.

For the Marimba River catchment a ratio S_b/S_{max} of 0.5 was assumed for S_{max} = 120 mm, implying that S_b = 60 mm based on soil characteristics around Harare and in the micro-catchment. The values adopted are indicative of predominantly shallow soils of granitic origin (refer to section 4.4 in this Chapter). From Equation 4.12 $\gamma°$ = 0.33, which implies that $B \approx 0.68$ from Equation 4.14. For $\gamma° \downarrow 0$ the slope B approaches 1, which means that every increase in effective rainfall is translated into transpiration (De Groen, 2002). Using the value of B = 0.68, the water balance of the Marimba catchment was simulated.

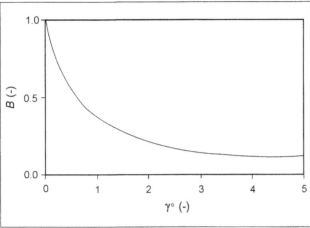

Figure 4-12 Relationship between and slope B, the slope between effective rainfall and transpiration
Source: De Groen (1992)

Estimation of groundwater storage and flow

Percolation (Q_p) is the process by which water moves downward through the soil under gravitational forces. The rate of percolation depends on the water content and the hydraulic conductivity of the soil. Hydraulic conductivity is a measure of the ability of a soil to conduct water in response to a gradient in water potential. During rainfall events when the soil is near saturation, if the hydraulic conductivity of the soil exceeds the precipitation rate, the water will move downward through the soil profile to the groundwater storage (S_g) and eventually emerge as stream flow (Q_g). Percolation (Q_p) can lead to nutrient losses from the soil profile as nutrients are leached below the rooting zone.

According to the water balance model used in this dissertation (see Figure 4.1) Q_p is given by Equation 4.15[5]:

$$Q_p = (1 - B - C)R_{eff} \qquad \text{(mm/month)} \qquad \text{Eq. 4.15}$$

Percolated water that moves to deeper aquifers generally follows streamlines that are much longer and the residence time of the water can be several years before the water emerges on the surface or as stream flow.

---

[5] Percolation (Q_p) of water from the rooting zone to the subsoil takes place when between the phreatic level and the boundary of the rooting zone is equal o greater than the soil suction at the root zone. Capillary rise (-Q_p) is the upward vertical flow of water from the phreatic level to the lower boundary of the rooting zone. There is no capillary rise if there is percolation, and vice versa (If there is equilibrium, there is neither capillary rise nor percolation). It is assumed that equation 4.15 gives the net percolation and that on a monthly time scale and in view of the land uses in the Marimba catchment $Q_p \gg (-Q_p)$.

Ground water flow (Q_g) is the movement of water under gravitational forces in the direction of a drain. When percolating water reaches a layer of low conductivity, a zone of saturation can occur (perched water table) and water can move down slope through the more permeable layer. Ground water flow results in lateral transport of nutrients from high elevation sites to lower elevation sites or directly to the stream channel.

The way this outflow (Q_g) behaves is generally described as a linear reservoir, where outflow is considered proportional to the amount of storage:

$$Q_g = {S_g}\big/{K} \qquad\qquad \text{(mm/month)} \qquad \text{Eq. 4.16}$$

where K is the reservoir constant, the '*time scale*' or the '*residence time*' of the water in S_g with a dimension of time (month). Equation 4.16 is an empirical formula, which has some similarity with the Darcy equation of flow of a liquid through a porous medium (Savenije, 2003).

In this dissertation change of ground water storage (dS_g/dt) for an '*average year*' was assumed to be equal to zero and hence for nutrient transport Q_p can be assumed to be equal to Q_g. This is because the observed hydrograph (Figure 4.13) at C24 does not demonstrate a recession at monthly time scale, K is the order of magnitude of one month or less. This is not surprising since the catchment consists of hard-rock with just a small layer of eroded rock with limited storage capacity. The underlying geology is granitic and it is assumed that in an *average year* the ground water storage is negligible as all ground stored water flows out within a period of 4 months (Amos *et al.*, 2003). As a result the sum of B (0.68) and C (0.24) are almost equal to 1.

Using the values of B, and C determined for the Marimba catchment the rainfall water balance for the micro-catchment was calculated using the average rainfall (R) for three stations for year 2000 (see Figure 4.4). Figure 4.14 shows the rainfall water balance for the micro-catchment. The values of R_{eff}, I, T, Q_s and Q_g are used in subsequent calculations for the municipal water balance, and P balances for the household and agriculture subsystems.

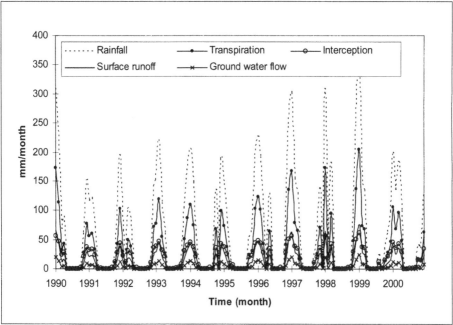

Figure 4-13 Observed monthly rainfall (*R*) and simulated monthly transpiration (*T*), interception (*I*), surface runoff (*Q_s*) and ground water flow (*Q_g*) for the Marimba catchment (1990-2000) (mm/month)

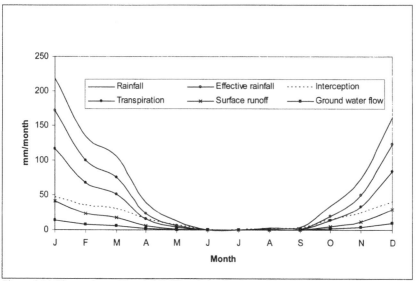

Figure 4-14 Observed monthly rainfall (*R*) and simulated monthly effective rainfall (*R_eff*), transpiration (*T*), interception (*I*), surface runoff (*Q_s*) and ground water flow (*Q_g*) for the micro-catchment in year 2000 (mm/month)

Municipal water balance

Attempts have been made to incorporate the interference of man in the hydrological cycle through the introduction of the water diversion cycle, which includes water withdrawal and water drainage. This diversion cycle exerts a significant influence on the terrestrial water cycle, especially in highly economically developed regions with a dense population (Savenije, 2003).

The municipal water system (Figure 4.15) defined in this dissertation consists of water supplied per unit area per month (W) through a piped network system, and mainly used for cleaning, nourishing and for local on-plot irrigation within the household subsystem[6] (refer to Table 3.19). The source in relation to the study area is a series of reservoirs described in Chapter 3 and hence W can be regarded as part of the natural '*blue water*' flux and stock system.

Accurate measurement of water flow and quality determinands for a number of dwellings or housing units is difficult due to diurnal and seasonal variations in water demand and use[7]. Based on survey data of the characteristics of domestic water usage together with metered consumption reasonable figures for water supplied (W) and components of the various water using activities can be established. From the studies carried out by Hoko (1999), and Gumbo (1998; 2000) water use fractions in the household subsystem were established (refer to Chapter 3 and Table 3.19). Assuming that storage of W is negligible for this subsystem a water balance equation as below can be assumed:

$$W = aW + bW + cW + dW \qquad\qquad \text{(mm/month)} \qquad \text{Eq. 4.17}$$

Where: aW is the component of municipal water used for on-plot irrigation
            bW is the component of municipal water used for cleaning that gives rise to the '*grey water*' flux consisting of three activities (bathing, laundry and non-consumptive uses in the kitchen i.e. washing dishes and food)
            cW is the component of municipal water used for toilet flushing thus giving rise to the '*brown water*' flux
            dW is the component of municipal water consumed either directly or as water contained in food
            a, b, c and d are respective fractions for water using activities obtained through observation (-) ($a + b + c + d = 1$). These are defined in detail with the relevant values in Table 4.3.

---

[6] Although the micro-study catchment considered in this dissertation consists of commercial activities which have a demand on water, these are considered to have an insignificant proportion of the total water demand in the area i.e. the area is predominantly residential.

[7] Usually a more detailed diurnal variation of water consumption, sewage flows and storm runoff is desirable to obtain a more dynamic picture of the water fluxes but for the purposes of this dissertation this is not necessary since a monthly time step is being considered and hence hourly or daily peaks and troughs are evened out.

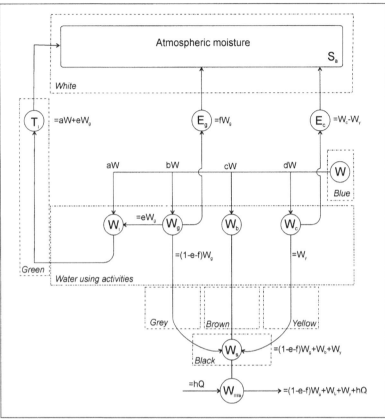

a, b, c, and d – water use fractions determined through observation (-) (see Table 4.3)

e – fraction of grey water diverted onto land for garden irrigation purposes (-)

E_c – water excreted by the human population mainly through respiration and perspiration $(W_c - W_y)$

E_g – evaporation from drying-out of laundry material and kitchen utensils

f – fraction of grey water evaporating either directly from laundry or after dripping to the ground surface (-)

h - fraction of storm water entering the foul sewer system conveyed as part of municipal sewage (-)

Q – discharge $(Q = Q_s + Q_g)$

S_a – storage in the atmosphere

T_i – transpiration arising from garden irrigation using municipal water

W – municipal water supply normalised to the catchment area

W_b – brown water generated from household activity related to toilet flushing after defaecation

W_c – municipal water consumed by population either directly or contained in ingested food products

W_g – grey water generated from activities related to nourishing and cleaning (kitchen, bathroom and laundry)

W_i – municipal water used for garden irrigation $(W_i = aW + eW_g)$

W_{ms} – municipal sewage water, which is a combination of yellow, black, and proportion of grey and storm water conveyed through a pipe to a sewage treatment plant $(W_{ms} = (1-e-f)W_g + W_b + W_y + hQ)$

W_s – foul sewage or '*black water*' which is a combination of yellow, brown, and a proportion of grey water $(W_s = W_{ms} - hQ)$. This corresponds to dry weather flow.

W_y – volume of yellow water excreted by an equivalent adult population per month normalised to the micro-catchment area (A)

Figure 4-15 The monthly municipal water balance model for the micro-catchment (mm/month)

W_i forms part of the '*green water*' system and it is assumed that it is entirely transpired (T) and enters the atmospheric '*white water*' system as moisture[8]. The values of the water use and diversion fractions (a, b, c, d, e, f, and h) for the micro-study catchment were primarily determined from a water use survey (see Chapter 3) (Hoko, 1999; Gumbo, 2000b). The values shown in Table 4.3 were verified against values from various literature sources listed below the table. The values are based on an average water use of 60 l/p.d as determined in Chapter 3 (compare also with Table 1.9 in Chapter 1).

Water use data record (metered consumption) for 1995 to 2000 was analysed and the variation of W values during this period are shown in Figure 4.16a. The water use is also affected by the cost of water, water use restrictions imposed by the local authority in drought situations or when the water system is approaching its design capacity. During good rainy seasons the water use restrictions are normally relaxed leading to increased consumption as evidenced in the year 1998 to 1999 in Figure 4.16a.

In year 2000 the average monthly water consumption as recorded by the Mufakose Housing Office was about 175 000 m^3/month[9] (W = 175 000/A = 175 000 × 10^3/6.5 × 10^6 = 27 mm/month) which translates to a total annual water consumption of 2.1 Mm^3/a (324 mm/a). The peak consumption of 31.4 mm/month was recorded in the month of October whilst the minimum consumption of 22.9 mm/month was recorded in the month of March (Figure 4.16b).

Table 4-3 Water use and diversion fractions for the micro-study area

Parameter	Value	Comment
a	0.20	Determines the volume of water used for on-plot garden irrigation
b	0.47	Determines the grey water volume generated from the kitchen b_1 = 0.08; bathroom b_2 = 0.25 and laundry b_3 = 0.15 ($b = b_1 + b_2 + b_3$)
c	0.30	Determines the volume of black water arising from toilet flushing. The average toilet cistern size is 10 litres. The water contained in faecal matter is assumed to be negligible when compared to the flush water volume[10].
d	0.03	Based on assumption that an average adult person in the study area consumes about 2 litres of water per day either directly or indirectly
e	0.15	Fraction of grey water applied on land for garden irrigation purposes
f	0.10	Fraction of grey water which evaporates from laundry fabrics, kitchen utensils and human body after washing and bathing
h	0.12	Fraction of storm-water entering the foul sewer system

Sources: MEWRD (1978); HDS & BCHOD (1982); MoLG & SALA (1990a, b, & c); Gleick (1996); Hoko (1999); Gumbo (2000b)

---

[8] It is assumed that interception (I), runoff ($Q = Q_s + Q_g$) arising from irrigation and the component of water used in the metabolic functions of the plant during photosynthesis or retained in plant tissues are all negligible i.e. $W_i = T_i$.

[9] This is the total volume of water billed and did not include losses or Unaccounted For Water (UFW). In Harare UFW is estimated at about 30% of the water produced (GKW *et al.*, 1996; O'Dwyer *et al.*, 1997).

[10] The upper limit normal for faecal excretion of water is approximately 1 l/p.d for adults. Large volumes are excreted in diarrhoea up to 10-20 litres/p.d in cases of Asiatic cholera. In congenial chloride diarrhoea, which persists throughout life; infants 0.3 l/pd; 3.0 l/p.d for adults (Lentner *et al.*, 1981).

The volume of urine excreted by an adult person per day (y)[11] was determined by observing the excretion rates of four students (one female and three male) from the Civil Engineering Department at the University of Zimbabwe in 1999. The students were aged between 22 and 24 years and had weight range of 55 to 65 kg. The observation was carried out in two-week periods in the month of June (winter time) and the other two weeks in the month of October (summer time) so as to establish the maximum and minimum rates of excretion respectively. The results obtained are summarised in Table 4.4.

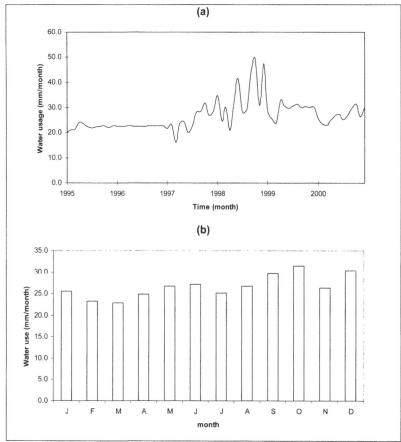

Figure 4-16 Variation of monthly water use 1995 to 2000 (a) Year 2000 monthly water use (b) in the micro-study catchment (mm/month)

The results of y obtained did not indicate any significant variation between male and female, age or body weight. The median values were adopted as an estimate for y values and it was assumed that these values were applicable for the adult population in the micro-study catchment. By comparison the y values observed were less than

---

[11] The physical properties and the chemical composition of urine are highly variable and are determined in large measure by the quantity and type of food consumed; more over the excretion of products of the endogenous metabolism is dependent on body mass. The composition of individual urine specimens is not entirely identical to that of the 24-hour urine; the excretion of many urine constituents is subject to a day and night rhythm (Lentner *et al.*, 1981).

recorded excretion rates in Europe (Feachem *et al.*, 1993; Del Porto & Steinfeld, 1999; Johansson *et al.*, 2000; Hoglund, 2001; Vinneras, 2002) (see Table 1.9 in Chapter 1).

Table 4-4 Urine excretion rates (*y*) for four students during winter and summer time (1999)

Parameter	Summer values	Winter values [a]
Average weight of students (kg)	60	60
Average age (years)	23	23
No of samples	56	56
Minimum (l/pd)	0.22	0.20
Maximum (l/pd)	0.60	0.76
Standard deviation (l/pd)	0.08	0.12
Average (l/pd)	0.38	0.47
Median (l/pd)	0.36	0.45

a) Winter time was assumed to be 4 months in a year i.e. the months of May, June, July and August

According to Lentner *et al.*, (1981) the urinary volume rate is increased at high water and salt intakes and a protein rich diet; it is decreased at low water intake, diet rich in carbohydrates, and in case of profuse sweating (see Table 4.5). Oliguria condition is regarded as the normal excretion rate. The volume rate is lower during the night than during the day (this rhythm is not observed in pregnant women).

Table 4-5 Evaluation of urinary volume rate

Condition	Volume rate (l/pd)
Severe polyuria	6-15
Moderate polyuria	2-6
Oliguria	<0.5
Anuria	<0.1
Total anuria	0

Source: Lentner *et al.*, (1981)

The total volumetric rate excreted by an equivalent adult population per month in the micro-catchment was computed from the following expression:

$$W_y = \frac{0.77H}{A}[30 \times y] \qquad \text{(mm/month)} \qquad \text{Eq. 4.18}$$

Where: *H* is the estimated population of the study area in year 2000
A is the area of catchment (m^2)
y is the urinary volumetric rate of excretion in l/p.d in winter or summer
0.77 is a factor for adjusting *y* according to the age structure of the population[12]. It was assumed that the age group 0-14 years and >65 years excrete 0.5*y*. Hence from Chapter 3 (Table 3.24) *(0.43 + 0.3)0.5 + (0.54) = 0.77*

The volume of domestic sewage (*W_s*) water depends on the population inhabiting the houses provided with water supply (*W*) and sewerage, as well as the per capita rate of domestic wastewater production. Field measurements of municipal sewage flows

---

[12] Johansson *et al.*, (2000) assumes an excretion rate of between 60-80% of the adult urinary volumetric for children and the elderly.

(W_{ms}) during wet weather indicated that 10-15% of rain water falling on roofs and other impervious surfaces enter the foul sewerage system (hQ_s), also component of the ground water flow (hQ_g) because of either design limitations or illegal connections by residents to the foul sewer system in a bid to avoid on-plot flooding.   This inevitably leads to overflows.   Although the dilution ratio of storm water and foul sewage (W_s) at the end-of-the-pipe conform to the Zimbabwean guidelines for the disposal of sewage effluent during wet weather (MEWRD, 1978), local malfunction of the sewerage system is experienced (see Chapter 3 and 5).

The municipal water balance for year 2000 was computed using respective values (Equation 4.17, Figure 4.15 and 4.16 and Table 4.3) and is depicted in Figure 4.17a for the major fluxes and Figure 4.17b for the minor fluxes.

The average return flow factor i.e. W_s/W obtained is about 0.66 (66%), which is less than 85% the recommended value for sewerage system design. This confirms that in general the sewerage network is design standards are conservative (MoLG & SALA, 1990a; HDS & BCHOD, 1982).

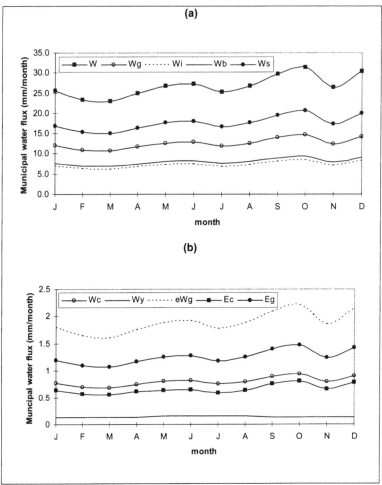

Figure 4-17 Municipal water balance-major fluxes (a) Municipal water balance-minor fluxes (b) in the micro-study catchment (mm/month)

4.3   The household subsystem

Food, beverage, soap and detergent P-flux

The two main activities being focused on with regard to P-bearing materials and connected to household activities are to '*nourish*' and to '*clean*'. To nourish comprises all processes and goods to produce solid and liquid food for man and to clean comprises all processes to maintain human health and to provide environmental protection from pollution (Baccini & Brunner, 1991). The food and beverage system consists of the food that residents in the micro-study area consume, the places where it is produced and the often-complex process by which it gets from producers to consumers.

P inflows into the "*household*" subsystem are predominantly food and beverage (P_{fb}) and P-bearing soaps and detergents (P_{sd}). Storage change ($dP/dt = 0$) is taken as negligible by assuming that as soon as household P-bearing goods are purchased they are immediately consumed or used within a month. The main P out-fluxes of concern included '*yellow water*' (P_y), '*grey water*' (P_g), '*brown water*' (P_b) (combination of the three components '*black water*' (P_s) and finally solid waste stream (P_{sw}) (see Figure 4.18).

The P-balance equation is hence defined as follows:

$$P_{fb} + P_{sd} = P_y + P_b + P_g + P_{sw} \qquad \text{(kg/month)} \qquad \text{Eq. 4.19}$$

The amount of P in food and beverage was established through mapping of weekly diet (then converted to monthly diet) of the inhabitants based on a national nutrition survey (CSO, 1998) and a local solid waste study (Mukonoweshuro, 2000; Madimutsa, 2000).   The most common food and beverage items consumed in considerably amounts and at least periodically were inventoried and the total P-content was determined in the laboratory. The P-content of maize meal and the most common leafy vegetable called 'rape' was determined in the laboratory using the ascorbic acid spectrophotometric method. The samples were first digested using the H_2SO_4/Se/salicyclic acid and H_2O_2 procedure (Vark & Van der Lee, 1989; APHO, 1992; Suess, 1995). Four test runs were conducted at the UNESCO-IHE Institute of Water Education Laboratory in Delft, the Netherlands and the set of results are shown in Table 4.6. Literature values[13] were used to compliment and confirm laboratory results (FAO, 1968; Matshalaga, 1997; Chitsiku, 1991; CSO, 1998; FAO, 2001) (Refer to Appendix B, Table B.4 for the full list).

---

[13] FAO (1968) and Chitsiku (1991) provide a compilation of data on nutritive values of foods indigenous to different regions of Africa and Zimbabwe in particular, respectively. More than 2000 food items were selected to represent the most commonly used food items in Africa. The tables provide basic and useful data and analytical information on proximate composition for use by those involved in the evaluation and improvements of the diets consumed in Africa. Nutritionists making food consumption surveys and scientists analysing foods could use the tables to identify systematically and scientifically the food products in terms of accepted scientific nomenclature (see Appendix B, Table B.5).

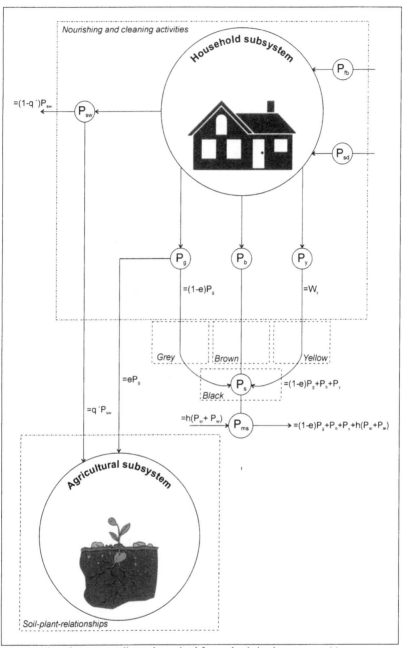

e – proportion of grey water diverted onto land for garden irrigation purposes (-)

P – phosphorus flux as P in kg/month

P_b – brown water P-flux emanating from toilet flushing of human faecal material

P_{fb} – food and beverage P-flux reaching the household subsystem i.e. imported from outside micro-catchment and some produced within

P_g – grey water P-flux emanating from activities related to nourishing and cleaning, taken as equal to P_{sd}

P_{le} - leaching P-flux due to percolation and groundwater flow (elaborated in the next section)

P_{ms} – municipal sewage P-flux, which is a combination of yellow, black, and proportion of grey and storm water P-fluxes conveyed through a pipe to a sewage treatment plant ($P_{ms} = (1-e)P_g + P_b + P_y + h(P_{sr} + P_{le})$)

P_s – foul sewage or '*black water*' P-flux which is a combination of yellow, brown, and proportion of grey water P-fluxes ($P_s = P_{ms} - h(P_{sr} + P_{le})$). This corresponds to the P-flux during dry weather

P_{sd} – soap and detergent P-flux reaching the household subsystem

P_{sr} - surface runoff P-flux dissolved in storm water (elaborated in the next section)

P_{sw} – organic solid waste P-flux derived from household activities and local vegetation growth and die-off derived from the agricultural subsystem

P_y – yellow water P-flux derived from urinary excretion

q' – proportion of organic solid waste which is either deliberately composted or is uncollected and end up being manure on agricultural land

Figure 4-18 Monthly household P balance model for the micro-catchment (kg/month as P)

Table 4-6 P-content of maize meal and leafy vegetable

Parameter	Maize meal	Leafy vegetable
No of samples	4	4
Minimum (g/kg as P)	1.8	0.9
Maximum (g/kg as P)	2.1	1.3
Standard deviation (g/kg as P)	0.1	0.2
Average (g/kg as P)	2.0	1.1
Median (g/kg as P)	2.1	1.1

The major food items (Table 4.7) were grouped according to the FAOSTAT (2001) database food classes used in Chapter 2 and Appendix A. The results of the national nutritional survey (CSO, 1998) provided a guideline to determine the average figures of consumption rate per person per annum for the micro-study area population for each food item. The combined P-content of identified food items is shown also in Table 4.7.

Using the P-content (σ) of the individual food and beverage (from laboratory results and Appendix B) the corresponding P-influx per month (P_{fb}) was calculated from the following expression:

$$P_{fb} = 0.77H\sum_{i=1}^{n} M_{fb,i} \times \sigma_{fb,i} \qquad \text{(kg/month)} \qquad \text{Eq. 4.20}$$

Where: H is the estimated population of the study area in year 2000

M_{fb} quantity of food and beverage per food group consumed per person per month (kg/p.month)

σ_{fb} is the phosphorus content as P in food or beverage material expressed as a ratio (kg/kg)

0.77 is a population adjustment factor according to the age structure

From Table 4.7 it can be deducted that an average adult in the micro-catchment consumes about 42 g/p.month of P which translates to 1.4 g/p.d as P (500 g/p.a as P). Phosphorus is the second most abundant mineral element in the 'human body[14] (the first is calcium), accounting for more than 20% of the body's minerals. Calcium

---

[14] Mean human mass worldwide = 50 kg, contains 0.9 kg phosphorus, most of it in bones. Accumulation of P in body mass is negligible for adults (i.e. > 14 years).

phosphates, for example, are the major constituent of the skeletal bones and teeth and contain 85% of the body's total phosphorus. Lack of phosphorus not only affects bone structure, but also appetite, growth and fertility. Recommended values for daily intake of phosphorus as P are shown in Table 4.8 (Johnston & Steen, 2000).

Table 4-7 Main components of the food flux in the household subsystem

Food group	kg/p month	P (σ) g/kg	P g/p month
Cereals (a)	13.3	1.85	24.6
Meat and fish (b)	1.4	1.55	2.2
Milk and eggs (c)	1.3	0.93	1.2
Vegetables (d)	7.3	0.85	6.2
Nuts (e)	0.7	4.05	2.8
Starchy roots (f)	0.6	0.46	0.3
Pulses (g)	0.9	4.62	4.2
Beverages (h)	0.9	0.13	0.1
Total	**26.4**	**-**	**41.6**

a)   The most common food is sadza (thick mealie meal porridge with the consistency of mashed potatoes made from white maize meal). Mealie meal has a P-content (σ) of 2.0 g/kg and it is estimated that an average household (9.9 individuals) consume about 20 kg of mealie meal per week (i.e. ~ 0.3 kg/pd). Bread is the second largest quantity of cereal based food consumed. It was estimated from the solid waste study that an average household consumes about two loaves of bread per day, each loaf weighing about 0.7 kg (i.e. ~ 0.07 kg/pd). Other cereals consumed include, rice, sorghum and barley, bringing the total consumption to about 0.43 kg/pd).

b)   Sadza is predominantly taken with meat, fish (mostly small fish known as 'kapenta') and some leafy vegetable. Meat products are inclusive of chicken, offal, beef or pork. From a random questionnaire confirmed by the quantity of bones in solid waste it was estimated that each household consumes approximately 3.5 kg per week of meat and fish (0.05 kg/pd).

c)   Popular milk products are 'fresh' milk and 'Lacto' also called 'sour or curdled milk'. The latter is normally consumed with sadza as relish. On average five 600 ml sachets of fresh milk were identified for each household and about three for 'Lacto' in the weekly solid waste generated (i.e. consumption rate of about 70 ml/p.d of milk products). It was assumed that each person consumes at least three eggs in a week (80 g per person per week).

d)   The popular vegetable is called 'rape' and has a typical P-content (σ) of about 1.1 g/kg. It is estimated than more that about 60% of the vegetable requirements are grown by the householders (see Chapter 3, Table 3.17 and 3.18). Other important vegetable food items identified included, cabbage, onions, tomatoes and other local and traditional vegetables. It was estimated that an average household consumes about 6.3 kg of leafy vegetables per week (0.06 kg/pd).

e)   Groundnuts either cooked or roasted are consumed especially following the harvest period i.e. after the month of April. This was evident from observations of solid waste stream i.e. higher fraction of groundnut shells and crop residues.

f)   These predominantly consisted of potatoes and sweet potatoes. The estimated consumption rate was about 0.6 kg/p month.

g)   Mostly peas, soya beans and green beans usually taken as relish with sadza. Estimated consumption rate was 0.9 kg/p month.

h)   Opaque beer called 'Chibuku' is the most popular beverage with a significant P-content. The beer is made from maize extracts, sorghum and barley. An average adult (based on volumes delivered to the local beer gardens) consumes about 4 l/week or 30 l/p.month (refer to Appendix B, Table B.5).

Table 4-8 Recommended daily intake of phosphorus for humans

Group	g/p.d as P
Children	0.48
Adults	0.70
Pregnant and lactating mothers	0.92

Source: Modified from Johnston & Steen (2000)

The type of soaps and detergents used were studied in a survey (Table 4.9) by observing weekly usage per household (Hoko, 1999; Mukonoweshuro, 2000). Chemical analysis for P-content (σ) of the most popular soaps was carried out using the ascorbic acid spectrophotometric method after digestion as with maize meal and leafy vegetable samples described previously. Table 4.10 shows the average P-content of soaps and detergents analysed together with the value range for the four test runs per sample. The quantities used per month multiplied by the average P-content of (Equation 4.21) in soap and detergents produced the related P-flux as soap and detergents (P_{sd}).

$$P_{sd} = H\sum_{i=1}^{n} M_{sd,i} \times \sigma_{sd,i} \qquad \text{(kg/month)} \qquad \text{Eq. 4.21}$$

Where: H is the estimated population of the study area in year 2000
M_{sd} quantity of soap and detergent used per person per month (kg/p month)
σ_{sd} is the phosphorus content as P in soap or detergent material expressed as a ratio (kg/kg)

Table 4-9 Common soaps, detergents and dishwashers in frequent use in the study area

Manufacturer	Washing soap	Bathing soap	Detergents	Dishwasher
Lever Brothers	Sunlight, Brilliant	Geisha, Lifebouy	Surf, Skip	Vim, Handy Andy, Sunlight liquid
Olivine Industries	Perfection	Romance, Jade	-	Lustro
Wallace Laboratories	-	Dolphin	Bymo	-
Colgate Pamolive	-	Choice, Protex	Cold Power	-
United refineries	Impala	Image	-	Sparkle
Other	Big Ben, King	Unity	-	Bubbles, Dish-brite

Table 4-10 Soap and detergent flux in the household subsystem

Group	kg/p month	P (σ) g/kg	P g/p month
Washing soap (a)	0.45	1.10 ±0.3	0.50
Bathing soap (b)	0.40	0.45 ±0.2	0.18
Detergent (c)	0.32	2.50 ±0.3	0.80
Dishwasher (d)	-	-	-
Total	**0.81**	**-**	**1.48**

a)   Washing soap is normally available in 750 g bars and an average household uses approximately 1.5 bars per week i.e. 4.5 kg/month or 0.45 kg/p.month

b)   Bathing soaps are usually packaged in 250 g (e.g. Choice soap) or 500 g tablets (e.g. Geisha). An average household uses about two tablets per week for the larger soaps i.e. 4 kg/month per household

c)   Detergents are available in powdered form with Surf and Cold Power being the most popular. Detergents are used interchangeable with washing soap. A weekly household load of laundry uses about 0.8 kg of detergent i.e. 3.2 kg/month

d)   Most of the dishwashers identified had a very low P-content. The most popular dishwasher is Vim which is a chlorine based compound. Hence dishwashers were not considered in the analysis of P_{sd}

Sewage and solid waste P-flux

The main components of the foul sewage P-flux (P_s) or '*black water*' are the yellow, brown, and a proportion of grey water P-fluxes (refer to Figure 4.18). Laboratory tests on the P-content[15] of '*fresh*' urine (one week old) were conducted during 1999 (refer to section 4.2). The sixteen tested samples (eight each for summer and winter urine) were obtained from a one week composite sample for each of the four students. A median value of 2.67 g/l was obtained for the sixteen samples tested (Table 4.11). There was no discernible difference observed between the P-content values for summer and winter urine samples. This could largely be due to the fact that composite samples were used. The expected results were that winter samples would have a lower P-concentration (dilute) than summer urine samples (concentrated).

Stale urine samples (more than six months old) were also tested in order to establish whether the P-content of urine diminished with time. From the results of four samples tested there was no clear indication that the P-content varied with time as compared to values obtained for the 'fresh' samples (see Table 4.11).

Table 4-11 P-content in urine samples (σ_y) as P

Parameter	Fresh samples	Stale samples [a]
No of samples tested	16	4
Minimum (g/l)	1.44	1.30
Maximum (g/l)	2.34	2.20
Standard deviation (g/l)	0.30	0.38
Average (g/l)	1.90	1.78
Median (g/l)	1.89	1.81

a) Stale urine samples were obtained from an ecological sanitation consultant and researcher Dr Peter Morgan of Aquamor in Harare. The samples had been stored in 2 litre plastic containers for a period of more than six months.

By combining the average volumetric urine excretion rates in Table 4.4 (0.36 + 0.45)/2 = 0.40 l/p.d and the median value of P-content for fresh urine of 1.89 g/l (Table 4.10) the P excretion rate for an average adult in the study catchment was calculated as 0.76 g/p.d (22.7 g/p month).

Normal urine generally reacts acid owing to the sulphuric and phosphoric acid derived from the degradation of proteins and phospholipids (mean pH of 6.17 for adults). On a vegetarian diet it can turn alkaline since organic acids of fruits and vegetables are broken down to bicarbonate. The urine pH is subject to a diurnal rhythm; the urine is least acid (sometimes alkaline) after awakening, and most acid toward midnight.

---

[15] About 95-100% of P in urine is in the form of phosphate ions (primary, secondary and pyrophosphate). Urine contains 60-80% of P intake. Most of it comes from food; a smaller amount originates from the endogenous metabolism of organic phosphates. The rate of excretion reaches a maximum during the night and a minimum in the forenoon. It is affected by parathyroid hormone and dependent on renal function (Lentner *et al.*, 1981). For <14 and >74 age group a 50% excretion rate of the adults can be assumed. Similar volumetric excretion rate for diet containing meat and vegetarian has been observed (Larsen & Gujer, 1996).

Under extreme conditions the urine can be acidified to a pH as low as 4.0 (extreme range of values 5.3 to 7.2) (Lenter et al., 1981).

The magnitude of the P-flux due to '*brown water*' (P_b)[16] was derived indirectly from wastewater characterisation studies and calculation (Inter-consult & GFJ, 1997). In 1997 a consortium of consulting engineers conducted a comprehensive study of the characteristics of wastewater emanating from Mabvuku and Tafara suburbs east of Harare city centre (see Figure 3.25). The two suburbs have a similar socio-economic profile as that of Mufakose suburb. Mabvuku and Tafara in 1995 had a population of about 60 000 inhabitants residing in 7 010 houses i.e. occupancy rate of about 8.5 persons per household.

Monthly water consumption and the corresponding sewage flows were monitored between 1990 and 1995, the results are shown in Figure 4.19a. Samples of influent raw sewage at Donnybrook sewage treatment works were collected randomly for period of 24 months (January 1994 to December 1995), the concentration of phosphorus as ortho-phosphate (P-PO_4) was analysed and the results obtained indicated a variation of TP shown in Figure 4.19b. From the data presented in Figure 4.19a it was deducted that the average return flow factor (W_{ms}/W) was about 0.75 or 75%. The average monthly sewage flow was calculated as 3 500 m³/month. This flow rate included the wet weather component i.e. hQ value.

Mass load calculations [Volume (m³/month) × P concentration (mg/l)] for the period 1994-1995 were performed and the resultant P loads kg/month divided by the population of Mabvuku and Tafara suburbs. The average dry weather flow was 3 500 m³/day (105 500 m³/month) at an average P concentration of 16.3 mg/l as ortho-P. To convert ortho-P values to total P it was assumed that 70-80% of the P in sewage is in dissolved ortho-phosphate form (see footnote 15 and 16) i.e. average sewage concentration of 21.1 mg/ as P.

---

[16] The major portion of P in faeces is the form of calcium phosphate (80%), organic substances (15%), and little phosphate ions. In adults 14-30% or more of faecal solids consists of dead bacteria and 25-40% of food residues (cellulose, muscle fibre, etc.); about one-third are inorganic substances one third-nitrogenous substances, one-sixth lipids, and one-sixth cellulose and like substances (Lentner et al., 1981; Larsen & Gujer, 1996).

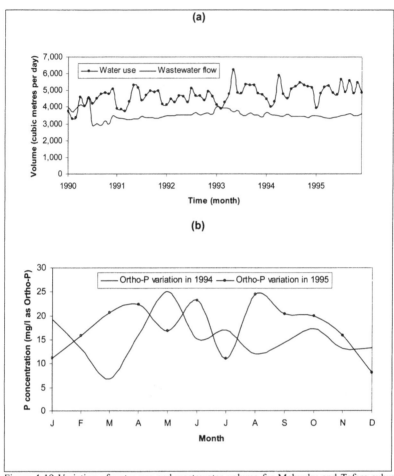

Figure 4-19 Variation of water use and wastewater volume for Mabvuku and Tafara suburbs 1990 to 1995 (m³/month) (a) Ortho-phosphate ion concentration of raw sewage measured at Donnybrook sewage works 1994 to 1995 (mg/l) (b)

Population equivalent values of P in municipal sewage i.e. the contribution per person per month (g/p month) as shown in Figure 4.20 were calculated for Mabvuku and Tafara suburbs. An average value of 37.3 g/p.month was obtained (straight line in Figure 4.20). From this value, and, assuming that the P-flux in surface runoff and leaching components which enter the foul sewer system i.e. $h(P_{sr} + P_{le})$ are small compared to that of the foul sewage P-flux ($P_s \approx P_{ms}$), and that the grey water P-flux (Pg) consists only of the soap and detergent[17] inputs, then, P_b can be calculated as $37.3 - (1-e)P_g - P_y = 37.3 - (1 - 0.15)1.48 - 22.7 = 13.2$ g/p month. The value of P_b calculated in this way shows that brown water P-flux makes up about 35% of the municipal sewage (P_{ms}).

---

[17] Grey water P-flux consists also of food residues due to cleaning activity of kitchen utensils.

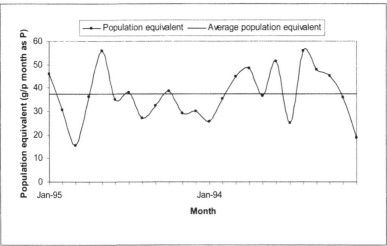

Figure 4-20 Calculated population equivalent of P in municipal sewage (g/p month) for Mabvuku and Tafara suburbs 1990 to 1995

For '*brown water*' the daily stool mass and transit time are affected by the composition of the diet (see Table 4.12). After ingestion of a diet low in indigestible carbohydrates ('low in fibre content') the stool mass is small and the transit time prolonged; the reverse is true for diet of 'high fibre content'. After prolonged fasting the stool mass decreases to about 10-22 g/pd. In adults on the customary western European diet, daily mass in excess of 200 g is pathological. It may range up to 1 000 g or more (extreme values of 6 000 g have been recorded) in cases of chronic pancreatic insufficiency, celiac disease, regional enteritis, and small bowel resection.

Table 4-12 Mouth-to-anus transit time and daily stool mass in relation to diet for adults

Diet	Transit time (hrs) [a]		Stool mass (g/pd)	
	Mean	Extreme range	Mean	Extreme range
Diet of high fibre content (vegetarians)	42	18-100	225	150-350
Mixed diet	44	23-64	155	120-260
European diet	83	44-144	104	39-223

a) The mouth-to-anus transit time was determined as the time within which 20 out of 25 ingested plastic cubes appear in the faeces
Source: Lentner *et al.*, (1981)

Solid waste in this dissertation is considered as unwanted material from households, which is not discharged through the pipe. The study of solid waste went beyond finding the composition of the solid waste and the disposal techniques but was also used as a way of finding out the socio-economic status of the community in Mufakose and Marimba suburbs. As discussed previously the solid waste analysis was used to establish the diet in the study area and to confirm the type of detergents used. Different solid waste categories contains different P-content hence by finding out the quantities of these categories produced one can establish the mount of P exported from the study area as solid waste.

The research recorded the weight of solid waste generated and analysed its composition, drawing on 100 samples from 50 households (Madimutsa, 2000;

Mukonoweshuro, 2000). The household level sampling was useful for determining the weekly volumes generated per household as well as for gauging the composition of the waste (see Figure 4.21). The households were selected using random, non-probabilistic sampling, working only with those households who agreed to participate. Based on the catalogue of suburbs in Harare inferences were made for suburbs with similar economic and social characteristics (refer to Chapter 3).

From the analysis of observed results, the solid waste generation rate in the micro-study catchment varied significantly with the time of the year, from 0.15 kg/p.d to about 0.3 kg/pd. The higher solid waste generation rates were recorded during the months coinciding with the harvesting period i.e. March, April and May. The values observed were comparable to the values quoted in Chapter 3. It was observed that a significant amount of crop residue from off-plot cultivation ended up at the household and reflected as an increase in the bio-degradable fraction of the solid waste generated (Madimutsa, 2000; Mukonoweshuro, 2000).

Up to 65% of the waste was found to be potentially compostable organic material consisting of mostly food scrap, agricultural residues and garden waste (Table 4.13). The variability in the composition for the 100 samples was below 10% suggesting general uniformity in the socio-economic status of the households. The solid waste composition figures obtained (Table 4.13) compared well to those listed in Table 3.23 in Chapter 3. An average generation rate 0.23 kg/p.d was adopted implying an organic solid waste fraction generation rate (λ) of 0.15 kg/p.d or 4.5 kg/p month. The moisture content of the different fractions was determined by measuring the dry mass of each solid waste component after heating in an oven for 48 hours at a constant temperature of 105 $^{\circ}$C.

Figure 4-21 Sorting and analysing solid waste from the micro-catchment, determination of bulk weight (a) and waste fraction weight (b)

Table 4-13 Solid waste characteristics as discarded in the micro-study catchment

Solid waste	Composition (%)	Moisture Content (%)
Paper and cardboard	11 ±8	8
Organic material	65 ±8	70
Plastics	8 ±8	2
Glass	5 ±8	2
Metals	3 ±8	1
Wood, leather and rubber	2 ±8	15
Textiles	4 ±8	10
Miscellaneous inert material	2 ±8	15

The calculated weighted average P-content of the main components of the food flux of 1.6 g/kg (see Table 4.7) was used as the representative P-content value of the solid waste. With suitable adjustment for moisture content of the organic waste (70%), the solid waste P-flux (P_{sw}) was calculated using Equation 4.22.

$$P_{sw} = \lambda H \sigma_{sw} \qquad \text{(kg/month)} \qquad \text{Eq. 4.22}$$

Where:　　H is the estimated population of the study area in year 2000
λ quantity of dry organic or biodegradable solid waste fraction generated per person per month (1.35 kg/p.month)
σ_{sw} is the phosphorus content as P in the organic fraction of solid waste expressed as a ratio (0.0016 kg/kg)

The average solid waste P-flux per person (P_{sw}/person) calculated from equation 4.22 is 2.16 g/p.month as P[18]. From the survey it was established that about 10% of the organic solid waste by weight was either deliberately composted on-site or due to collection problems dumped into open land where it decomposed and got incorporated into the soil matrix (i.e. $q' = 0.10$). The various components of the household water and P-flux per person per month for the micro-study catchment are compared and summarised in Figure 4.22a and 4.22b.

The household P-flux balance per person (Figure 4.22b) shows for that there is a discrepancy or error of 8.2% between the input and output P-flux. This is due to a number of assumptions made in the calculation of P_b and P_{sw} in particular. However the error margin is deemed to be acceptable for this kind of analysis.

---

[18] For a Swedish community 21% of P-content of food stuffs ends up in solid organic waste (household level), 65% in urine and faeces, 4 % waste from trade (market) and 10% in industrial and restaurant waste (processing) (Jonsson, 1999).

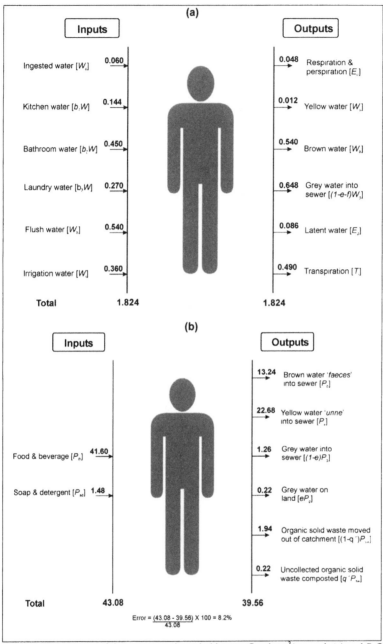

Figure 4-22 Summary of the per person household water flux in m³/p.month (a) and P-flux balance in g/p.month as P (b) for the micro-study area in year 2000[19]

---

[19] The values are per adult person per month (p month) and are based on an average adult weight of 60 kg.

4.4   The agricultural subsystem

Soil deficiencies are a major constraint on African agriculture. Soil in the savanna are extremely weathered, the chemical status is poor due to deficiencies in phosphate and organic nitrogen, and the amounts of phosphorus and sulphur mineralised annually are often below requirements of high crop yields. At the same time highly acidic and weathered soils have high capacities for phosphorus fixation (Fernandes, 1989; Smaling, 1993; Rockstrom, 1997).

The optimum environment for phosphorus availability requires that the soil climate is just slightly acid to slightly alkaline. This requires the soil reaction to be between pH 6.3 to pH 7.0. The degree of aliveness of the soil also plays a major role in P availability. Maintaining or increasing the amount of active soil organic matter (humus) can best assist this function. For example, a land use system with a phosphorus-fixing soil, application of low dose of P-fertiliser may not result in a measurable yield increase at all because P added is quickly immobilised. Application of a higher dose is needed to saturate the immediate P fixing capacity of the soil and bring about the desired increase in production.

In this dissertation P-fluxes indicated in Figure 4.21 for urban agriculture activity extending over an area of about 2.9 Mm^2 (A_c = both on-plot and off-plot) are characterised and estimated. The storage change in the monthly P balance model (dP/dt) consists of two components, P stored in the soil (P_{ss}) and P stored in the biomass (P_{bs}). It assumed that during land preparation in the months of September and October of each year all the remaining P_{bs} reverts to P_{ss} i.e. the crop residues are ploughed into the soil. During the growing season part of P_{ss} is taken up by the crops and incorporated into the plant biomass as P_{bs}. As explained in Chapter 3, land preparation also involves burning of agricultural residues, grass and weeds (see Figure 3.24f). A variety of crops are grown in the micro-catchment (see Chapter 3), but the major ones are maize, sweet potatoes and vegetables (rape, tomatoes and onions). For simplicity the maize crop is used in this dissertation for P-flux and stocks calculations for this subsystem.

Animal food and feed droppings were ignored. The only potential P-flux in the study area that was recognised but not included in the balance in Figure 4.21 is bird droppings (mostly chickens, ducks and pigeons), which was omitted because no estimate of bird numbers could be obtained.

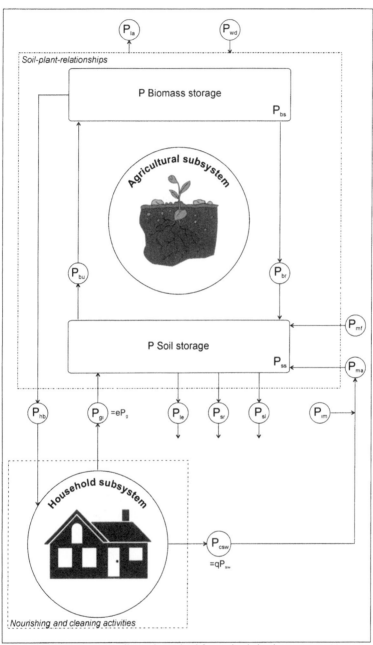

e – proportion of grey water diverted onto land for garden irrigation purposes (-)

P – phosphorus flux as P in kg/month

P_{br} – biomass residue, which remains in the agricultural subsystem after the harvest and is subsequently ploughed back into the soil during land preparation

P_{bs} – P biomass storage, with an initial value set at zero and then accumulates P during the plant growth period and is depleted to zero after the land preparation activity

P_{bu} – P biomass uptake from the soil storage occurring in the agricultural subsystem during the plant growth phase

P_{csw} – composted organic solid waste P-flux emanating from household activities ($P_{csw} = q'P_{sw}$ as defined in the household subsystem)

P_{gi} - grey water irrigation P-flux arising from grey water application on on-plot agricultural land ($P_{gi} = eP_g$)

P_{hb} – harvested biomass or economic yield P-flux of the crop also contains the inedible (mainly leaves, the core of a maize cob and other waste derived during post-harvest processing) biomass advertently or inadvertently transported from the agricultural subsystem to the household subsystem

P_{im} – imported manure P-flux from outside the micro-study catchment, mostly waste from wood furniture manufacturing and tobacco residue normally provided at no charge to the urban farmers ($P_{im} = M_{im} \times \sigma_{im}$), where M is the mass imported per month and σ in the P-content

P_{la} – land to atmosphere P-flux either in gaseous form or attached to fine particles of organic and inorganic matter. Main processes include losses of P due to slash and burn practices during land preparation and soil losses due to wind erosion (refer to P_{wd} below)

P_{le} - leaching P-flux due to percolation and groundwater flow

P_{ma} – manure application P-flux, which is a combination of imported manure (P_{im}) and the proportion of organic solid waste which is composted and deposited on agricultural land (P_{csw}) i.e. $P_{ma} = P_{im} + P_{csw}$

P_{mf} – mineral P-based fertiliser applied on agricultural land

P_{sl} – soil loss P-flux arising from soil erosion phenomena. Represents the particulate P (PP) losses from the agricultural subsystem (refer to Chapter 2)

P_{sr} - surface runoff P-flux dissolved in storm water. Represents the dissolved P (DP) losses from the agricultural subsystem (refer to Chapter 2)

P_{ss} – P soil storage, which contains both the plant available (*labile*) and unavailable (*non-labile*) P fractions. Repreesnt the most critical storage in P-supply, uptake and biomass and crop yields. P_{ss} in this dissertation is calculated for the top 0.2 m of soil i.e. the so-called plough layer and only labile P is considered (see Figure 1.13 in Chapter 1)

P_{wd} – wet and dry deposition P-flux from the atmosphere to the land mass (see Chapter 2). P_{wd} and P_{la} are assumed to be of the same order of magnitude and hence the interchanges cancel out in the mass balance analysis. On a micro-scale in relation to agricultural activities they can be assumed to be insignificant.

q' – proportion of organic solid waste which is either deliberately composted or is uncollected and end up being manure on agricultural land

Figure 4-23 Monthly agriculture P balance model for the micro-catchment (kg/month as P)

Atmospheric deposition and land to atmosphere P-fluxes

Data on P inputs through atmospheric deposition (P_{wd}) and losses from land to atmosphere in gaseous or particulate form (P_{la}) is generally limited globally (refer to Chapter 2). The scarce documented measurements of atmospheric deposition in Africa are summarised by Poels (1987) and Stoorvogel & Smaling (1990). Approximate ranges quoted by Smaling (1993) and White (1979) for P are 0.2-2 kg/ha.a i.e. 0.002-0.02 g/m^2. month. For this dissertation no values of either P_{wd} or P_{la} were measured or estimated. It is assumed that their relative magnitudes are very small compared to other P-fluxes and hence are not included in the subsequent analysis[20].

Slash-and-burn operations are common during land preparation in the months of September to November within the study area. Few studies have investigated changes in the total P reserves immediately following a slash-and-burn operation, but the general view is that P is not mobile in highly fixing soils (Sanchez, 1976; FAO, 1980a; 1980b; Kellman, 1984; Uhl, 1987; Jordan, 1989). However, some results from

---

[20] Atmospheric inputs of P to water bodies are generally low, although could be significant at a local scale with regard to nutrient enrichment. Increases may result from wind erosion of cultivated land. The drier the soil the more susceptible it is to wind erosion and the airborne material increases the P content of subsequent rainfall and dry deposition (Sharpley *et al.*, 1995).

Hands *et al.*, (1995) show that the entire ash P content derived from the burnt forest biomass may pass through the top 0.20 m of soil within a few weeks of the burn.

P stocks in the soil

The primary reservoirs or stocks of P in a soil are:
- Soil solution
- Exchangeable cations and anions
- Sorbed cations and anions
- Organic matter
- Primary and secondary minerals

The plant availability of P varies greatly between the different pools (refer to Chapter 2). P contained in the soil solution are readily available for plant uptake while P contained in organic matter and primary minerals must first undergo mineralisation and chemical weathering, respectively. The scope of this dissertation does not entail the distinction of the various pools of P as in the CENTURY model described in Chapter 2, but rather focuses on bio-available or *labile* soil P stock.

The soil P stock (P_{ss}) is generally reported in terms of kg/ha or g/m^2 or kg/m^3 of soil. Chemical analyses of solid-phase P concentrations are determined on a weight basis (g/kg as or mg/kg of soil as P) for each soil horizon. To convert P concentrations on a weight basis to a unit area basis (i.e. kg/ha or g/m^2), the effective volume of the rooting zone and the mass of soil contained within this zone must be determined. The bulk density (ρ_b) can be obtained by the coring method, paraffin-coated clod method, or by digging quantitative pits. Field results for the micro-catchment indicated an average typical density of about 1 400 kg/m^3 (Bennet, 1985).

Soil analyses are typically performed on the <2 mm soil fraction and it is assumed that the >2 mm fraction has a negligible P supplying capacity. Thus, to determine the effective P storage volume (V_e) of each horizon, the coarse fragment volume (>2 mm) is subtracted from the total volume of soil within a given horizon. Based on the nutrient concentrations (σ) and the weight of a particular horizon, P_{ss} associated with a given horizon can be calculated. The top 0.3 m (*plough layer*) of the soil is deemed to be more important for P stocks. In Zimbabwe the *plough layer* is assumed to be 0.2 m and P_{ss} is normally quoted for this layer[21].

---

[21] Little attention is paid to the nomenclature of horizonation in Zimbabwe because generally poor horizon differentiation, and because the Zimbabwe classification system tales no account of diagnostic horizons (Bennet, 1985).

Equation 4.23 can be used to calculate the P-stock in the plough layer for the area under cultivation within the micro-catchment:

$$P_{ss} = \rho_b V_e \sigma_{ss} \qquad\qquad \text{(kg)} \qquad\qquad \text{Eq. 4.23}$$

Where: ρ_b is bulk density (kg/m^3)

V_e is the effective volume for a specified soil horizon (i.e. total soil volume less >2 mm fraction) (m^3). V_e = factor × 0.2A_c (A_c is the total area under cultivation in m^2)

σ_{ss} is the total phosphorus concentration (kg/kg)

Because of the great spatial variability of soil properties on a given landscape, a large number of replicate samples are required to estimate P pools on a landscape scale. Thus, while it may appear a simple matter to estimate P_{ss}, it realistically requires a tremendous effort due to the large number of replicate samples and analytical measurements required to produce reasonably rigorous estimates.

Eight principal sites were identified within the micro-study catchment for soil sampling based on a fair geographical spread and ensuring most of the soil types were sampled (see Figure 4.24). Soil samples were collected after the harvesting period in year 2000 and a range of mechanical and fertility tests were conducted at the Chemistry and Soil Research Institute in Harare. A hand auger shown in Figure 4.25 was used at each site for soil depths of 0.2, 0.6, 0.9 and 1.2 m as advised by the Chemistry and Soil Research Institute.

The results of the physical and mechanical analysis of the soils shown in Table 4.14 show that the hilly terraces and upper slopes have reddish fine sand clay loam soils and heavily leached brownish and pale coloured sand loam soils. The stream bank soils are greyish sandy clay loamy soils. Fine clay sand loam and fine sand loam covered 67% and 33% of the micro-catchment area, respectively.

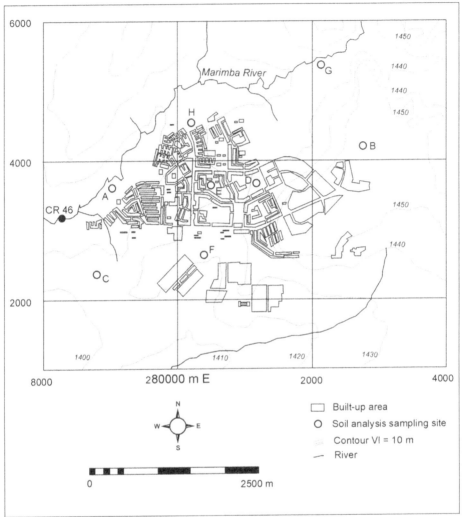

Figure 4-24 Location of soil sampling sites within the micro-catchment

Figure 4-25 Top soil and depth sampling in the micro-catchment, using a hand auger after the harvest period (May 2000) (a) using a spade after top-dressing (December 2000) (b)
Photographs: Gumbo (2000)

Table 4-14 Summary of physical and mechanical soil test results for the agricultural subsystem

Sample or Site	Colour Code [a]	(%), size grades in mm[22]					
		Clay (<0.002)	Silt (0.002-0.02)	Fine Sand (0.02-0.2)	Medium and Coarse Sand (0.2-2.0)	>2 mm fraction	Textural Class [b]
A	RB	23	11	47	12	7	MG/SCL
B	GB	28	15	40	10	7	MG/SC
C	B	25	14	44	10	7	MG/SCL
D	P/B	14	6	61	10	9	MG/SL
E	P/B	6	6	53	22	13	MG/S
F	P/B	15	6	38	25	16	MG/S
G	GB	9	7	51	16	17	MG/SL
H	B	11	8	57	19	17	MG/SL
Average	-	16	9	49	16	12	-

a) Colour key: R = Red/Reddish, B = Brown/Brownish, Y = Yellow/Yellowish, G = Grey/Greyish, BK = Black/Blackish, S = Strong, D = Dark, L = Light, V = Very, P = Pale.
b) Texture key: S = Sand/Sandy, L = Loam/Loamy, C = Clay, HC = Heavy Clay, FG = Fine Grained, MG = Medium Grained, CG = Course Grained

---

[22] The texture of the soil is an assessment of the particle size distribution determined using the Hydrometer (Bouyoucos) and sieve method of the mineral fraction having particle diameters of no greater than 2 mm. This, known as the fine earth, is divided into sand, silt and clay. The relative proportions of sand, silt and clay place a soil into a particle size or textural class based on the separates, or particle size grades, and a triangular diagram used in Zimbabwe (Bennet, 1985; Nyamapfene, 1991).

The available P content was determined by the Chemistry and Soil Research Institute in Harare, Zimbabwe using a procedure developed in their research laboratory (Table 4.15). The soil is shaken for 16 hours with an '*anion exchange resin*' which progressively absorbs the phosphate as it comes into solution, thus tending to simulate what occurs in the filed. Experiments have shown that this method provides a much better indication of the availability of P in Zimbabwean soils than any other method, especially on fertilised soils i.e. it is more sensitive to the residual effect of phosphate fertiliser (refer to Table 2.1 in Chapter 2).

Table 4-15 Summary of chemical and soil fertility test results for the agricultural subsystem

Sample or Site	pH [a]	Mineral Nitrogen (mg/kg) [b]		Available P (mg/kg) as P_2O_5	Exchangeable cations (me/100 g)				
		Initial [c]	Al [d]		K [e]	Ca	Al	Mg	Total
A	4.8	25.0	41.0	11.0	0.34	3.96	0.35	1.40	6.05
B	5.0	10.0	38.0	12.0	0.50	12.41	0.50	3.72	17.13
C	4.6	9.0	19.0	7.0	0.39	7.84	0.40	2.72	11.35
D	4.2	18.0	19.0	3.0	0.17	1.79	0.60	0.58	3.14
E	4.0	4.0	19.0	3.0	0.13	0.80	0.90	0.47	2.30
F	4.9	15.0	16.0	2.0	0.15	1.75	0.50	0.51	2.91
G	6.4	24.0	52.0	45.0	0.43	8.44	0.30	1.20	10.37
H	4.9	8.0	20.0	18.0	0.35	1.69	0.30	0.57	2.91
Average	4.9	14.1	28.0	12.6	0.31	4.84	0.48	1.40	7.02

a) Calcium –chloride method which employs a dilute solution of M/100 $CaCl_2$ was used instead of the conventional distilled water method. The calcium-chloride method gives much more accurate laboratory results and more importantly is more indicative of what the soil acidity will be under filed conditions during the growing season.
b) The total nitrogen content of a soil reflects its organic matter content, abnormally low figures for a particular soil type ma indicate severe erosion.
c) The Initial mineral nitrogen is the amount of nitrate-nitrogen plus ammonia-nitrogen, expressed in mg/kg, actually present as received. Determined using the potassium-chloride extraction and Nessler's method.
d) The mineral nitrogen after incubation (AI) is the total of nitrate-nitrogen plus ammonia-nitrogen present in the sample after it has been maintained in an incubator for two weeks at the optimum moisture and temperature ($35°$ C) for microbial decomposition of organic matter. The value is closely related to the amount on nitrogen that becomes mineralised in the filed during the growing season.
e) The exchangeable potassium is the fraction of the total soil potassium that is held on the surface of clay particles in a form in which it can become available to plants measured in mg equivalent K per 100 g of soil (me/100 g). While exchangeable K status of the top soil is a useful guide to the potash-supply power of the soil, there are other factors that may profoundly affect it.

The Chemistry and Soil Research Institute has drawn up a scale of *desirable* and *critical* values of various chemical and soil fertility parameters. This scale is used to interpret values obtained from laboratory results. Table 4.16 presents the average values obtained and their standard deviations in the micro-study catchment and these are compared and interpreted for the various parameters measured.

Table 4-16 Summary of soil fertility test results for the agricultural subsystem

Parameter	Value	Desirable	Interpretation
pH	4.9 ±0.7	5.2-7.8	Strongly acid; there is a progressive risk of fertility being adversely affected and lime addition is required.
Total P mg/kg as P	130 ±5	-	-
Available P mg/kg as P_2O_5	13 ±14	> 30	Deficient; large yield increases may be expected with adequate dressings of phosphatic fertiliser
Potassium me/100g	0.3 ±0.1	> 0.3	Rich; no potash required
Mineral nitrogen (AI) mg/kg	28 ±14	> 40	Low; arable lands cropped with little fertiliser or manure
Organic carbon % [a]	0.8 ±0.4	3	Normally should be above 1%

a) Determined using the Walkley-Black Method of digestion using potassium di-chromate ($K_2Cr_2O_7$) solution.

In Zimbabwe a soil is considered P deficient if the concentration of available P in the plough layer is less than 30 mg/kg as P_2O_5 (Cooper & Fenner, 1981; Grant, 1981). General fertiliser requirements are usually made based on such soil fertility tests. The average ratio of available P to total P calculated for the micro-catchment is 1:10. This demonstrates that a large percentage of P could either be in organic non-labile form or be strongly absorbed in or precipitated by the soil and not immediately available to the crop.

Using Equation 4.23 the bio-available P_{ss} was calculated for available P using the average value of σ_{ss} in Table 4.15, a r_b value of 1 400 kg/m^3, A_c = 2.9 Mm^2 and an effective volume factor of 0.88 (Note from Table 4.14 course material >2 mm is about 12% of the soil sample). Hence P_{ss} = 1 400 × 0.88 × 2.9 × 10^6 × 0.2 × 13 × 10^{-6} = 9 289 kg of bio-available P as P_2O_5 or 4 053 kg as P. P_{ss} can also be represented per unit area of land under cultivation for the depth of the *plough layer* i.e. P_{ss} = 2.65 g/m^2 available P as P_2O_5 or 1.16 g/m^2 as P.

Figure 4.26 shows the variation with depth of pH, available P as P_2O_5 (mg/kg) and total P as P mg/kg) for four sampling sites A, B, C and D where depth sampling and analysis was conducted. Generally the pH increased with depth (soil became more alkaline i.e. less acidic), whilst as expected the available P, total P and organic carbon content decreased with increasing depth (refer to Figure 2.6 Chapter 2).

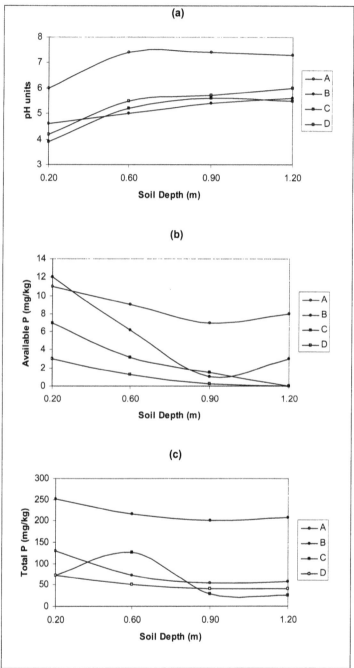

Figure 4-26 Variation with depth of pH (a), available P as P_2O_5 (mg/kg) (b), total P as P (mg/kg) (c) for soil sampled at sites A, B, C and D in year 2000.

P-export processes in the agricultural subsystem

The processes that control P-fluxes from terrestrial to aquatic systems are complex. They include a host of human and natural processes as detailed in Chapter 2. P-cycling in soils is greatly influenced by the water cycle. Soil surface conditions influence the water cycle (Q_s and Q_g), energy flow, nutrient cycling, and succession. Soil surface conditions have direct and indirect effects on water infiltration, soil organic matter, soil aeration, soil compaction, runoff, and evaporation.

Numerous comprehensive mathematical models have been developed to simulate the fate of P in soil and its transport to surface waters, with the purpose of aiding the selection of management practices capable of minimising water quality problems (Sharpley *et al.*, 1995). These include, but are not limited to ANSWERS (Areal Non-point Source Watershed Environment Response Simulation, Beasley *et al.*, 1985), AGNPS (Agricultural Non-point Pollution Model, Young *et al.*, 1989), CREAMS (Chemical, Runoff, and Erosion from Agricultural Management Systems, Knisel *et al.*, 1980), GAMES (Guelph Model for Evaluating the Effects of Agricultural Management Systems on Erosion and Sedimentation, Cook *et al.*, 1985), HSPF (Hydrologic Simulation Program - Fortran, Johanson *et al.*, 1984), ARM (Agricultural Runoff Model, Donigan *et al.*, 1977), and EPIC (Erosion-Productivity Impact Calculator, Sharpley & Williams, 1990). In general, these models describe the loss of P in soluble reactive and particulate forms (Sharpley *et al.*, 1995).

Water and wind erosion are mainly responsible for surface and air transport of P and strongly influence the seasonality of inputs to aquatic systems. Soluble as well as particulate forms of P compounds are mobilised by this process usually with a large contribution of P bound to inorganic and organic soil and plant derived particles. From Figure 4.23 the main P out-fluxes from the P soil storage (P_{ss}) are: soil loss P-flux arising from soil erosion phenomena (P_{sl}) which represents the particulate P (PP) losses from the agricultural subsystem; surface runoff P-flux dissolved in storm water (P_{sr}) which represents the dissolved soluble P (DP) losses from the agricultural subsystem and - leaching P-flux due to percolation and groundwater flow (P_{le}).

Leaching (P_{le}) is believed to be of minor importance. It is responsible for subsurface and underground transport of organic and inorganic soluble P compounds. These processes are strongly site specific depending on land cover and land use, topography, P input with fertilisers, rainfall, sorption capacity of soils, local hydrological discharge, meteorological conditions and many other factors (Sharpley *et al.*, 1995).

Regression equations relating overland flow (Q_s) and soil loss (Z_l) and hence P-loss due to erosion (P_{sl}) were established by Stocking (1986) and these relationships constituted the basis of the so-called Soil Loss Estimation Model for Southern Africa (SLEMSA). The research work done by Elwell (1975, 1977) and Elwell & Stocking (1982) is summarised in Box 4.E. The approach used in establishing the relationships is used in this dissertation for estimating P_{sl}. The characteristics of the four soil series studied by Elwell & Stocking (1982) are shown in Table 4.17 and they represent the major soil groupings in Zimbabwe.

For the four soil series investigated, series I to III i.e. all soils other than well drained granite sand behaved similarly, whilst series IV i.e. poor granite sandveld behaved differently. The best simplifying approximation of annual data was to take the regression equations for series I to III and the equations for series IV separately. The corresponding correlation coefficients by soil for each relationship indicate best-fit regressions. For simplicity the final regression equations for prediction and extrapolation of P_{sl}, and Z_l are shown in Table 4.18.

Box 4-E   Erosion and nutrient loss experiments in Zimbabwe

The most intensive soil conservation and erosion research programme in Africa was instituted during the period of Federation in Rhodesia (now Zambia and Zimbabwe) and Nyasaland (now Malawi) between the years 1953 to 1964. In the late 1960's and early 1970's most the soil loss and runoff data generated was classified, documented and analysed, storm-by-storm and annually, for the many treatments (slope ranging from 3 to 8% and grass, maize and tobacco rotated in various sequences) and for four soil types(see Table 4.16) used in the experiments (Elwell, 1975; 1977; Elwell & Stocking, 1982). The results of this work was published and a new model of soil loss estimation model for Southern Africa (SLEMSA), was constructed for practical field conditions in Zimbabwe (Elwell & Stocking, 1982). The data collected from the largest experimental site at Henderson Research Station, 20 km outside Harare also contained storm-by-storm records of nitrogen (N), phosphorus (P) and organic carbon (C) concentrations in the sediment samples.

Initial graphical analysis of the data from several plots showed a remarkable correlation between N, P and C losses and soil loss, thus opening up the exciting prospect of being able to predict nutrient losses and their economic impact under various land uses, soil types and rates of erosion (Stocking, 1986). The original raw data was checked for consistency and the well-known phenomenon of a 'flush' of nutrients in an early part of the storm was observed. Weaknesses in the data -not as a result of experimental errors- were apparent due to the original experimental organisation. Most importantly the nutrients contained in the runoff and the nutrients directly leached through the soil were not considered.

The equations in Table 4.18 are a simplification[23] of the SLEMSA Model (Elwell, 1977; Elwell & Stocking, 1982). From the micro-catchment soil analysis results Table 4.14 the soils tend to fall into series I-III category[24] and hence using the corresponding P_{sl} expression the P-flux due to soil erosion was calculated using the *'average year'* Q_s values determined in section 4.2 rainfall water balance; $Q_s = 0.24 \times R_{eff}$ and $Q_g = 0.08 \times R_{eff}$, where R_{eff} is the effective rainfall derived from the year 2000 average rainfall (see Figure 4.4(c)). The P_{sl} values estimated in this way could be underestimates because of poor soils in the micro-catchment with limited humic substances, which act as a soil conditioner for P retention, steep slopes and lack of conservation tillage knowledge, by the urban farmers.

---

[23] It should be noted that these estimated rates of erosion hide very considerable spatial and temporal variation, and are based on mean values of parameters which themselves also have great variation. experience suggests that these estimates are of the correct order of magnitude for filed erosion rates under stated conditions in Zimbabwe (Stocking, 1986).

[24] Urban agriculture in Harare mainly occurs on vleis, clay and sandy vleis (ENDA, 1995; Mbiba, 1995).

Table 4-17 Summary soil description for the four series investigated at Henderson Research Station

Description	Series			
	I	**II**	**III**	**IV**
Zimbabwe classification [a]	Fersiallitic, Mazowe 5SE (S1)	Fersiallitic, Mazowe 5E (E1)	Fersiallitic, Mazowe 5G (G3)	Fersiallitic, Mazowe 5G (G1)
Soil taxonomy	Aquultic Haplustalf	Oxic Rhodustalf	Typic Psammaquent	Arevic Haplustalf
FAO (1988) [b]	Ferric luvisol	Ferric luvisol or Eutric nitosol	Eutric gleysol	Chromic luvisol
Depth	Moderately deep	Deep	Shallow	Deep
Texture	Sand clay loam to clay loam	Clay to heavy clay	Sand to loamy sand	Loamy sand to sandy loam
Colour	Greyish-brown to yellow	Reddish-brown to red	Light yellow to pale brown	Grey-brown to yellow
Drainage	Imperfect	Well drained	Poorly drained	Good to moderate
Parent material	Argillaceous meta-sediments	Epidiorite and dolerite	Granite	Granite
Fertility and use	Reasonably fertile for commercial farming	Highly prized for their fertility and have a high tendency for P fixation	Poor soils, leached and unproductive, only viable for grazing	Require fertilisers and organic manures for cropping

a) Refer to Chapter 3, Figure 3.9. Each soil, in its way, reflects the major soil types in Zimbabwe in the areas of highest rural population and agricultural significance.
b) FAO-UNESCO Soil Map of the World (1988)
Source: Stocking (1986)

Table 4-18 Final regression equations for prediction and extrapolation of mean soil and P-losses

Parameter	Series I-III (All soils other than well-drained granite sands) [a]	Series IV (Well-drained granite sands)
Soil loss (g/m^2)	$Z_l = 3.225 \times Q_s$	$Z_l = 1.690 \times Q_s$
P-loss due to soil erosion (g/m^2)	$P_{sl} = 0.000155 \times Z_l$ $= 0.0005\, Q_s$	$P_{sl} = 0.000031 \times Z_l^{1.6}$ $= 0.00007\, Q_s^{1.6}$

a) Surface runoff (Q_s) is in mm/month or mm/a for annual losses
Source: Adapted from Stocking (1986)

Loss of P in storm and ground water flow (P_{sr} and P_{le}) are some of the primary processes of its export from a system. Usually P fluxes are calculated by coupling P concentrations (σ) with surface runoff (Q_s) or ground water flow (Q_g). Storm water P concentrations (σ_{sr}) may display large fluctuations in concentrations during a single storm event. Thus, several water samples need to be collected for P-flux quantification during the rising and falling limbs of the hydrograph to determine accurate fluxes from the watershed.

Water quality data from station CR46 located on the Marimba River on the periphery of the micro-study catchment (see Figure 3.21 and 3.27 in Chapter 3) was collected and recorded by the City of Harare for a period of four years. In order to calculate the value of P_{sr} and P_{le} the concentration of P (σ_{sr}) from the river water samples is used in this dissertation. Figure 4.27a shows the monthly variation of recorded values of σ_{sr} in mg/l as P_2O_5 from 1995 to 1998. It is assumed that the concentration of P in surface runoff and in ground water is the same i.e. $\sigma_{sr} = \sigma_{le}$.

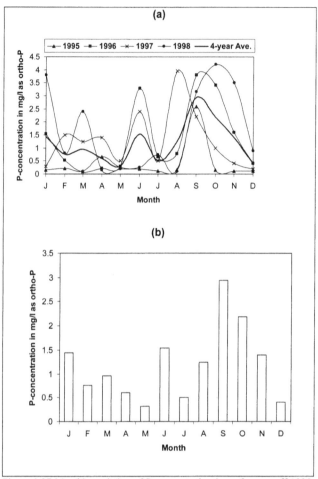

Figure 4-27 Monthly variation of P concentration in surface runoff 1995 to 1998 (a), average monthly P concentration variation (1995-1998) (b), (mg/l) as ortho-P measured at water quality station CR46 on the Marimba River.

Hence the general expression for P_{sr} as P is given by Equation 4.24:

$$P_{sr} = \frac{A_c Q_s \sigma_{sr}}{2.29} \qquad \text{(kg/month)} \qquad \text{Eq. 4.24}$$

Where: Q_s is the monthly surface runoff generated in mm/month
           σ_{sr} is the phosphorus concentration in the runoff as P_2O_5 in mg/l
           A_c is the total area under cultivation in m^2
           2.29 is the conversion factor of P_2O_5 to elemental P (see Chapter 2, footnote 14)

P_{le} is calculated using the same expression by substituting Q_s with Q_g. By using Equation 4.24 P_{sr} and P_{le} fluxes were calculated and these are shown graphically in Figure 4.28.

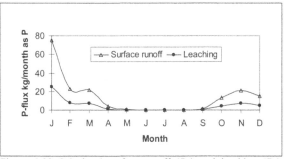

Figure 4-28 Calculated surface runoff ($\hat{P}_{sr}$) and leaching ($\hat{P}_{le}$) P-fluxes for the micro-catchment (kg/month)

Mineral fertiliser and manure application P-fluxes

The general fertiliser requirements of a range of field crops on soils of high, medium, and low nutrient status have been published widely in Zimbabwe (Cooper & Fenner, 1981; Bratton & Truscott, 1985; Fernades, 1989; ENDA, 1995). Whilst the recommended values provide a useful guide, it is emphasised that the only means of assessing the nutrient status of a soil, and therefore, the most economic use of fertiliser is analysis of a representative sample of the soil by methods that have been correlated with crop responses in field experiments.

Table 4.19 is adapted from Cooper & Fenner (1981) and it provides suggested rates of fertiliser application for various crops. In Zimbabwe the general recommended mineral fertiliser application rate for commercial maize production is 420 kg/ha.a and is commonly administered as Compound D fertiliser (see Table 3.4 Chapter 3) also called *maizefert* (8:14:7; $N:P_2O_5:K_2O$). This application rate translates to 58 kg/ha.a as P_2O_5 (5.8 g/m^2.a). In year 2000 the price of Compound D fertiliser was about US$ 0.50/kg.

Table 4-19 Suggested rates for P-based fertiliser requirements for various crops in Zimbabwe

Crop	Probable available P status (g/m^2 as P_2O_5)		
	Very deficient [a]	Deficient to marginal [b]	Adequate [c]
Grain crops, cotton, beets, lettuce, cucurbits and sweet potatoes	9-6 [e]	6-3	3-2
Potatoes, tomatoes [d], brassicas, carrots, celery, onions and peas	18-12	12-6	6-4
Field legumes and sunflowers	6-4	4-2	Nil
Intensive pastures in high rainfall areas	16-12	12-8	8-4
Lucerne	14-10	10-6	6-2

a) Very deficient: Virgin or reverted land, and cultivated land on which little or no phosphate has been applied n the past
b) Deficient to marginal: Cultivated lands on which a moderate total amount of phosphate has been applied in the past
c) Adequate: Cultivated lands on which a large amount of phosphate has been applied in the past by means of regular, uninterrupted, annual dressings
d) For potatoes and tomatoes, where heavy yields are expected, the rates may be increased with advantage

e) It is probably unwise to use the lowest rates suggested unless soil analysis has confirmed that the available phosphate status is indeed adequate

Source: Cooper & Fenner (1981)

From previous studies and field surveys conducted in year 1999 to 2001 it was established that the majority of the urban farmers in the micro-study area apply *maizefert* as basal fertiliser once a year or season at application rates of about 10-20% of the recommended commercial farming value i.e. 0.58 to 1.16 g/m^2.a as P_2O_5 (ENDA, 1995; Mbiba, 1995; Nyamangara & Mzezwa, 1996; 1999). A median value of 15% of the recommended application rate is assumed i.e. urban farmers apply 0.87 g/m^2.a as P_2O_5. This implies that basal fertiliser is applied mostly during the months of October, November and December. Ammonium nitrate is applied as "top-dressing" usually twice because of its susceptibility to leaching and washout, firstly just after germination of the maize crop (early December) and finally immediately after weeding when the maize crop is about 0.5 m high (late January). The study by ENDA (1995) also established that on-plot gardening relied on organic fertiliser whilst off-plot cropping used significant amount of synthetic (inorganic) or mineral fertiliser. The general expression for the mineral fertiliser P-flux is (P_{mf}) is given by Equation 4.25:

$$P_{mf} = \frac{\theta A_c M_{mf} \sigma_{mf}}{2.29} \qquad\qquad \text{(kg/month)} \qquad \text{Eq. 4.25}$$

Where: θ is the proportion of the recommended commercial fertiliser application rate applied by urban farmers as Compound D (-)

M_{mf} is the recommended mass of mineral fertiliser applied per month as '*maizefert*' or Compound D (kg/month)

σ_{mf} is the total phosphorus concentration in the fertiliser as P_2O_5 = 0.14 (kg/kg)

2.29 is the conversion factor of P_2O_5 to elemental P (see Chapter 2, footnote 14)

Using Equation 4.25 and a once off application rate or total of monthly application rates of (P_{mf}) = 0.87 g/m^2.a, the total mass M_{mf} of fertiliser used per year in the agricultural subsystem can be calculated i.e. = 0.87 × 10^{-3} × A_c = 0.87 × 10^{-3}× 2.9 × 10^6 = 2 520 kg/a of phosphorus as P_2O_5. Therefore urban agriculture accounts for the importation of about 2 520/2.29 = 1 100 kg/a of mineral fertiliser as elemental P.

Most urban farmers in Harare practice low-input agriculture that depends on organic matter in the soil to sustain production (ENDA, 1995, Mbiba, 1995). Soil organic matter therefore plays an important part in establishing the intrinsic properties of a soil, which make plant growth possible. Soil organic matter helps sustain soil fertility by improving retention of mineral nutrients, increasing the water-holding capacity of soils, and increasing the amount of soil flora and fauna. Continuous cropping, burning of harvest residues and erosion reduce the level of soil organic matter which judging from the values of organic carbon in Table 4.16 is very low for the soils encountered in the micro-study area.

Figure 4.23 shows that the manure application P-flux consists of two components P_{csw} and P_{im}. The magnitude of P_{csw} is calculated from the household subsystem. From the ENDA report (1995), 60% of the cultivators use imported organic manure on gardens in the form of sawdust, mulch, and tobacco residue (refer to Chapter 3). P-content of these materials (σ) is generally regarded to be very low. Sawdust is the least effective as a fertiliser as it does not decompose easily. From the household survey results it was noted that several households used significant amount of imported organic matter as manure. An estimate of at least 10 kg/household.a of imported manure was made for the 10 100 stands in Mufakose and Marimba suburbs i.e. total mass of imported manure per annum (M_{im}) of 101 000 kg/a. Laboratory test on the P-content (σ_{im}) of sawdust and tobacco residue indicated an average value of 0.0006 kg/kg (for most samples it was undetectable). Hence using a similar expression to Equation 4.24, the annual P-flux due to imported manure was calculated to be 61 kg/a as P. This is small compared to the total annual value of composted solid waste fraction $P_{csw} = 0.22 \times 10^{-3} \times 100\ 000 \times 12 = 264$ kg/a as P (refer to Figure 4.22b).

Biomass up-take, harvested and residual biomass P-fluxes

The capacity of soils to be productive depends on more than just plant nutrients. The physical, biological, and chemical characteristics of a soil - for example its organic matter content, acidity, texture, depth, and water-retention capacity - all influence fertility (Gruhn *et al.*, 2000). Because these attributes differ among soils, soils differ in their quality. Some soils, because of their texture or depth, for example, are inherently productive because they can store and make available large amounts of water and nutrients to plants. Conversely, other soils have such poor nutrient and organic matter content that they are virtually infertile.

The amount of P taken into a plant by the root, economic organ and straw during growth on a monthly time step is difficult to measure. The total monthly biomass uptake of P (or P-demand, P_{bu}) changes greatly over the course of development of a crop from establishment to maturity. During the period of early establishment, P-uptake will increase with increasing gross ecosystem production. P_{bu} rates will peak at approximately the time of canopy closure. Following canopy closure, gross production slows due to competition and mortality and more carbon is allocated to woody materials having a lower nutrient content (e.g. the stem). Thus, uptake rates will decline from their maximum level at canopy closure and maintain a relatively constant value.

The P biomass stock (P_{bs}) is the quantity of nutrients contained in the biomass of an ecosystem. The value of P_{bs} vary widely between plant species. P_{bs} is typically segregated by individual species within an ecosystem and within a given species into various components: e.g. foliage or straw, economic harvest or yield, woody tissue, bark, and roots (White, 1979; Baccini & Brunner, 1991). In some ecosystems, P stored within the biomass pool may represent a large fraction of the extractable P (assumed to be plant available). It is important to note that in this subsystem the total P stored is $P_{bs} + P_{ss}$. The primary components of the post-harvest biomass residue (P_{br}) in the case of maize, for example are the straw, roots, and inedible reproductive tissues.

Although a variety of crops are grown in the micro-catchment (see Chapter 3) for simplifying model computations, maize (*Zea mays L.*) is selected in this dissertation for P-flux and stock analysis. Maize requires a fertile soil and is sensitive to water logging. Varieties grown require a frost-free growing period of 80 to 210 days (Table 4.20) depending on the cultivar selected. Germination is best between 18 and 21 °C, greatly reduced below 13 C and failing at 10 °C. The optimum temperature for tasseling is between 21 and 30 °C. The crop does not tolerate much salt: yield reductions are 0%, 50% and 100% at electrical conductivity of <0.03 and 0.10 mS/m, respectively. Maize is sensitive to sodicity to the extent that yield reductions of up to 50% occur at sodium saturation values of 15% or less. The crop grows in soils with a pH between 5.0 and 8.0; the optimum is between 6.0 and 7.0 (IFA, 1992; Driessen & Konijn, 1992).

Table 4-20 Indicative values for the duration of the various developments stages, initial rooting depth and the maximum rooting depth of maize

Development stages	Duration (days)	Rooting depth (m)
Initial	15-30	0.10
Vegetative	30-45	-
Mid-season	30-45	-
Late-season	10-30	0.1-1.70 (max)

The nutrient status of maize is at times judged by the levels of nutrient elements in the economic produce or '*yield*' (normally the storage organ), and the crop residue also called '*straw or stover*'. Pot trials and field experiments have shown that plants cannot grow normally if they cannot maintain specific minimum concentrations of P in yield and straw (IFA, 1992; Driessen & Konijn, 1992; Tiessen, 1995). Indicative minimum concentrations on P are given in Table 4.21 for four groups of crops. The World Fertiliser Use Manual (IFA, 1992) provides typical P removal rates by maize (grain and straw) for some values of the economic yield (Y_{hb}). These values are given in Table 4.22.

Table 4-21 Indicative minimum concentrations on phosphorus in economic yield and straw for type of crops

Crop	Yield	Straw
	(kg/kg as P)	
Grain crops	0.0011	0.0005
Oils seeds	0.0045	0.0007
Root crops	0.0013	0.0011
Tuber crops	0.0005	0.0019

Source: Adapted from Driessen & Konijn (1992)

Table 4-22 P-removal rates based on observed yield and straw production for maize

Yield	Grain	Straw
(kg/m^2)	(g/m^2 as P)	
0.95	3.10	0.79
0.63	1.75	1.00

Source: Adapted from IFA (1992)

For maize, the water-limited dry mass economic yield (Y'_{hb}) and combined dry biomass production potential (Y'_{bu}) are quoted by Driessen & Konijn (1992) as 0.790 kg/m^2 and 1.867 kg/m^2 respectively[25]. This implies that the water-limited dry biomass production potential of straw only (Y'_{sb}) would be Y'_{bu} - Y'_{hb} = 1.867 – 0.790 = 1.077 kg/m^2. The P-uptake requirement (P_{bu}) for maize can be calculated by multiplying the dry masses of yield (Y_{hb}) and straw (Y_{bu}) by their respective minimum P concentrations given in Table 4.20. For example, for a scenario with the water-limited yield potential of 0.790 kg/m^2 and a corresponding potential biomass production of 1.867 kg/m^2 the total biomass P-uptake (P'_{bu}) would be:

$$P'_{bu} = [0.790 \times 0.0011 + (1.867 – 0.790) \times 0.0005] \times 10^3 = 1.41 \text{ g/m}^2$$

The calculated P-uptake requirement is the minimum requirement. A crop of maize could take up more than 1.41 g/m^2 but this would not result in more biomass production or economic yield. It could possibly improve the quality of the product (Driessen & Konijn, 1992). Assuming that the proportion of Y'_{sb} and Y'_{hb} (1.077/0.790 = 1.36) is the same regardless of the economic yield of maize or harvested biomass (Y_{hb}), a general expression for the aggregate P_{bu} can be developed as follows:

$$P_{bu} = A_c[Y_{hb}\sigma_{hb} + Y_{sb}\sigma_{sb}] \qquad \text{(kg)} \qquad \text{Eq. 4.26}$$

Where: Y_{hb} is the economic yield or harvested biomass of maize (kg/m^2)
Y_{sb} is the yield of maize straw biomass (kg/m^2)
σ_{hb} is the P-content in the harvested biomass as P (0.0011 kg/kg from Table 4.21)
σ_{sb} is the P-content in the maize straw biomass as P (0.0005 kg/kg from Table 4.21)
A_c is the total area under cultivation (m^2)

Since Y_{sb} = 1.36Y_{hb} and substituting for known values in Equation 4.26 the aggregate P_{bu} for the entire growing season for maize = $A_c[Y_{hb} \times 0.0011 + 1.36Y_{hb} \times 0.0005]$ = 0.0018A_c Y_{hb}. The harvested biomass or economic yield P-flux (P_{hb}) can be expressed simply as in Equation 4.27.

$$P_{hb} = 0.00114_c Y_{hb} \qquad \text{(kg)} \qquad \text{Eq. 4.27}$$

Where: Y_{hb} is the economic yield or harvested biomass of maize (kg/m^2)
A_c is the total area under cultivation (m^2)

---

[25] It is assumed that fertiliser trials are conducted under the same conditions i.e. all plants in an experiment grow under the same temperature, solar radiation and water supply, and weeding, plant protection and harvesting are optimum.

Estimated average maize yield from the surveys conducted in the year 1999 and 2000 indicated an average value of 0.20 kg/m^2. This value of Y_{hb} compared well with reported figures from Mbiba (1995) and ENDA (1995). The value is comparably higher than the less than 0.10 kg/m^2 observed in communal areas of Zimbabwe (Grant, 1981; Bratton & Truscott, 1985). Using this value in Equation 4.27 gives P_{hb} value of 638 kg/a as P.

The residual biomass is what remains within the agricultural subsystem after the harvest period. It is the difference between the biomass P-uptake and the harvested P-flux $P_{br} = P_{bu} - P_{hb}$. The biomass stock is assumed to be equal to zero in the month of October each year as it is incorporated into the soil matrix after land preparation.

So far, P_{mf}, P_{ma}, P_{bu}, P_{hb} and P_{br} have been calculated as annual or growing season totals i.e. kg/a. In order to convert these P-fluxes to monthly values there is need to establish disaggregating factors based on observation of the urban agricultural dynamics in the micro-study area. The factors corresponding to the agricultural activity within each month are listed in Table 4.23.

Table 4-23 Disaggregating factors for converting annual or seasonal to monthly P-fluxes

P-flux	Disaggregating factor for each month											
	J	F	M	A	M	J	J	A	S	O	N	D
P_{mf} [a]	0.0	0.0	0.0	0.0	0.0	0.0	0.0	0.0	0.0	0.5	0.3	0.2
P_{ma} [b]	0.0	0.1	0.2	0.3	0.2	0.0	0.0	0.0	0.0	0.1	0.1	0.0
P_{bu} [c]	0.2	0.1	0.1	0.0	0.0	0.0	0.0	0.0	0.0	0.1	0.2	0.3
P_{hb} [d]	0.0	0.0	0.4	0.3	0.3	0.0	0.0	0.0	0.0	0.0	0.0	0.0
P_{br} [e]	0.0	0.0	0.4	0.3	0.3	0.0	0.0	0.0	0.0	0.0	0.0	0.0

a) Mineral fertiliser as P is applied as basal fertiliser compound D before sowing in the months of October and November and a small portion for the late maize crop in December.
b) Imported manure application (P_{im}) is applied in a similar pattern as mineral fertiliser i.e. during the months of October to December. However composted solid waste applications (P_{csw}) increase during the month of March to May during and immediately after the harvesting period.
c) Biomass P-uptake pattern for maize was derived from Leigh & Johnston (1986) and Barber & Olson (1968) for a situation where P is not a limiting nutrient. Sowing is assumed to take place in the first week of October in each growing season. P-uptake for maize peaks at about 60 days after sowing and then recedes until it reaches about zero when the crop matures and is harvested (see Figure 4.28).
d) The harvesting period for the maize crop is normally between the months of March to May. Part of the crop is harvested and consumed before it dries out whilst remainder is left to dry in the field.
e) The biomass residue starts accumulating at a steady rate in the field at the same time when harvesting begins in the month of March each year. During land preparation in September and October the accumulated biomass is burnt or ploughed into the soil.

The biomass P-uptake (P_{bu}) changes during maize growth. Daily uptake values after sowing were modified from uptake patterns provided by Leigh & Johnston (1986) and Barber & Olson (1968). The derived biomass P-uptake pattern for maize is shown in Figure 4.29.

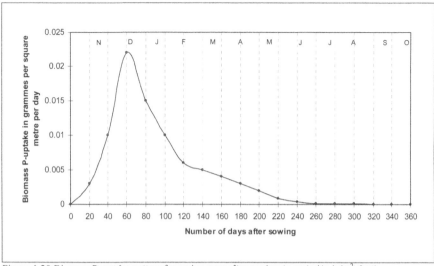

Figure 4-29 Biomass P-uptake pattern for maize crop after sowing to maturity (g/m².day)
Source: Adapted from Barber & Olson (1968) and Leigh & Johnston (1986).

The monthly agricultural subsystem P-balance for year 2000 was calculated using the disaggregating factors listed in Table 4.23 for those P-fluxes calculated as annual or seasonal totals. Figure 4.30 shows graphically the calculated monthly variation of P-fluxes for the agricultural subsystem in the micro-study area.

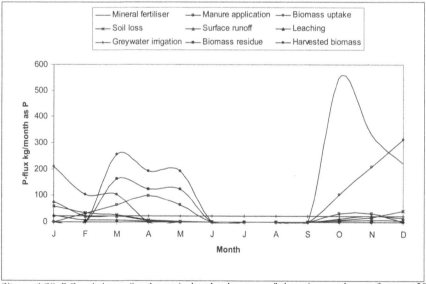

Figure 4-30 P-flux balance for the agricultural subsystem of the micro-catchment for year 2000 (kg/month as P)

It is apparent from Figure 4.30 that the various fluxes have peaks and troughs at different time of the year depending on the agricultural activity. In proposing options for short-cutting the P-cycle it is important to observe these trends and design suitable engineering and management systems which match the agricultural activities. The limitation in this case is that maize (seasonal crop) has been used for the analysis. The ideal scenario should involve a variety of crops (perennial and seasonal) with different P-uptake patterns.

So far the four sub-systems have treated and analysed more or less independently and to provide flexibility and changes of the assumed and measured input values there is a need to build a model which combines both subsystems.

4.5 Modelling monthly flows and stocks

Introduction

The use of STELLA as a systems analysis model in this dissertation was critical to piece in the four subsystems presented and to generate insight and hence create a picture showing the linkages and interdependent relationships in the study of P and water fluxes (see Chapter 1).

It is not the intention of this dissertation to provide details on the use of STELLA software, menus, controls, building blocks, tools and objects. Most of the 'how to' information is provided in the various technical manuals provided by High Performance Systems (HPS), Inc. (HPS, 1992, 1993, 1994, 1996). STELLA model provides a multi-level, hierarchical environment for construction and interaction with the model. The environment consists of two major layers: the High-Level Mapping layer and the Model Construction layer. An Equations View is provided to view the entities on the Model Construction layer in a list format, for rapid modification of variable definitions, and for easy exporting of equations from the model (HPS, 1996).

The Model building blocks of STELLA, stocks, flows, converters and connectors have been introduced in Chapter 1. The controls of the Model Construction layer are shown in Figure 4.31. The specific menu items and mechanical aspects of using the building blocks are fully described in the Technical Documentation of STELLA (HPS, 1996).

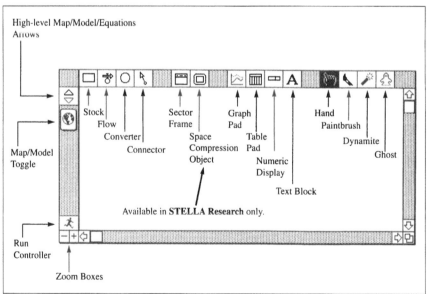

Figure 4-31 Overview of the STELLA operating environment (Model Construction layer)
Source: HPS (1994)

The advantage of layering in STELLA is to manage complexity. The environment fosters a top-down approach to model development by using the tools and objects on the High-Level Mapping layer to create a high-level systems map. Then, on the Model Construction layer a more detailed representation of the relevant processes can be developed. Sub-models allow detailed structures to be embedded into a single icon (HPS, 1992; 1996). The interactive environment provides opportunities for other users of the model to experience the dynamics of the system modelled.

STELLA allows selection of units of measurements to make sure that variables are combined in an internally consistent manner. This facility is invaluable in system dynamics models and MFSA where usually a range of units are preferred for certain stocks or flows.

The P-calculator for the micro-catchment

The High-Level Map in Figure 4.32 shows the four process frames representing the subsystems used in this dissertation in the analysis of flows and stocks, with Bundled connectors linking the frames, indicating that relationships exist between them. The connectors and Process Frames facilitate a 'top-down' approach to model construction and provide navigational capabilities to the stock/flow structure on the Model Construction layer.

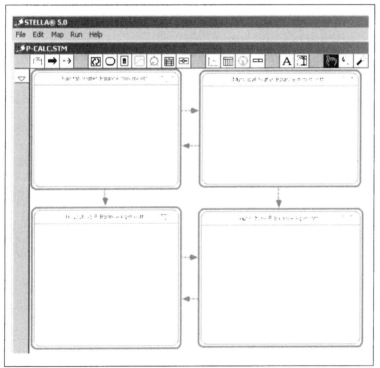

Figure 4-32 Overview of the process frames and connectors of the P-calculator (Map layer)

In the Model Construction layer by using stocks, flows and converters the four sub-models were constructed. Input data like monthly rainfall (*R*) from Microsoft Excel spreadsheet was entered into the model using the graphical input function in STELLA. Using the paste link command which is supported under Dynamic Data Exchange (DDE) in the Windows environment the corresponding converters were represented as graphical functions with monthly time scale and the simulation time-step (d*t*) set at one (see Figure 4.33). The data links are continually updated as changes are made in either application.

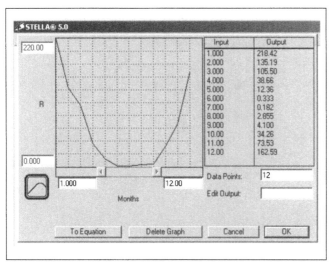

Figure 4-33 Representation of monthly rainfall of the mirco-catchment using the graphic input function in STELLA

The constructed sub-models are shown in Figure 4.34, 4.35, 4.36 and 4.37. For each sub-model a defined Graph and Table Pad icon is shown. The input and output variables were viewed using either in tabular format or graphically after launching the Run command.

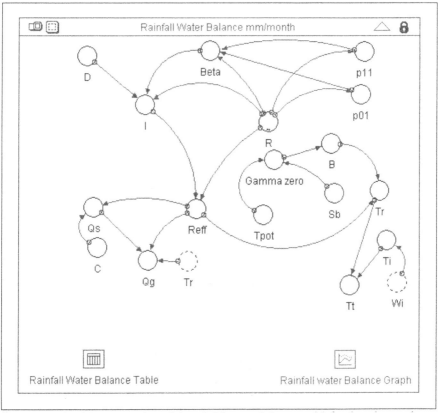

Figure 4-34 Pictorial representation of the rainfall water balance model for the micro-catchment (mm/month)

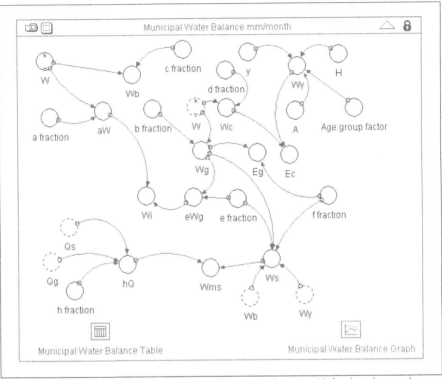

Figure 4-35 Pictorial representation of the municipal water balance model for the micro-catchment (mm/month)

In the Equation View mode, the equations for the P-calculator are listed by sector or subsystem and full documentation is included as part of the equation listing (see Appendix C). The Ghost tool is used to make replicas or aliases of individual stocks, flows, and converters. Ghosting is an antidote to spaghetti models which can be visually overwhelming. Its real value is mainly for cosmetic purposes. A Ghost adds no real structure to a model. In Figure 4.34 to 4.37 several ghosts have been created in the sub-models. These are represented as converters with drawn with a discontinuous line.

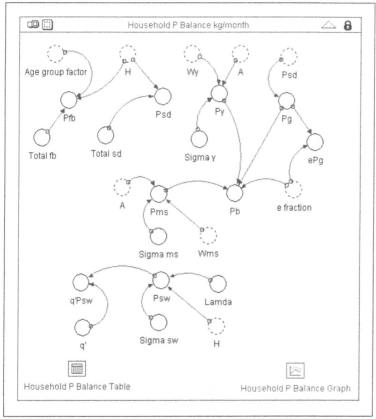

Figure 4-36 Pictorial representation of the household P balance model for the micro-catchment (kg/month as P)

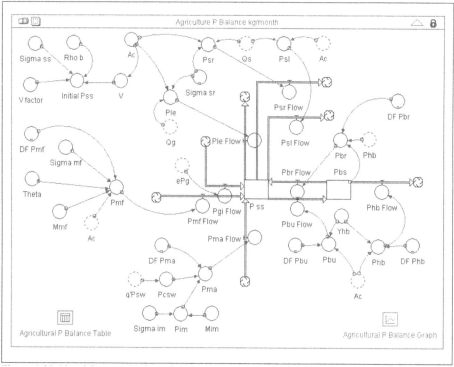

Figure 4-37 Pictorial representation of the agricultural P balance model for the micro-catchment (kg/month as P)

The Table Pad was used to display the numerical output from the simulations, and as a locus for exporting data to Microsoft Excel spreadsheet. For each subsystem the water and P-fluxes were added to the table using the Table Pad Dialogue box in STELLA. The Graph Pad was used as a repository for plotting water and P-fluxes and stocks generated by the model simulation runs. Figure 4.38 provides an example of a Table and Graph Pad simulation outputs for the agricultural subsystem.

Simulated output was exported to Microsoft Excel for ease of data handling, executing certain algebraic procedures and plotting of graphs where STELLA has limited capabilities. This was done via the Table Pad which is the only locus in STELLA software for exporting data via a link.

The output from STELLA is used in the following Chapter to create P-balances after aggregation of monthly P-fluxes and stocks in to annual values. The P-balances are essential in visualising the flows and stocks within the urban-shed and assist in decision-making.

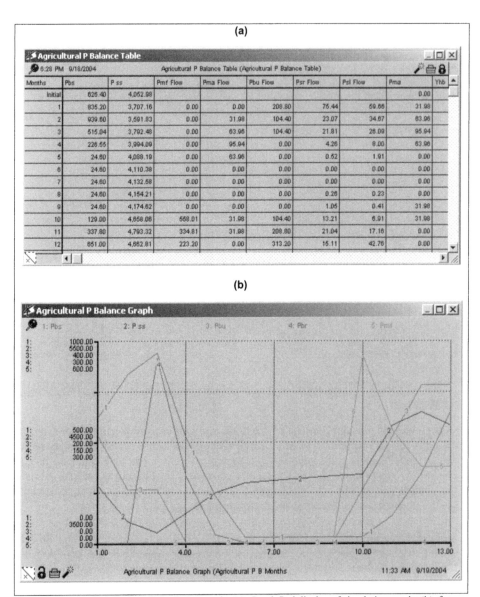

Figure 4-38 Table Pad display of simulation results (a), Graph Pad display of simulation results (b), for the agricultural subsystem in STELLA.

4.6   Conclusions

Establishing water and P flows and stocks in the urban-shed necessitated making a number of assumptions which is a common procedure in MFSA (see Chapter 1). This approach to problem solving involved the stripping away of unnecessary detail, so that only essentials remain. The raw data (primary, secondary and tertiary) that has been used to create the subsystems has varying levels of uncertainty and in combination creates a system which may not be well balanced. The year 2000 defined earlier as an *'average year'* only provides a snap-shot of the dynamics of each of the subsystems and hence provides limited understanding of the long-term variation and trends of the water and P flows and stocks within the micro-catchment.

An attempt has been made in Appendix D to look at the sensitivity and reliability of the various input parameters used in defining the models and to try and ascertain the extent to which each parameter influences the final output. Since the focus of this dissertation is on the increasing food productivity in urban agriculture by ensuring adequate bio-available P for crop uptake the value of bio-available P-stock in the soil (P_{ss}) in relation to the biomass uptake (P_{bu}) for a certain target economic yield (Y_{hb}) is critical.

Calculations for year 2000 for the micro-catchment indicate a value of bio-available P soil storage (P_{ss}) of 4 053 kg as P whilst the corresponding yield 0.2 kg/m^2 for maize only requires a total P-uptake of 1 044 kg as P. The P_{ss} value can ideally provide adequate P to attain the water-limited dry mass economic yield (Y'_{hb}) of 0.79 kg/m^2 (when calculated for the maize crop in the micro-catchment $P'_{bu} = A_c \times 1.41$ g/m^2 = 2 900 000 $\times$ 1.41 $\times$ 10^{-3} = 4 089 kg as P).

This confirms that the capacity of soils to be productive depends on more than just one plant nutrient. The physical, biological, and chemical characteristics of a soil all influence fertility and hence the crop yields. In addition the biochemical transformations of P in the soil are a fascinating event though incompletely understood in soil science, despite a great deal of study devoted to the chemical and biological behaviour of P in soils by soil chemists, agronomists, and horticulturists. The chemistry of soil P is complex owing to the ability of each phosphate ion to form a multiplicity of compounds of different composition and variable solubility (refer to Chapter 2).

For example, in a land use system with a P-fixing soil, application of low dose of P-fertiliser may not result in a measurable yield increase at all because P added is quickly immobilised. Application a higher dose is needed to saturate the immediate P-fixing capacity of the soil and bring about the desired increase in production. The determination of bio-available P-content of a soil on its own is of small agronomic value in the evaluation of the fertility potential and crop yields.

Plants on P-fertilised soils generally recover from 20 to 30% of the applied phosphate in the first year and progressively decreasing amounts in succeeding years. Agronomists commonly accept the research findings that the remaining portion of the added P is converted into forms of less availability, or 'fixed forms'. There is some evidence favouring the concept that organic matter in mineral soils helps to maintain soil phosphorus in the available form, the mechanisms being possibly an anion exchange which prevents the chemical combination of phosphorus with iron and aluminium or by replacing combined phosphate. However the fixing capacity of soil maybe seen as an asset rather than a detriment to long-term needs in soil fertility.

The P-calculator developed in this dissertation is a STELLA and Excel-based P-balance analysis programme designed to help planners, designers and implementers of ecological sanitation systems which aim at closing or short-cutting the P-cycle in urban ecosystems. STELLA software enhances the systems thinking approach and help to generate insight and understanding of webs of interdependent relationships in the study of P-fluxes and stocks in urban ecosystems. STELLA also enabled the superimposition or combination of the water flux and the P-flux to create a model, which can be adjusted accordingly to satisfy different boundary conditions.

The P-calculator can be used to simulate P-fluxes and stocks from the household and agricultural subsystem and hence providing an indication of where interventions can be introduced to ensure sustainable use of P within the urban environment.

5 Options for short-cutting the P-Cycle

5.1 Introduction

The difficulty of guaranteeing the quality of wastewater effluent and sludge, and the limited possibilities to recycle nutrients such as P within the traditional end-of-pipe waterborne system could be regarded as incitements to develop localised source separation systems (refer to Bellagio Principles in Chapter 1), i.e. systems designed to separate different waste fluxes such as faeces, urine, biodegradable organic waste, rain water and grey water (Van der Ryn, 1995; Hellstrom, 1998).

Currently research and development has yielded innovative water-conserving appliances, composting toilets and accessories, urine-diverting toilet facilities, vacuum-flush and micro-flush toilets, grey water system components, and rainwater harvesting systems (refer to Appendix F for internet sites describing such systems). This Chapter looks at some of these options in relation to the micro-study catchment discussed in Chapter 3 and 4.

Phosphorus is an example of a limited resource as shown in Chapter 2. In general, farmers make use of different P sources, namely mineral P fertilisers, other P sources which include organic matter, human and animal manure. From an economic point of view the different sources should be used efficiently to arrive at sustainable land use systems (Hermans, 1999). Here, sustainability will be defined by two aspects: meeting the growing food demand i.e. increasing crop yields, and maintenance or enhancement of environmental quality. Environmental quality comprises two important issues, the maintenance of soil fertility and the avoidance of environmental pollution.

Three options are considered; with option zero being the year 2000 setup and assuming that a business-as-usual approach is maintained. This option is devoted to an environment in which there is an excess of P transferred to Lake Chivero via sewage treatment works and sanitary landfills (P_{ms} and P_{sw}) or ground and surface water flow (P_{le}, P_{sr} and P_{sl}). At the same time there is a need for P-based fertiliser to boost the yields in the urban agriculture activities. Obviously this is a case in which recycling of nutrients back to the fields is of primary interest.

Two other options considered as far as short-cutting the P cycle is concerned are; urine or *yellow water* separation, collection, storage and application on agricultural land (diversion of P_y) and; composting of organic residues from the household subsystem (faeces or *brown water* and the organic solid waste fraction i.e. P_b and P_{sw}) which ideally should be applied to agricultural land to increase the available P-status of the soil and as manure to improve the soil condition. The diversion of P_b onto arable land is however considered to be technically, economically and socio-culturally challenging in terms of adaptability of the existing situation, and is therefore not included in the analysis. The two promising selected options are compared to the reference or existing conventional system in relation to P-recycling (see Figure 5.1).

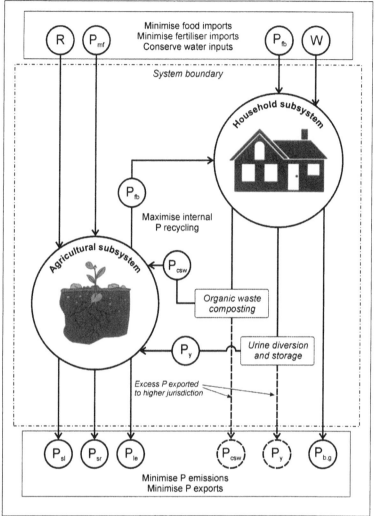

Figure 5-1 Conceptual model of the options for short-cutting the P-cycle in urban ecosystems

From Chapter 4 it has been shown that P contained in detergents or *grey water* (P_g) is insignificant in terms of supplying adequate quantities required for crop-uptake (maize in particular). Recycling of grey water in therefore only attractive in closing the water cycle and not the P-cycle and is subsequently not given further consideration in this dissertation.

Crop yields from rain fed urban agriculture can be significantly improved via the use of *grey water* as supplementary irrigation (i.e. increasing the *e* factor in the P-calculator, STELLA model in Chapter 4) to overcome short-term droughts which are critical to the crop and reduce yield considerably (FAO, 2000). It is important to note that besides systems for short-cutting the P-cycle, a combination with systems that focus on water conservation and closing the water cycle at local scale enhances an approach to the so-called life-support systems, i.e. small-scale systems that maintain an ecosystem supporting human life (refer to Chapter 1).

In accordance with the Bellagio principles Figure 5.1 shows that the optimum or desirable scenario for the selected promising options requires maximum internal P-recycling (P_y and P_{csw}), whilst at the same time; 1. Limiting, minimising and conserving any external resource input (water and food); 2. Minimising P-emissions from both subsystems and transfer or export of P beyond the level of jurisdiction determined by the system boundary.

5.2 Sustainability indicators

The definition of sustainability as espoused in *Our Common Future* (WCED, 1987) dictates that any water supply or sanitation system should '*meet the needs of the present generation without compromising the ability of future generations to meet their own needs*'. Even if such a definition is unquantifiable and unverifiable unambiguously, it implies that the sustainability of any water or sanitation system should include an efficient use of physical resources and a limited use of non-renewable resources (Pearce & Turner, 1990; Beck *et al.*, 1994; Holmberg, 1995, Azar *et al.*, 1996, Butler & Parkinson 1997; Larsen, 1997, Otterpohl *et al.*, 1997, Hellstrom, 1998; Lundin *et al.*, 1999, Mels *et al.*, 1999; Foxon 2000, Karrman, 2000). Another implication of the definition of a sustainable sanitation system is the concern about the conditions in the recipient, i.e. '*the physical conditions of production and diversity within the ecosphere must not systematically deteriorate*' (Holmberg, 1995).

Literature is replete with methodologies to assess sustainability of innovative urban water systems against traditional systems and attempts have been made to capture sustainability in a single indicator (Loetscher, 1999; Balkema *et al.*, 2001; Muhleck *et al.*, 2003; Hellstrom *et al.*, 2003; Van der Vleuten-Balkema, 2003). These methodologies have mostly involved Energy Analysis also called *Exergy* (Hellstrom, 1998), Cost-Benefit Analysis (CBA) (De Groot, 1992), Life Cycle Assessment (LCA) (ISO 14040-14043), Environmental Impact Assessment (EIA), Carrying Capacity (CC) concept and its land area-based indicator the Ecological Footprint (EF), (Wakernagel & Rees, 1996; Chambers *et al.*, 2000).

Multi-Criteria Analysis (MCA) has also been used extensively to establish preferences between options by reference to an explicit set of objectives that the decision maker has identified, and for which established measurable criteria to assess the extent to which the objectives have been achieved (Shoemaker, 1991; Keeney, 1992; Philips & Phillips, 1993). MCA is about techniques which do not necessarily rely on monetary valuations like CBA. This provides an improved way of presenting monetised and non-monetised impacts of water and sanitation options to decision makers (Golub, 1997, Hammond *et al.*, 1999).

A limitation of most of the methodologies outlined is that they cannot show that an action adds more to welfare than it detracts. The subjectivity in terms such as '*human needs*' and '*quality of life*' make it impossible to define and measure sustainability unambiguously. Scores and weights, when used in e.g. MCA, LCA and EIA, are also explicit and are developed according to established techniques. Eventually the end result or aggregated indicator is hard to communicate, and difficult to relate to natural limits. Another major drawback of these approaches is that assessment requires lots of data, and aggregating data into the standardised environmental impact categories means loss of insight in recovery of valuable finite resources like P, a subject specifically interesting from an ecological sanitation point of view.

A set of sustainability criteria has been proposed by various researchers and has generally been divided into economic, environmental, social-cultural and functional criteria as summarised in Box 5.A (Holmberg, 1995; Lundin *et al.*, 1999; Bijleveld, 2003; Van der Vleuten-Balkema, 2003). A cautionary approach is necessary when using these indicators since sustainability is by definition impossible to measure quantitatively. As knowledge proceeds, ideas change, situations and priorities shift, and hence there is no guarantee that the solutions identified as sustainable today will prove to be sustainable in the far future. Although, one may expect that a system with 100% reuse and powered by solar energy could stand the test of time (Van der Vleuten-Balkema, 2003).

Both social and cultural criteria are hard to quantify and are therefore often not addressed. However, these criteria play an important role in the implementation of water and sanitation technology. This is the case especially when the end user is directly involved, like in ecological sanitation (refer to Box 1.J in Chapter 1).

5.3 Indicators considered

The scope of this dissertation is limited to analysing the potential for short-cutting the P-cycle in urban ecosystems as outlined in Chapter 1. The sustainability criteria of most relevance are those listed under the environmental category or more specifically those related to utilisation of natural resources, environmental impact and potential to recycle nutrients in agriculture. As acknowledged in Chapter 1 this presents a limitation in the approach and methodology. A detailed analysis of functionality, economic and socio-cultural criteria is essential and when the methodology used in this dissertation is combined with other approaches it could provide insights to the bigger picture which is amenable MCA. As indicated in the previous section there is a lot of research and literature, which has been generated, in the last decade focusing on some aspects not fully considered in this dissertation.

Box 5-A   Sustainability criteria for urban water systems

Adapted from Lundin *et al.* (1999) and Van der Vleuten-Balkema (2003)

Criteria	Description
Functional	
Adaptability	Indication of flexibility of the process with respect to the implementation on different scales, increasing or decreasing capacity, and anticipate on changes in legislation for example.
Durability	Lifetime of installation.
Maintenance	Indication of maintenance required: frequency or costs and time needed for maintenance.
Performance	Expressed in removal of BOD or COD, heavy metals, organic micro-pollutants, pathogens, and nutrients.
Reliability	Indication of sensitivity of the process concerning malfunctioning equipment and instrumentation.
Robustness	Indication of sensitivity of the process concerning toxic substances, shock loads, and seasonal effects.
Economic	
Affordability	Costs and foreign exchange required relative to national or regional budget. The amount of money spent by users on water and sanitation relative to their total budget.
Cost effectiveness	Performance relative to costs.
Costs Net Present Value	Investment costs (specified for: land, materials, equipment and labour), maintenance costs and costs of dismantling the installation at the end of its life time.
Labour	Number of employees needed for operation and maintenance.
Willingness to pay	Indication of the amount of money the user is willing to pay for (improved) water and sanitation service.
Environmental	
Biotic	Mineral material depletion potential, biodiversity, global warming, nitrification.
Emissions	Untreated or treated wastewater and sludge.
Energy	Energy used, produced.
Land	Land area required.
Nutrients	Amount of nutrients suitable for reuse, indication of nutrient quality, organic matter recycled through sludge reuse and biogas production.
Water	Total water use, discharge, amount of water suitable for reuse, indication of water quality.
Socio-Cultural	
Acceptance	Indication of the cultural changes and impacts: convenience and correspondence with local ethics.
Expertise	Indication whether a system can be designed and built or can be repaired, replicated and improved locally or in the country or only by specialised manufacturers.
Institutional requirements	Indication of the efforts needed to control and enforce the existing regulations and of embedding of technology in policymaking.
Participation	Indication of the possibilities for participation of the end user.
Sustainable behaviour	Indication of stimulation for sustainable behaviour.

The main assumption in evaluating the proposed options is that P is the limiting nutrient in attaining the maximum water-limited cop yield (Y'_{hb} as maize) of 0.79 kg/m^2. As explained in the conclusions of Chapter 4 there is no direct relationship between bio-available P-content in the soil storage (P_{ss}) and the biomass production potential (refer to Figure 2.3 in Chapter 2). What is evident though is that a certain threshold P_{ss} is critical to ensure that the desirable economic yields are achieved. As stated in Chapters 2 and 4, maize only draws a small part of P dressing (P_{bu}) applied at sowing during the course of its growth. The application rate of P consists of a *build-up* or *corrective* application as well as an additional amount to replace that removed by the crop, the so-called *maintenance* application.

A number of studies have shown that excessive P applications (within the usual economic range) have no negative dramatic influence on crop yields due to toxicity (Rothamsted Experimental Station, 1991; Sharpley *et al.*, 1995; Withers *et al.*, 2001). However the continuous accumulation of surplus P in the soil increases the risk of P-losses through leaching to ground water sources (P_{le}), surface runoff and soil erosion to surface waters (P_{sr} and P_{sl}).

Available space for urban agriculture (A_c) to absorb P arising from the household subsystem could be a limiting factor (refer to Figure 1.4 Chapter 1). The practicality of the Bellagio principles and the possibility of localising urban human demands on the hinterland (ecological footprints) are tested (Chapter 1). If claims by Dr Gus Nilson that an area of 1000 m^2 including a house can feed a family (Esrey *et al.*, 1998), then urban housing plots should be planned with this minimum land area requirement. The proximity of production to consumption i.e. shortening the food cycle in particular determines the sustainability of the actions that might be taken in the urban environment.

5.4 Existing situation: option zero

The existing situation has been fully described in Chapter 1, 3 and 4. In this section a number of observations are elaborated so as to make effective comparisons with the proposed options. Since one of the main concepts being challenged in this dissertation is that of waterborne sanitation with its end-of-the-pipe management, it is important to focus on the design features and performance of such a system within the micro-catchment. The summary features of traditional waterborne sanitation and ecological sanitation are compared in Box 1.J in Chapter 1.

All houses in the micro-catchment have access to the conventional waterborne sanitation facilities centred on the water closet (WC), the sewer system and central wastewater treatment. The WC is connected to a sewerage system which collects the sewage for subsequent treatment. Some of the positive features of the conventional WC systems are that it is easy to clean, is almost odourless, is indoors, and has considerable health benefits. The legal system and technical expertise with regard to urban sanitation in Zimbabwe is geared towards centralised conventional WCs and sewer systems and contain minimal reference to or acknowledgement of the existence of decentralised (composting, urine-diverting WCs and anaerobic digestion) systems.

Generally the sewerage infrastructure in Zimbabwe is over-designed (Box 5.B). This is demonstrated by the continued satisfactory performance even as the numbers of users increase above the initial design figures (Chapter 3). Malfunction of the sewerage system is related sewage leakage and misuse of sewers by introduction of stones, large masses of rags, cotton wool (commonly used by women in menstrual periods) and other waste material[1]. Ingress of storm flow during wet weather, although difficult to quantify precisely, also increases the hydraulic load on the system (see Chapter 4). Despite this waterborne sanitation remains the preferred option for urban sanitation in Zimbabwe particularly in densely populated areas because of its ability in maintaining sanitary conditions and minimising waterborne or water related diseases.

Box 5-B   Design features of sewerage in Zimbabwe
Adapted from MEWRD (1978); MoLG & SALA (1990a);
HDS & BCHOD (1982)

Design Manuals: Coverage is about 96% in cities the difference usually being on septic tanks. Sanitation manual design procedures Vol. 5 (MoLG & SALA, 1990a) is the official design guideline recognised by the Ministry of Local Government. The manual is used in conjunction with two additional guidelines namely; the Model Building By-Laws guided by section 183 of the Urban Councils Act [Chapter 214] that specifies the standard materials for fixtures and fittings for water, sewerage and stormwater infrastructure; secondly there is the Guidelines for the Disposal of Sewage Effluent During Wet Weather (MEWRD, 1978). The former also specifies the use and protection of public storm water drains and sewers.

Design Parameters: 85% return as sewage is assumed for average water demand of 850 l/day per stand. This gives 720 l/day per stand as the Annual Dry Weather Flow (ADWF) (MoLG & SALA, 1990a; HDS & BCHOD, 1982). Peak factors for wet weather flow range from 5.25 to 2.70 (MEWRD, 1978). The factors decrease with increasing sewage flows because of attenuation with large area of reticulation. The Colebrook-White formula is generally used for sizing the pipes (velocity of flow and discharge). Minimum self-cleansing velocity 0.6m/s and maximum of 3 m/s to avoid scour damage to pipe. Grade should not be flatter than 1:100 until flow reaches 0.75 l/s especially in low cost housing areas where misuse of sewers by introduction of stones, large masses of rags and other waste.

Construction Details: Sewer lines usually located at the back common mid-block boundary. Minimum depth of about 1.0 m and 7.0 m probably represents the extreme practical limit in Zimbabwe. Maximum manhole spacing for 100 mm diameter. Pipe maximum spacing 30 m and for 150 mm, 200 to 250 is 75 m and 100 m respectively.

Materials: Generally 150 mm sewer pipes are used, house connections are almost always 100 mm. All sewer pipes are flexible and this is achieved by laying rigid pipes with flexible joints. Glazed earthenware (clay) pipes, concrete, asbestos cement, steel pipes are commonly used and recently also PVC pipes.

The generated sewage (W_s = mixture of yellow, grey and brown water) is conveyed though a pipeline to a centralised location at Crowborough sewage treatment plant which employs BNR technology to remove nutrients from the effluent (see Chapter 3). Although treatment at the end of the pipe is usually regarded as satisfactory the trend is that there has been a slow build-up of nutrients, heavy metals, pathogens and other residuals in receiving media due to discharge of effluent and sludge leading to water quality problems in Lake Chivero.

---

[1] The majority of the people use old newspapers for anal cleansing, these do not disintegrate easily within the sewer system and the high rate of blockages can be attributed to this.

Using the results of the P-calculator developed in Chapter 4 annual water fluxes through the micro-catchment area can be represented as in Figure 5.2. The *source* and *sink* of the water fluxes are basically the atmosphere and Lake Chivero. These are not shown in Figure 5.1 as they fall outside the system boundary. The advantage of presenting the results pictorially as annual totals is that the relative magnitudes are more visible as opposed to graphical representations shown in Figure 4.14 and 4.17.

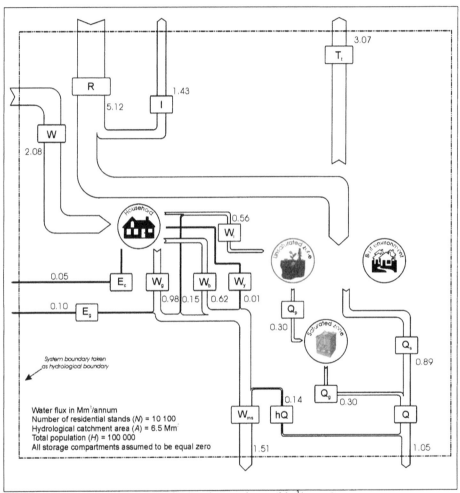

Figure 5-2 The annual urban water flux for the micro-catchment (Mm³/annum)

The bulk of the food and beverage consumed is imported from outside the micro-catchment. Only a few of the 10 100 households (N) rely on produce from the urban agricultural activities. The quantity of P contained in biodegradable solid waste material recycled is also small (P_{csw}). Hence the local recycling of P from the household to the agricultural subsystem is limited and the system resembles an open-ended P-flow described in Chapter 1 and 2 (Figure 5.3). From Figure 5.3 it is apparent that efforts have to focus on short-cutting the P_y, P_b and P_{sw} fluxes such that P_{hb} is maximised and hence less imports of food.

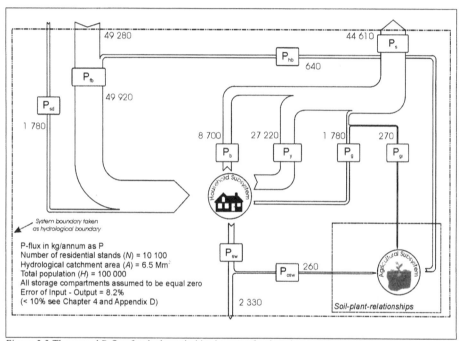

Figure 5-3 The annual P-flux for the household subsystem for the micro-catchment (kg/annum)

The calculated total area under cultivation per household (A_c/N) is 287 m^2 or alternatively 29 m^2/person. This figure falls shot of the 100 to 800 m^2 range required to produce grain crops to feed an adult person each year (refer to Chapter 1). Further, this available area is insignificant compared to the area required to grow all the crops (including non-edible) of 5 300 m^2 quoted by Wackernagel *et al.*, (2002a) as the average area exerted on earth's bio-capacity per person (refer to Table 1.1 in Chapter 1). Figure 5.4 illustrates the annual P-fluxes and stocks for the agricultural subsystem for an *average year* (taken as year 2000) in the micro-catchment.

Although there appears to be an apparent net accumulation of bio-available P soil storage of about 610 kg for the year under consideration (refer also to Figure 4.38a STELLA model output table) there is greater possibility that bio-available P in the soil is declining taking into account the error margins of the P-calculator. Other P-fluxes like clearing of land through slash-and-burn of maize straw and other plant residues could be significant even though not included in the model.

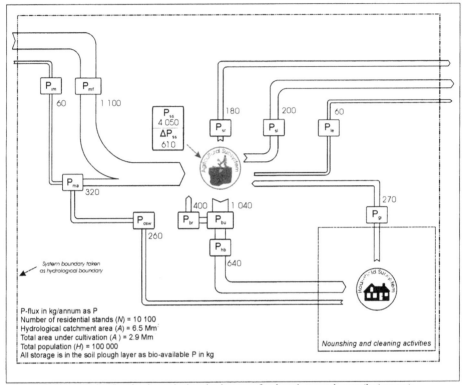

Figure 5-4 The annual P-flux for the agricultural subsystem for the micro-catchment (kg/annum)

The existing conventional system is relatively reliable, although its flexibility is low because of the high investment costs and the below ground construction. Adaptation and anticipation on new technologies or developments is almost not possible because of the costs and expertise required.

5.5 Urine diversion, storage and application on arable land: option one

The favourability of diverting urine, storing it for a specified time period and applying it on arable land has been explained in Chapter 1 and summarised in Box 1.G. The principle of the *modern* urine-diverting toilet is simple; urine and faecal matter are never mixed. The appearance of the toilet itself includes a range of materials and design (refer to Figure 1.2). The management may involve only a single household, or may comprise several households connected to a communal system. . The highest concentration of P is obtained if the urine and faeces are collected without flush water. However, it is possible to use a water flushed toilet and collect P from the urine and faeces in relatively concentrated form.

Urine in itself presents virtually no risk of infection. But it can be contaminated with the pathogens in faeces (Hoglund, 2001). Human health aspects have been evaluated regarding risk of transmission of infectious diseases (see Box 5.C), however research on the possible ingestion of excreted pharmaceuticals, hormones (estrogens and xeno-estrogens), and other dioxins is limited (Arnbjerg-Nielsen *et al.*, 2003).

Box 5-C   Hygiene and protection against infection

In Sweden recommendations have been developed to minimise the risks-barriers on the use of human excreta particularly urine (Hoglund *et al.*, 1998; Johansson *et al.*, 2000; Hoglund, 2001; Vinneras, 2002; Jonsson *et al.*, 2004). These recommendations compliment the World Health Organisation (WHO) published guidelines for the safe use of wastewater and excreta in agriculture and aquaculture (Mara & Cairncross, 1989; Blumenthal *et al.*, 2000; 2001) The recommendations and guidelines were developed to protect the health of both workers and consumers of agricultural or aquaculture produce (refer to Chapter 1).

These Swedish recommendations are presented as a framework with prescribed measures to reduce infection risk: 1. *Pre-treatment* through storage eliminates or reduces the number of pathogens in urine; 2. Application techniques should reduce the formation of aerosols and the exposure of the farmer; 3. *Restrictions with respect to crops* (see Table 5.1). As a rule of thumb urine can be used for all crops if it has been stored at not les that 20 °C for at least six months; 4. *Site selection* helps to minimise the exposure people and animals in the area; 5. *Avoiding risk groups* where excretion of pathogens and medicines is known to be higher than normal; 6. *Adjusting the size of the pathogen cycle* in relation to storage routines for large and small private household systems and harvest times.

Table 5-1 The relationship between storage conditions, the pathogen content of urine mixture and recommendations for crops in larger systems [a]

Storage temperature	Storage period	Presence of pathogens in the urine mixture [b]	Recommended crops
4 °C	≥ 1 month	Viruses, protozoa	Forage and Food crops that are to be processed
4 °C	≥ 6 months	Viruses	Food crops that are to be processed, forage crops [c]
20 °C	≥ 1 month	Viruses	Food crops that are to be processed, forage crops [c]
20 °C	≥ 6 months	Probably none	All crops [d]

a) 'Larger systems' in this case means that human urine mixture is used to fertilise crops that are consumed by persons other than the members of the household from which the urine was collected.
b) Gram-positive and sporulating bacteria are not included
c) Except grasslands for production of animal feed.
d) In the case of food crops consumed raw it is recommended that fertilisation with urine be discontinued at least one month prior to harvesting and that the urine be incorporated into the soil.
e) For small systems no storage is required, but the urine should be incorporated into the soil and the crop should nor be harvested within one month of urine fertilisation.
Source: Johansson *et al.*, (2000); Jonsson et al., (2004); Schonning & Stenstrom (2004)

This toilet may use some water[2] (low-volume flush systems of 0.2 l/p.d) or be waterless. Urine is collected in a bowl or container, and the faecal matter and paper could either be flushed separately with grey water (as with traditional WCs) or collected in a net or bucket in a chamber below the floor (for dry systems). There are numerous reports on user appreciation of the various toilets in the house or yard and collection efficiency of urine. A urine-diverting toilet is odourless and for an indoor installation has the same positive features as the WC when it comes to convenience and hygienic safety (Johansson *et al.*, 2000; Drangert, 2003; Vinneras & Jonsson, 2003).

Potential drawbacks to urine handling systems are the risk of ammonia evaporation and the relatively large volumes that need to be handled particularly when low-volume flush is preferred. The amount of urine solution may be reduced by using drying techniques provided that ammonia evaporation is avoided. One method to reduce this risk is to prevent the decomposition or urea to ammoniacal nitrogen by adding small amounts of acid to the fresh urine before storage (Hellstrom, 1998, Vinneras, 2002). The P-content of urine is normally assumed to remain constant during sanitisation because of minimal gaseous losses.

The low-flush urine diverting toilet scenario is very similar to the conventional system. The only difference is the urine diversion. The most probable scenario within the micro-catchment is a house with a urine-diverting toilet installed and the urine collected undiluted or with minimal dilution in a semi-central collection tank and after a desirable period of storage applied to arable land. The brown water or faeces are flushed with some water into the existing sewer system and transported, together with the grey water, to the existing central wastewater treatment plant. Because the yellow water is not delivered to Crowborough treatment plant it is assumed that there would be less efforts on P-removal, less energy inputs, and possibly treatment costs would be reduced.

This scenario can be represented in the P-calculator in STELLA as an additional flow into the agricultural subsystem. The magnitude of this flow is taken as equal to the P-flux due to yellow water from the household subsystem (P_y). Figure 5.5a shows the agricultural system sub-model with P_y added (compare with figure 4.37). The results of the simulation are shown in the graphical output in Figure 5.6a. In essence this scenario results in a total annual diversion of 27 220 kg (see Figure 5.3) into the soil storage compartment (P_{ss}). Assuming that P is the limiting input in attaining the maximum desirable harvested biomass or yield (Y'_{hb} as maize) of 0.79 kg/m[2], then by setting $Y_{hb} = 0.79$ kg/m[2] in the agricultural sub-model the depletion P can be simulated (Figure 5.6b).

As calculated in Chapter 4 the maximum yield (Y'_{hb}) and for the available agricultural space (A_c) the maize crop can only remove (P'_{bu}) 4 089 kg/a as P. This implies that 100% diversion of P_y results in excess P in the plough layer which cannot be absorbed by the maize. In this case export of P_y would be necessary. The total land area can be calculated using P-calculator by adjusting the value of A_c, Y_{hb} and P_{bu}. A land area of 6.6 $\times$ A_c (19.30 Mm[2]) would be required to absorb all the P contained in urine

---

[2] For reasons of acceptation and rinsing the toilet bowl, a number of toilet seat designs allow for little water for urine flushing.

produced within the mico-catchment. The per capita land requirement for maximum maize production would therefore be 19.30 $Mm^2/100\ 000 = 193\ m^2$/person.

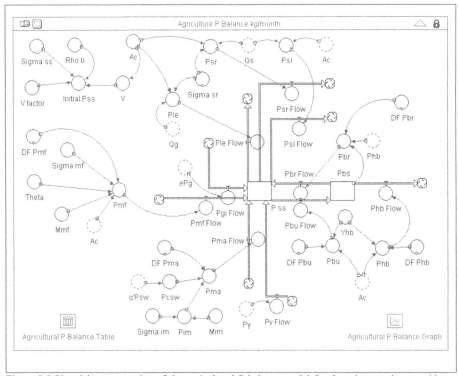

Figure 5-5 Pictorial representation of the agricultural P balance model for the micro-catchment with yellow water application in STELLA (kg/month as P)

The land area calculation assumes that the all the biomass produced is harvested from the field. In the case where biomass residue (P_{br}) remains and is incorporated into the soil storage during land preparation, the P_y diversion rate in theory has to be reduced in such a way that it only supplies the amount of P equivalent to the P harvested (P_{hb}) i.e. *maintenance* application. For maximum biomass production this would be equivalent to 2 520 kg/a as P i.e. approximately 10% of P_y produced. As discussed in Chapter 2 and further in section 5.7 of this Chapter, a decision to apply enough bio-available P to match the amount of P removed in the harvest $(P_{hb} = P_y)$ can only be made after exhaustive field experiments which include, the P holding capacity or saturation point, the rate P mineralisation (from organic to inorganic, refer to Chapter 2), and the forms of P in the various compartments described in Chapter 2 and 4.

Environmental sustainability criteria calls for efficient use of P as any excess or undue accumulations would eventually cause degradation of land, water and air. Export of P_y beyond the micro-catchment is necessary in this case once the *build-up* or *corrective* application component of P has been satisfied.

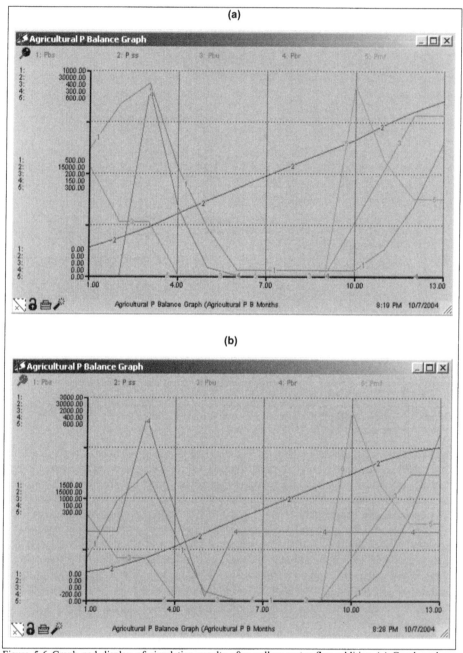

Figure 5-6 Graph pad display of simulation results after yellow water flux addition (a) Graph pad display of simulation results after setting the yield to the maximum value(b), in STELLA (kg/month as P)

From an engineering feasibility perspective the source-separated urine has to be collected, transported in a pipe network to a central storage point serving a number of households or neighbourhoods. The volume of urine generated is of practical significance given that it has to be stored for a minimum period of six months at 20 °C to inactivate all pathogens for unrestricted crop application (see Box 5C). The six month storage recommendation is convenient for annual (seasonal) cropping pattern occurring in the study area and timing of P application. P is normally applied as basal fertiliser (see Chapter 3 and 4) i.e. before sowing i.e. in the months of October and November of each year. This means that by this time of the year, each year, sufficient urine mixture last collected around April-May should be available

Assuming that the households in the micro-catchment are divided into ten neighbourhoods each having 1 010 households, then from the P-calculator the total volume of urine generated per neighbourhood (W_y) in six months would be [(0.14 mm/month × 6 months × 6.5 × 10^6 m^2/(1000 × 10 neighbourhoods)] equal to 546 m^3 per neighbourhood. Ideally, storage facilities should be provided in duplicate to ensure that fresh and stabilised urine does not mix and also for enabling maintenance of the system. If rectangular below ground concrete tanks are used with a maximum depth of say 2 m then the plan area of each tank would be 273 m^2 i.e. 16 × 16 m. The storage tank sizes appear to be practical and can be located strategically by considering accessibility and location of agricultural where the yellow water has to be applied.

Transportation of excess yellow water could be done using specialised tanker trucks, whilst application on agricultural land can be done by tractor drawn tankers with hoses fed from a centrally located distributor for larger areas. But since most of the gardens and off-plot fields are small, small volume tanks of say 0.1 m^3 could be designed and fitted into a wheel-barrow configuration and urine applied to the field using a hose pipe[3].

The introduction of urine separation, storage and transport facilities can be achieved with minimal retrofitting of the existing conventional system. The costs can be weighed against the cost of P-based fertiliser (US$ 0.50/kg in year 2000), or the market price of the produce i.e. cost of maize per unit weight (US$/kg). From the cost of P-based fertiliser yellow water produced per annum within the micro-catchment is worth some US$ 13 610.00. In other words the value of urine excreted each year by each adult person within the micro-catchment area is worth US$ 0.14 as P-based fertiliser equivalent. A holistic CBA analysis could include the benefits derived from reduced P-fluxes reaching the sewage treatment plant at Crowborough (less energy and infrastructure required for P-removal) and other external environmental costs like the cost of P-pollution of Lake Chivero water.

The P-calculator assumes 100% efficiency in separation, collection, storage and field application of P contained in yellow water. In practice the toilet to field continuum has a number of inefficiencies. Separation and collection efficiencies of 40-95% of

---

[3] A variety of transport systems and technology for the storage of urine are available on the market. In Sweden the Stockholm water Company uses special tankers to transport urine from holding tanks within the neighbourhood to water tight inflatable storage '*balloon tanks*'. Stratification of nutrient concentration has been observed during storage and therefore occasional stirring is recommended (Johansson *et al.*, 2000).

the theoretical potential quantities of yellow water have been reported in studies conducted in Sweden (Johansson et al., 2000; Vinneras, 2002). It has been observed that the higher the motivation of the residents the higher the quantity and the better the quality of the yellow water product.

The flexibility of the proposed system is similar to the conventional system, mainly because of the high investment costs and the underground construction infrastructure required. Adaptation and anticipation on new technologies or developments is limited or too expensive for built-up areas especially when the brown water and grey water still have to be transported via a sewer system. But once the urine diverting WCs have been introduced together with the necessary collection pipe work and storage facilities, new techniques for recycling (short-cutting) of yellow water at a local scale can be introduced more easily. The reliability of such a system could only be a little bit lower than the conventional system, because of the additional yellow water system.

5.6 Stabilisation of organic residues and application on arable land: option two

The degree of aliveness of the soil plays a major role in P availability. Maintaining or increasing the amount of active soil organic matter (humus) can best assist this function. At the same time soils that reveal abnormally high concentration of humus usually have low levels of microbial activity. This is due to the facts that in these soils, conditions prevail that are more conducive to preservation than decomposition. In order to have an active organic base there must be an equitable mineral, physical and biological environment.

Organic P-flux residues from the micro-catchment which are of interest in relation to short-cutting the P-cycle are brown water P-flux (P_b) and the biodegradable solid waste fraction (P_{sw}). There are three possible scenarios within this option: 1. separate collection and stabilisation of brown water; 2. separation collection and stabilisation of the biodegradable solid waste fraction and; 3. separate collection and combined processing of brown water and organic solid waste (Obeng & Wright, 1987). The principles of ecological sanitation (refer to Chapter 1) dictate that that '*non mixing*' and '*non-diluting*' systems are preferable. This suggests that dry systems are the logical starting point implying that brown water has to be separately collected using dry-systems like composting toilets or low-volume and vacuum flush toilets. Similarly organic solid waste has to be as dry as possible for ease of transport but having sufficient moisture for stabilisation processes to occur.

As indicated in the introduction of this Chapter adapting the existing conventional system for a separate brown water collection system could be technically and financially impractical and it is likely not to be easily accepted by the users in the short-term. In new residential areas probably the composting toilet scenario could be implemented. The composting container must be accessible, to control the composting process and to empty the container after some period to retrieve the compost. The rest of the wastewater produced in the house, mostly grey water, after collection in smaller sewer lines could be treated in a compact facility consisting of a constructed wetland which can function as a green area or an educational facility.

This could be done in a centralised (neighbourhood scale and upwards) or semi-centralised system (household or a number of households).

Like organic manure, organic solid waste is a source of P, albeit a relatively poor source in comparison with commercial fertilisers, human urine or faeces as shown in Chapter 4. It helps improve soil structure by adding organic matter to the soil. It is also a source of the secondary nutrients and micronutrients that are necessary in small quantities for proper plant growth. In addition, urban waste transforms material that would otherwise be slated for costly disposal into a useful farm product (see Chapter 1).

In the previous section it was demonstrated that yellow water separation, stabilisation and application on arable land can provide the requisite P for maximum maize yield production. At the same time if the biomass residues are managed expeditiously (refraining from slash and burn activities) they could also provide the much needed organic manure. Possibly from a perspective of the soil organic carbon content there is no urgent requirement for stabilisation of solid waste and application on arable land. The organic carbon content of the investigated soils in the micro-catchment has been shown to be below the desirable value of 3% (see Table 4.15) hence the need to increase it by increasing the amount of or recycled solid waste (P_{csw}).

Urban solid waste needs to be treated carefully because it may contain heavy metals, parasites, and other pathogens. The build-up of heavy metal concentrations in the soil can be a cause for concern. While trace amounts of some heavy metals play a critical role in plant metabolism, excessive amounts have reduced crop yields and could be dangerous to public and grazing livestock (Conway & Pretty, 1991; Kirchmann & Widen, 1994). To minimise these risks the continuous application of urban solid waste needs to be monitored in order to ensure that heavy metal and overall nutrient concentrations do not reach toxic levels and do not damage the environment through leaching and eutrophication.

Urban organic solid waste may also contain organic compounds such as dyes, inks, pesticides, and solvents that are often found in commercial and industrial waste. These contaminants have been shown to cause genetic damage, while others, such as bacteria, protozoa, and viruses can cause salmonellosis, amoebic dysentery, and infectious hepatitis (Gotaas, 1956; Tchobanoglous *et al.*, 1993; Rosenburg & Furedy, 1996).

Biological processes are preferred in the transformation of the organic fraction of solid waste into gaseous, liquid, and solid conversion products. In essence these can be classified into two categories aerobic (composting) or anaerobic (digestion) processes (see Box 5.D).

Box 5-D Biological conversion technologies

Aerobic composting
Composting is a stabilisation process used for organic wastes such as sewage sludge, animal and agricultural wastes, and household organic waste. Composting systems are aerobic; therefore, the main products are stabilised sludge, water, carbon dioxide, and heat (Gotaas, 1956; Tchobanoglous *et al.*, 1993). The most important operational factors are aeration, detention time, temperature, moisture, carbon-nitrogen (C:N) ratio, and pH. For rapid composting, the composting material should have moisture content between 40-65% (Tchobanoglous *et al.*, 1993; Eklind & Kirchmann, 2000).

The overall volume reduction is 30-35% and because of the water loss, the weight can be reduced by 40-80% (Eklind & Kirchmann, 2000). The P in organic matter is not lost during composting. Heat inactivation is one of the main factors in pathogen removal in composting, therefore, the main challenge is to make sure that all material is exposed to significant high temperatures (>50 °C) for certain time (>10 days) during the composting process. Composting can reduce virus concentration to below their detection limit (Haug, 1993; Karrman 2000). Also pathogenic bacteria, *Ascaris* eggs, and other *Helminths* eggs can be eliminated effectively in composting. However, re-growth of bacteria such as faecal coliforms and *Salmonella* may occur.

Due to the long retention times (7 to 14 weeks and up to one year for some systems) relatively large amounts of space may be required.

Anaerobic digestion
Anaerobic digestion is a biological fermentation process (in the absence of molecular oxygen) that can be used to treat organic solid waste. The main advantages of anaerobic digestion are: methane production, low sludge production, highly stabilised sludge, low operational costs, and low nutrient requirement due to the lower growth rate of anaerobes, and the production of mineralized compounds (NH_4^+, PO_4^{3-}, S^{2-}) (Gray, 1999). This methane gas can be converted to electricity and heat by cogeneration.

The main disadvantages are: low sludge activity, low reactor capacity, high capital costs, energy use for heating, long retention times, the need for post-treatment (high residual BOD and COD), inhibitory effects, and the production of malodorous compounds (H_2S) (Gray, 1999).

The Upflow Anaerobic Sludge Bioreactor (UASB) has received a lot of international acclaim as the technology of choice in stabilisation of domestic sewage and organic waste (Zeeman, 1991; Zeeman & Lettinga, 1999; IWA, 2002). The UASB reactor requires skilled operation and management and a higher capital investment that composting.

Conversion of organic solid wastes into compost can be done within the micro-catchment in two stages; one at the domestic level by each household and the other at a centralised neighbourhood composting unit. The first important step for both stages is separation at the source of the different solid waste fractions. Organic wastes from kitchen (food residues), yard wastes and non-edible biomass from the gardens and agricultural subsystem should be separated from recyclable waste like plastic, paper, glass, leather and inert debris, and should be placed in separate containers.

In the agricultural subsystem of the P-calculator the maximum recycling potential of P_{sw} can be realised by adjusting the value of q' from 0.10 to 1.0. When the value is set at one, this implies that 100% of the organic biodegradable waste is stabilised and applied onto arable land. This translates to an addition of 2 590 kg/a as P. This value is comparable to the amount of P-fertiliser contained in the maximum water-limited yield (P_{hb}) of 0.79 kg/m^2. The limitation could be that a smaller fraction of P in stabilised organic material is available in ionic form which is readily taken up by plants as compared to yellow water. The diversion of P_{sw} onto arable land could be important during the soil P *build-up* or *corrective* application phase.

The weight of organic solid waste that requires stabilisation is calculated from the generation rate per person in the P-calculator of λ = 1.35 kg/p.month and moisture content of 70% (1.35 × 100 000 × 12/0.7) to obtain a value of 23.14 × 10^6 kg/a. Assuming an un-compacted density of about 150 kg/m^3 (refer to Chapter 1) the volume of organic solid waste generated would be 154 × 10^3 m^3. If for example a windrow[4] composting process is adopted, with a three month stabilisation period (including curing of compost) an area of no less than 0.1 Mm^2 would be required for the processing plant. The land area requirement although representing less than 5% of the land area currently under cultivation the location of such a facility in proximity to a residential area can pose a number of nuisances like fly breeding, rodent and odour problems.

The alternative would be to go for in-vessel (composting accomplished inside an enclosed container) aerobic or anaerobic composting systems. The popularity of in-vessel composting systems in recent years has been due to better odour control, faster throughput, lower labour costs, and smaller area requirements.

The value of P contained in organic solid waste as per P-based fertiliser can similarly be computed as for yellow water P generated per annum. The value translates to about US$ 1 295.00. This value converted to per person organic solid waste generated is approximately US$ 0.01/a. The total value of organic solid was can also be calculated holistically by taking into account the costs of collection, transportation and disposal (typical elements of a solid waste management process). In addition, external environmental costs, which include cost of landfill disposal, leachate handling and treatment and impacts of possible contamination of surface and ground water sources as is the case of Golden Quarry in Harare.

The reliability of on-site aerobic or anaerobic stabilisation processes for organic waste handling could be lower than the conventional system of collection and disposal at a centralised landfill plant. The sensitivity of such systems is also higher than the conventional system because they require some level of expertise for good operation and maintenance.

---

[4] Windrow composting is one of the oldest methods of composting. A high-rate windrow composting employs windrows with typical cross section of 1.5 m high and 5 m wide. The actual dimensions depend on the type of equipment used to turn the composting waste (Tchobanoglous *et al.*, 1993).

5.7 Enhancing P availability to plants

The recovery fraction of P can be defined as the ratio of applied P and P taken up by the crop. In theory the element recovery fraction can assume any value between 0 and 1. In practice it varies from less than 0.1 to, say, 0.8. Since the uptake requirement has to be met by applying P-based fertiliser the mass fraction of P in fertiliser is important. The skill and motivation of the individual farmer largely determine the efficiency of fertiliser use. Improving uptake of P by adapting management attributes is a basic characteristic of agriculture.

However, P residues accumulated in the soil are not entirely lost - but this is not a reason for accumulating these residues unnecessarily. There are critical values of P below which yield decreases appreciably and which represent a financial loss to the farmer. To accumulate P in the soil above these critical levels is an unnecessary cost for the farmer. It may also pose an environmental risk, in that soil lost by water or wind erosion to streams, rivers and lakes takes its nutrient load with it (FAO, 1980a; Smaling & Braun, 1996; Johnston, 1995; 1997a; 1997b).

In fact, improvements in fertiliser use efficiency can be detected in most agriculturally advanced regions, but this can be attributed to improvements in cultivation practices, techniques of fertiliser application and crop varieties. Apart from some developments in coated, controlled release fertilisers and nitrification inhibitors, there has been little significant change in the fundamental nature of the main fertiliser products for many years, or even decades (see Box 5.E). There is little incentive to invest in the research and development of a bulky, low-priced commodity, which offers little scope for product differentiation (IFA & UNEP, 1998; 2000).

Various soil conservation technologies have been promoted in the developing world to prevent the physical loss of soil and P through leaching and erosion. These fall under three general categories: 1. Practices such as terracing, alley cropping, and low-till farming, which alter the local physical environment of the field and thereby prevent soil and P from being carried away; 2. Mulch application, cover crops, and intercropping, act as physical barriers to wind and water erosion and help to improve soil characteristics and structure; 3. Organic manures such as solid waste, animal and green manures also aid soil conservation by improving soil structure and replenishing secondary P and micronutrients (FAO, 1980b).

Improved application and targeting of inorganic and organic fertiliser not only conserves P in the soil, but also makes P-uptake more efficient. Most crops make inefficient use of P. Inefficiencies in fertiliser use represent both an environmental hazard and a substantial economic loss. Often less than 15% of applied P is found in the harvest crop (UNIDO, 1998). More efficient fertiliser uptake by plants has two major benefits. Farmers can apply less fertiliser while maintaining increased yields, thus saving money. More efficient absorption also reduces the quantity of plant nutrients, which might otherwise leach into water systems (IFA & UNEP, 1998; 2002).

Box 5-E   Methods of improving the efficiency of P use

Integrated nutrient management: The relief of a nutrient deficiency by applying fertiliser will increase plant growth and alter the requirement for other nutrients. For example, the yield increases gained from use of N-fertilisers (Tandon, 1987; De Datta *et al.*, 1990; Gruhn *et al.*, 2000) have led to inefficient N use because of induced P deficiency. Similarly, increased soil P status decreases the plant uptake of other nutrients, which then become key factors in limiting yield and efficiency of P fertiliser use. Further, interactions between soil acidity, lime and phosphate availability are also important in managing nutrients in a holistic fashion (Gerner & Baanante 1995; Mokwunye, 1995; Teboh, 1995). Balanced fertilisation is the nutrient mix which gives the optimum economic return in the sense that the proper balance assures that the plant absorbs each nutrient and does not waste any of what is applied (IFA & UNEP, 2002).

Soil organic matter management: Conservation tillage, use of mulches, manures and crop residues and rotations with restorative crops such as legumes and green manures plays a key role in efficient utilisation of soil and fertiliser P, especially on acid, P deficient soils of the tropics. Soil organic matter content interacts with most soil fertility parameters through improvement of soil structure, provision of N, S and P, increased cation exchange capacity, increased soil water holding capacity and alleviation of Al toxicity (Von Uexkull, 1987; Tandon, 1992; 1998). All these factors impact on plant root growth. Increased root length or mycorrhizal extension has the largest effect on plant P uptake efficiency (Hedley *et al.*, 1982, 1990; 1993; 1995; Maene, 2000).

Nutrient accounting: Nutrient accounting is a tool for reducing unwanted environmental impacts (CLM, 1988; IFA & UNEP, 2002). In brief, this management technique requires a farmer to budget the quantity of nutrients introduced to the farming system and the quantity escaping either through crop harvest or losses to the environment. Nutrient accounting was first developed in the Netherlands[5] to respond to concerns about nutrient pollution of water resources (CLM, 1988). Balanced fertilisation, as well as timely and appropriate application, can help achieve the correct nutrient budgets.

Precision agriculture: In developed countries, computers and satellite technology have paved the way for precision agriculture, which uses high-tech systems to map the specific nutrient needs of small plots within crop fields (Withers, 1999). This allows a farmer with the proper equipment to tailor the application of nutrients to suit the needs (Tandon, 1986; Dibb *et al.*, 1990, IFA & UNEP, 2002).

Fertigation: Fertigation enables farmers to maximise the use of water resources and to increase the efficiency of fertiliser use. It is particularly useful for high value crops under arid and semi-arid conditions. It involves adding soluble fertilisers into irrigation systems using a drip system, which allows uniform water distribution and feeding of the crop. However, as with any technique, it must be properly managed; use and maintenance require skilled labour. Initial investment costs are high (Anon, 1984; Kofoed *et al.*, 1986; Beaton & Murphy, 1988; IFA & UNEP, 2002).

Selection of P-efficient plant species: Recently more emphasis has been placed on choosing plants suited to P deficient soil conditions by selecting alternative plant species or genetically improving existing species. Plants vary widely in their abilities to survive and grow in soils of low P status (Barber, 1980;1984). Varietal differences in P uptake have been observed for many plant species, such as, maize (Nielsen & Barber, 1978). Species differences in P uptake characteristics can be used to introduce plants with greater P uptake efficiencies into soils of low P status. Varietal differences, induced by genetic manipulation, can be used in a similar manner.

---

[5] The Centre for Agriculture and Environment (CLM) is an independent, non-governmental organisation based in the Netherlands, founded in 1981. Its objective is to promote a sustainable and integrated agriculture. A simple tool known as the Nutrient Management Yardstick has been developed by CLM to track farm level nutrient flows.

The aim is to optimise production per unit of fertiliser while applying the required quantities to satisfy the world's agricultural requirements. Various management techniques summarised in Box 5.E have been employed in recent years mostly in the developed world to optimise fertiliser use and application (Headley *et al.*, 1995; Withers, 1999; Maene, 2000).

5.8 Conclusions

The existing system (WC, sewer and wastewater treatment plant or landfill site) of human and organic waste handling offers little opportunity for local recycling of P on agricultural land. Although the conventional setup is perceived to be functioning relatively well and is the preferred option, especially from the users' point of view it fails the environmental sustainability criteria since it causes immense pollution in the receiving waters of Lake Chivero. Option zero is therefore not desirable as it is wasteful of valuable nutrients such as P required in the popular urban agricultural activities.

Urine-diverting toilets, option one which involves collection and storage facilities before field application offer the most attractive option for short-cutting the P-cycle within the micro-catchment. Yellow water management is inherently inter-linked to closing the P-cycle locally and is the cornerstone of ecological sanitation. As little as 10% recycling of the potential P contained in yellow water can satisfy the biomass P-uptake requirements to produce a maximum yield of maize within the micro-catchment.

The amount of available agricultural land per capita is the major constraint in realising 100% recycling of P in yellow water. The calculations reveal that an adult person living within the micro-catchment requires a land area of about 190 m^2 for assimilation of P contained in his/her excreted annually through the yellow water fraction when growing maize. At present only 29 m^2/person is available for urban agriculture. The town planning guidelines in Zimbabwe provide up to about 2 000 m^2 for a plot in the so-called low density suburbs, whilst within the micro-catchment the average plot size is about 200 m^2 inclusive of land required for the superstructure and other land-uses. The alternative is to resort to crops with a higher P-uptake rate or export excess P contained in yellow water for use outside the micro-catchment. The feasibility of this requires further investigation.

This diverting toilet system with its ancillary infrastructure compares quite favourably over the traditional WC in terms of environmental sustainability. If installed indoors, it can also match the conventional WC for socio–cultural features. A crucial comparison between the two systems might deal with the affordability of investment and operation. The magnitude of savings of financial resources for the household depends on whether they prefer to handle the reuse of sanitised urine themselves or pay for the collection services or even sell the valuable liquid as fertiliser if there is vibrant market of local farmers. The market value of P in yellow water has been estimated at US\$ 13 610.00/a or US\$ 0.14/p.a.

The acceptance of the urine-diverting scenario is a little bit less than the conventional system, but better than the dry or composting scenario. People are a little bit reluctant for choosing a urine-diverting toilet, because it is unknown and it requires more discipline than the conventional toilet. More attention is needed for good separation. Some flushing with water is possible, like in the conventional system, and is perceived as hygienic and is very similar to the conventional system and therefore more acceptable.

Currently, effective use of urban solid waste (option two) could be hampered by its bulkiness, high water content of about 70% (see Chapter 4), and need for transportation of the manure product to the off-plot arable land. There is also a possibility of contamination of the organic waste with non-decomposable household items, and high handling, storage, transport, and application costs.

Social acceptance can play an important role in the decision to application of stabilised solid waste as compost. Community mobilisation to participate by separating biodegradable organic waste fraction is critical. From an environmental criterion requiring the closing of the P-cycle organic solid waste is less attractive as compared to yellow water diversion storage and sanitization before application in the fields. The P-supplying potential is less than 10% that of urine. Available space for stabilization of organic solid waste could also be a disadvantage.

However, given the inability of the formal municipal system to collect and safely dispose solid waste effectively, and the unavailability of organic manures (low organic carbon content of local soils), the relative abundance and benefit of solid waste as a soil amendment, and the rising cost of environmentally safe waste disposal, economies may make organic solid waste an appropriate manure of choice especially in view of urban agriculture being practised widely.

The full application of MCA, LCA and EIA techniques in this dissertation is possible given the data requirements to compare an existing and perceived solution. The score system used in most of these techniques is presents results which are hard to communicate, and difficult to relate to sustainability indicators. The solution o this would be to introduce pilot schemes (in new residential areas or through retrofitting residential areas having a conventional waterborne sewage system. Only in this way could meaningful comparisons be made. Costs and benefits, and social acceptance of the proposed options can only be measured with some degree of accuracy when pilot schemes are installed.

6 Conclusions

6.1 General

Conceptually, short-cutting the P-cycle in urban ecosystems is feasible, in accordance with the Bellagio principles of environmental sanitation, sustainable city and sustainable agriculture concepts at household and neighbourhood scale, with potential social, economic and environmental benefits.

The choice of P as an indicator element or to test the feasibility of short-cutting nutrient and water cycles in the urban environment has got several advantages and some limitations. The P-fluxes and stocks analysis correlates food, soap, fertiliser, and manure products and at the same time reflecting on biomass P-uptake, soil storage variation and potential impacts from P pollution. The two main limitations are the complexity of the biogeochemistry of P and the fact that natural ecosystems do not rely entirely on one element (there is a web of interrelationships between various elements and their cycles are synergistic).

In year 2000, the average global 'ecological footprint' for growing crops to meet human needs in area terms was estimated at 5 300 m^3/person. With rapid urbanisation and other humanity's area demand on biologically productive space this area is regarded to have exceeded the earth's capacity. The net import of P from the hinterland or other distant 'elsewheres' as food into the urban centres can be minimised if part of the food requirements can be grown as close as possible to their point of consumption or use.

Efficient use of P-based fertiliser, in particular, is fundamental to the sustained productivity and viability of agricultural systems worldwide. Modest calculations indicate that the current economic reserves would last at least 100 years at consumption rates of about 14 Mt P/a. Whilst there is no supply "crisis" for the phosphate industry's main raw material (phosphate rock), it is clear that reserves are finite. Recycling of P at a local scale is essential to conserve the finite resource and at the same time limit its dispersal into agricultural soil sinks which has led to excessive pollution of fresh water resources.

Many of the facets of 20th and 21st century human society have radical effects upon biogeochemical cycling. Activities such as clearing of forests, extensive cultivation and urban waste disposal and drainage systems have enhanced the transport of P from terrestrial to aquatic environments to an estimated 22 Mt P/a. As a consequence of these activities, concentrations of P in rivers and supply of P to lakes and estuaries have increased almost three-fold from an estimated 8 Mt P/a of pre-industrial and intensive agriculture era.

The contribution of P from sewage into the aquatic environment globally is small (estimated at less than 0.5 Mt P/a) as compared to P from agricultural lands. In year 2000 in excess of 2 Mt P/a was transported on a global basis in crop, animal and fertiliser products (imports and exports) enhancing the phenomenon of soil mining in some countries. These actions have decoupled patterns of supply, consumption and waste production from the natural P-cycle, and have made the cycle dependent on economics and trade.

At a global scale solutions leading to the short-cutting of the P-cycle might not have apparent or easily quantifiable impacts because of the many dynamics which affect the mobilisation and transportation of P from various *sources* to *sinks*. Most of the P-flows arise due to the need to feed the urban populations. With rapid urbanisation, particularly in the developing world, this would lead to increased use of P on land with the resulting increased P-flows through the city and potentially ending up being dumped into water bodies. Manageable solutions appear to be related to shortening the food supply line and hence the need to focus on food production within the city limits.

Views on urban sanitation are changing from getting rid of human excreta and organic solid residues (*waste*) to exploring beneficial use (*recovery and recycling the resources*). The concept of ecological sanitation has gained momentum and it is seen as one of the solutions for providing access to adequate sanitation at least to half of the 2.4 billion people who lack this service by 2015. The link between sanitation, food, agriculture, soil fertility has been established. At present sanitation experts, biologists, agriculturists, medical professionals, are involved in reshaping this concept. However, a number of barriers need to be surmounted and these are mostly related to negative perceptions about human excreta and organic solid waste, city planning and design, public health, social acceptance and technological constraints.

Urban agriculture needs to be elevated from its current status to a level where it is viewed as a legitimate land use activity for urban waste fluxes to be recycled at household or neighbourhood scale. There is better scope for efficiency in urban agriculture because it relies on existing urban fabric like transport and communication network, water supply, proximity to the market and pressures to use space effectively. The very shortest food supply line is therefore in need of massive promotion and this in turn will enable the short-cutting of nutrient and water cycles.

Systems analysis and Material Flow and Stock Accounting (MFSA) as applied in this dissertation, are essential tools of disaggregating a complex problem (processes, flows and stocks at various spatial and temporal scales), breaking the problem into more manageable pieces to allow data and judgements to be brought to bear on the pieces, and then of reassembling the pieces to present a coherent overall picture to decision makers. The purpose of this methodology has been to serve more as an aid to thinking and decision-making, rather than taking the decision.

Systems analysis and Material Flow and Stock Accounting (MFSA) are useful tools to for early recognition of resource depletion and environmental impacts of human activities at local, country, continental and global scales. They provide a methodology of linking the activities in the city, induced flows and stocks to the '*hinterland*' (source to sink). The approach reveals that a city, suburb neighbourhood or household is not an isolated system.

Material Flow and Stock Accounting (MFSA) requires numerous data (primary, secondary and tertiary) for establishing P-balances at various scales. Understanding the underlying processes of P transformation, mobilisation and transport within the different compartments of the biosphere is critical because of the level of uncertainty associated with all forms of data used in the analysis. As a consequence the P-balances of the some systems described in this dissertation do not balance because of the error margin (±20% or more) and lack data on some processes.

The assessment of the quality of data is an important problem. It is essential that the generation and composition of data are sufficiently transparent. In this dissertation data from administrative records and hybrid data from other studies done at various times has been used. In most cases it has been difficult or impossible to assess the quality of data.

By using the STELLA based P-calculator in the micro-catchment helps to identify sensitive input data which needs to be determined with more precision and hence partly overcome data quality concerns. In addition the systems analysis software, STELLA, allows for better data validation and consistency in handling data with different units of measurement.

Once the P-calculator is set, simulations of P-fluxes and stocks become much easier and linkages between the two subsystems investigated *viz.* household and agricultural subsystems become more lucid and helps to support better decision making in selecting suitable options for short-cutting the P-cycle.

6.2 Specific

At a global scale (macro P-flux analysis) P transfers and transformations do not reveal accurately and precisely the impact of man's activities on its natural cycle. But at smaller scales (micro P-flux analysis) e.g. river basins, the distortion of the natural P-cycle is evident and can be quantified with better certainty. It is such spatial scales where options for short-cutting the P-cycle can be considered and implemented with clear measurable outcomes.

The following conclusions are valid for Zimbabwe, Lake Chivero basin and the micro-catchment:

Initial estimates indicate that P-based fertiliser application rates for year 2000 in Zimbabwe of 18 940 Mg/a as P is exceeded by the total P losses from the soil (soil erosion, runoff and leaching losses) amounting to 57 160 Mg/a as P. Soil erosion and land degradation are the major environmental problems leading to loss of soil fertility. In year 2000 Zimbabwe was a net importer of P (190 Mg/a) in terms of trade in vegetal, animal and fertiliser products. However, year 2000 represents a below average scenario because of the reduced exports arising from a combination of drought and the land reform programme which disrupted most farming activities.

P imports in terms of P-based fertiliser are still low (10%) compared to the domestic consumption of about 18 940 Mg/a as P. In effect the Zimbabwe is still self-reliant in terms of P-based fertiliser.

The Lake Chivero case study provides a graphic example of the open-ended P-flows in urbanised catchments which effectively upset the natural P-cycle with the resultant water quality and biodiversity problems. The estimated P-fluxes and stocks for year 2000 indicate that the Lake received about 676 Mg/a as P compared with an outflow of 265 Mg/a as P. There is a steady accumulation of P in the lake with a large percentage of the P trapped in the sediments below the lake. The amount of P discharged into the Lake each year corresponds to about 4% of Zimbabwe's agricultural P usage in year 2000.

It is likely that in the short-term the recovery of the Lake from being eutrophic to a mesotrophic or oligotrophic condition would not be possible even if the input of P into the lake is reduced to zero i.e. there is already a high internal P-loading which exceeds the lake system's assimilative capacity. A likely short-term solution would be the removal of P from lake sediments e.g. by dredging. Such a physical intervention is very expensive and could also result in extensive environmental damage.

The micro-catchment consisting of Mufakose and Marimba suburbs consists of a thriving urban agricultural sector, which partly relies on expensive synthetic fertiliser as P to increase crop yields. The P in the waste fluxes (mainly sewage and solid waste) could be diverted on to land under cultivation and hence boost the crop yields whilst at the same time reducing the P-pollution load which ends up in Lake Chivero.

The micro-catchment provides a localised area where the possibilities of short-cutting the P-cycle can be demonstrated and quantified. The P-calculator for the micro-catchment captures the dynamics of P-fluxes and stocks on a monthly or annual time scale. The P-calculator helps to un-nest the black-boxes of the various processes which influence P mobilisation and transport from source to sink in urban ecosystems.

Calculations for year 2000 for the micro-catchment indicate a value of bio-available P soil storage (P_{ss}) of 4 050 kg as P whilst the corresponding yield 0.2 kg/m^2 for maize only requires a total P-uptake of 1 040 kg as P. The P_{ss} value can ideally provide adequate P to attain the water-limited dry mass economic yield (Y'_{hb}) of 0.79 kg/m^2 (when calculated for the maize crop in the micro-catchment P'_{bu} = 4 090 kg as P).

This confirms that the capacity of soils to be productive depends on more than just one plant nutrient. The physical, biological, and chemical characteristics of a soil all influence fertility and hence the crop yields.

Relationships between crop yields and the amount of bio-available P in the soil for crop uptake, as measured by standard soil tests indicate that there is a threshold or critical level of residual soil P. This critical level defines the point at which P fertilisation can switch between build-up and maintenance. Although often assumed to be the same, this critical level of residual soil P is not necessarily the level above which or below which yield responses to fresh P will not or will occur, respectively.

Two options are considered as far as short-cutting the P cycle is concerned *viz.* urine or *yellow water* separation, collection, storage and application on agricultural land (diversion of P_y) and; composting of organic residues from the household subsystem (faeces or *brown water* and the organic solid waste fraction i.e. P_b and P_{sw}) which ideally should be applied to agricultural land to increase the available P-status of the soil and as manure to improve the soil condition. The diversion of P_b onto arable land is however considered to be technically, economically and socio-culturally challenging in terms of adaptability of the existing situation, and is therefore not included in the analysis. The two promising selected options are compared to the reference or existing conventional system in relation to P-recycling.

Urine-diverting toilets, option one which involves collection and storage facilities before field application offer the most attractive option for short-cutting the P-cycle within the micro-catchment. Yellow water management is inherently inter-linked to closing the P-cycle locally and is the cornerstone of ecological sanitation. As little as 10% recycling of the potential P contained in yellow water can satisfy the biomass P-uptake requirements to produce a maximum yield of maize within the micro-catchment.

The amount of available agricultural land per capita is the major constraint in realising 100% recycling of P in yellow water. The calculations reveal that an adult person living within the micro-catchment requires a land area of about 190 m² for assimilation of P contained in his/her excreted annually through the yellow water fraction when growing maize. At present only 29 m²/person is available for urban agriculture. The town planning guidelines in Zimbabwe provide up to about 2 000 m² for a plot in the so-called low density suburbs, whilst within the micro-catchment the average plot size is about 200 m² inclusive of land required for the superstructure and other land-uses. The alternative is to resort to crops with a higher P-uptake rate or export excess P contained in yellow water for use outside the micro-catchment. The feasibility of this requires further investigation.

This diverting toilet system with its ancillary infrastructure compares quite favourably over the traditional water closet (WC) in terms of environmental sustainability. If installed indoors, it can also match the conventional WC for socio–cultural features. A crucial comparison between the two systems might deal with the affordability of investment and operation. The magnitude of savings of financial resources for the household depends on whether they prefer to handle the reuse of sanitised urine themselves or pay for the collection services or even sell the valuable liquid as fertiliser if there is vibrant market of local farmers. The market value of P in yellow water has been estimated at US$ 13 610.00/p.a or US$ 0.14/p.a.

The acceptance of the urine-diverting scenario is a little bit less than the conventional system, but better than the dry or composting scenario. People are a little bit reluctant for choosing a urine-diverting toilet, because it is unknown and it requires more discipline than the conventional toilet. More attention is needed for good separation. Some flushing with water is possible, like in the conventional system, and is perceived as hygienic and is very similar to the conventional system and therefore more acceptable.

Currently, effective use of urban solid waste (option two) could be hampered by its bulkiness, high water content of about 70%, and need for transportation of the manure product to the off-plot arable land. There is also a possibility of contamination of the organic waste with non-decomposable household items, and high handling, storage, transport, and application costs.

Social acceptance can play an important role in the decision to application of stabilised solid waste as compost.   Community mobilisation to participate by separating biodegradable organic waste fraction is critical.   From an environmental criterion requiring the closing of the P-cycle organic solid waste is less attractive as compared to yellow water diversion storage and sanitization before application in the fields.   The P-supplying potential is less than 10% that of urine.   Available space for stabilization of organic solid waste could also be a disadvantage.

Urban agriculture is not yet fully recognised as an important factor in development of Harare into a sustainable city.   Therefore, there is a need to for local authority officials and other decision makers, planners and engineers to integrate urban agriculture with other land uses.

6.3 Further research

It is clear that the incorporation of green zones, which include land for urban agriculture, is not yet fully understood by town planners, engineers and policy-makers.   City and town planning requires some rethinking so as to reduce the ecological footprint they exert on the hinterland and other distant areas.

The following research themes require further investigation:
- The land required to assimilate the waste emissions coming from cities needs to be determined (possibly per person) for each locality based on current or perceived developmental plans.   Once this land area has been determined a conscious decision has to be made whether it has to be incorporated within the city boundaries or elsewhere.
- Phosphorus dynamics in soils still remains a grey subject.   There is need for applied research and demonstration projects illustrating how and when P-based fertilisers (including organic and human waste) should be used.   The recommendations or guidelines should also indicate the advantages of using non-synthetic P-based fertilisers.
- Extraction of P from seawater could be the only option once the mined resource is depleted.   Research in the technology and processes for extraction need to be investigated further.

7 Epilogue

Being on a so-called sandwich construction to me was the best arrangement to undertake PhD research studies. But I did not realise that it would easily lead to insurmountable challenges and suffering from contradictory demands from supervisors, professors, family and colleagues. Many have been through this road before but a few have liked to share their agonies and triumphs as part of their dissertation. Behind every PhD dissertation there is always an untold story.

This is a summary of my story, although still, some things shall remain unsaid. From the day I started working on my PhD I was excited but frustrated by the requirement to undergo more formal training before being academically accepted whilst at the same time my employer (The University of Zimbabwe) and sponsor heavily encouraged me to build up other research and teaching experience for under and post-graduate courses (including publication and conference activities).

I thought I was young and energetic and could take most of these challenges in a stride and along four years had been prescribed as the maximum period. I had totally forgotten that my employer, family and friends still expected 100% of the time I afforded them before starting on the research. Financial support seemed not to be a big issue and I think remained so until the final year. Time management eluded me and towards the end caused quite some discomfort to me, my promoter and supervisor, family and friends. I had heard from other PhD fellows of their personality clashes with promoters and supervisors, lack of attention from an overworked promoter, the list is long. At UNESCO-IHE Institute of Water Education there were annual PhD seminars (still on-going), which addressed some of the problems and some obviously did not have ready answers. Luckily for me student-promoter-supervisor dynamics seemed to be the least of my concerns.

The year I embarked on my research coincided with the economic and political melt-down in Zimbabwe. Life and carrier as a researcher or university academic became uncertain. Professionals were on the move. The immediate family wanted to know how I was going to secure a living during period? Sacrifices had to be made and who said '*life is going to be easy*'? Suddenly I added a count on the statistic, '*one of the Zimbabweans in Diaspora*'.

Three years into my research, I was still struggling to find time to attend to the many activities I had un-intentionally committed myself to. I needed a character shift. I could not do everything and anywhere a few cared about all the other small things I did for them. You will be judged by what you deliver (dissertation on the table). My salvation came a few months before putting together this manuscript. I suddenly realised one has to revert to the Stone Age way of life (trust your instincts), forget about replying to those numerous e-mails, switch of the cell phone and the TV. But do not switch off the family. My eleven-year-old son once asked me '*Dad, what is this PhD you have been harping about for all these years, and when you get it, what difference would it bring to us*'? I thought about it and in my mind I told myself: probably more research and post-doctoral fellowships.

References

Abesinghe, D.H. Shanableh, A. & Rigden, B. (1996). Biofilters for water reuse in aquaculture. Wat. Sci. Tech. Vol. 34 No 11, pp 253-260.

Ackermand, W.C., Harmeson, R.H. and Sinclair, R.A. (1970). Some long-term trends in water quality of rivers and lakes. In EOS, 51, 6, pp 516-522.

Adeptu, J.A. and Corey, R.B. (1976). Organic phosphorus as a predictor of plant available phosphorus in soils of southern Nigeria. Soil Sci., Vol. 122, pp 159-164.

Adeptu, J.A. and Corey, R.B. (1977). Changes in N and P availability and P fractions in Iwo soil from Nigeria under intensive cultivation. Plant and Soil, Vol. 46, pp 309-316.

Admiraal, W., Jacobs, D.M.L.H.A., Breugem, P. and De Ruyter Van Stevenick, E.D. (1992). Effects of phytoplankton on the elemental composition (C, N, P) of suspended particulate material in the lower river Rhine. Hydrobiologia 235/236: 479-489.

Adriaanse, A. (1993). Environmental policy performance indicators. SDU Publishers, The Hague, 175 pp.

Alcamo, J. (ed.). (1994). IMAGE 2.0: Integrated modeling of global climate change. Kluwer Academic Publishers. Dordrecht. 318 pp.

Alexandratos, N. (ed.). (1988). World agriculture: Towards 2000. An FAO Study. Belhaven Press, London and New York University Press, New York.

Alexandratos, N. (ed.). (1995). World Agriculture: Towards 2010. An FAO Study. Food and Agriculture Organisation of the United Nations, Rome and Wiley and Sons, Chichester, 488 pp.

Allan, J.A. (1994). Mechanisms for reducing tension on water: substituting for water (in the Middle East). MEED, Vol 38, No 4, January 1994, pp 12-14.

Allard, M. (ed.). (1992). GEMS water operational guide. 3rd edition. GEMS/W. 92.1, NWRI, Burlington, Canada.

Allen, J. (2002). People challenges in biospheric systems for long-term habitation in remote areas: Space stations, Moon, and Mars expeditions. Life Support and Biosphere Science, Vol. 8, pp 67-70.

Alling, A., Nelson, M., and Silverstone, S. (1993b). Life under glass: The inside story of Biosphere 2. Synergetic Press, Santa Fe, New Mexico, pp 288. ISBN 1 88 248 07 2.

Alling, A., Nelson, M., Leigh, L., Frye, R., Alvarez-Romo, N., MacCallum, T. and Allen, J. (1993a). Experiments on the closed ecological system in the Biosphere 2 Test Module. Ecological Microcosms. Springer-Verlag, New York, pp 463-479.

Almeida, M.C., Butler, D. and Friedler, E. (1999). At-source domestic wastewater quality. Urban Water, Vol. 1 No. 1 pp 49-55.

Amer, F., Bouldin, D.R., Black, C.A. and Duke, F.R. (1955). Characterisation of soil phosphate by anion exchange resin adsorption and 32P equilibration. Plant and Soil, Vol. 60, pp391-408.

Amos, A. Hranova, R., and Love, D. (2003). Impacts of sewage sludge land disposal practice on groundwater in Harare, Zimbabwe. 1st IWA Regional Conference, "Water: Key to sustainable development in Africa", 14-17 September 2003, Cape Town, South Africa.

Amy, G., Pitt, R., Singh, R., Bradford, W.L., Lagraff, M.B. (1974). Water quality management planning for urban runoff. United States Environmental Protection Agency, Washington, DC, EPA 440/9-75-004. NTIS PB 241 689/AS.

Anderberg, S., Bauer, G., Ermoliev, Y. and Stigliani, W.M. (1993). Mathematical tools for studies of industrial metabolism. IIASA working paper WP-93-9, Laxenburg Austria, 28 pp.

Anderson, G. (1980). Assessing organic phosphorus in soils. In: Khasaweh, F.E., Sample, E.C. and Kamprath, E.J. (eds.). The role of phosphorus in agriculture. ASA, Madison, Wisconsin. pp. 411-431.

Anon. (1984). Fluid fertilisers. Bulletin Y-185. TVA-NFDC, Muscle Shoals, Alabama. 130 pp.

Anon. (1985). Fertiliser research programme for Africa. Rep. Int. Fund. Agric. Dev. IFDC Muscle Shoals, Alabama.

APHO. (1992). Standard methods for the examiniation of water and wastewater. 18th edition pp 4-108, 4-115 and 4-116. American Public Health Organisation.

Argenti, O. (2000). Food for the Cities. Urban food supply and distribution policies to reduce food insecurity. A briefing guide for mayors, city executives and urban planners of Cities in developing countries and countries in transition, Food into Cities Collection, DT/43-00E. FAO, Rome.

Argenti, O. (2001). Food for the cities. Food supply and distribution policies to reduce urban food insecurity. A briefing guide for mayors, city planners and urban planners in developing countries and countries in transition. URL: www.fao.org/ag/ags/agsm/sada/pages/dt/dt4300E.htm . Accessed July 2002.

Armar-Klemesu, M. (2000). Urban agriculture and food security, nutrition, and health pp 99-117. In: Bakker, N., Dubbeling, M., Guendel, S., Sabel-Koschella, U. and De Zeeuw, H. (eds.). Growing Cities, Growing Food. DSE: Feldafing.

Arnbjerg-Nielsen, K., Hansen, L., Kjolholt, J., Stuer-Lauridsen, F., Hasling, A.B., Stenstrom, T.A., Schonning, C., Westrell, T., Carlsen, A. and Halling-Sorensen, B. (2003). Risk assessment of local handling of human faeces with focus on pathogens and pharmaceuticals. Paper presented during the 2nd International Symposium on Ecological Sanitation, 7-11 April, 2003, Lubeck, Baltic Sea, Germany.

Asano, T. and Levine A.D. (1996). Wastewater reclamation, recycling, and reuse, past, present and future, Wat.Sci.Tech, Vol. 33, No. 10-11, pp 1-14.

Austin, A. (2003). Ecosan: An unsuccessful sanitation scheme at a rural school: lessons learnt from the project failure. Paper presented during the 2nd International Symposium on Ecological Sanitation, 7-11 April, 2003, Lubeck, Baltic Sea, Germany.

Ausubel, J.H. (1997). Industrial ecology: Reflections on a colloquium. Proc. National Academy of Sciences USA, 89, pp 879-884.

AWWA. (1975). Standard methods for the examination of water and wastewater. 14th Edition, pp 624, American Waterworks Association (AWWA).

Ayres, R.U. (1978). Resources, environment, and economics: Applications of the materials and energy balance principle. John Wiley & Sons, New York.

Ayres, R.U. (1989). Industrial metabolism. In: Technology and environment, Ausubel, J. and Sladovich, H. (eds). National Academy Press, Washington D.C.

Ayres, R.U. (1992). Industrial metabolism and the grand nutrient cycles. Working Paper 92/64/EPS, INSEAD, Fontainebleau.

Ayres, R.U. and Ayres, L.W. (1996). Industrial ecology: Towards closing the material cycle, pp 379. Edward Elgar Publishing, Inc. Cheltenham.

Ayres, R.U. and Simonis, U.E. (eds.). (1992). Industrial metabolism: Restructuring for sustainable Development. United Nations University Press, New York.

Ayres, R.U., and Ayres, L.W. (1998). Accounting for resources, 1; Economy-wide applications of mass-balance principles to materials and waste. Edward Elgar, Cheltenham.

Azar, C., Holmberg, J., and Lidgren, K. (1996). Socio-ecological indicators for sustainability. Ecological economics 18, pp 89-112.

Baccini, P. and Brunner, P.H. (1991). Metabolism of the anthroposhpere. Springer-Verlag, Berlin Heildelburg.

Baffaut, C., and Delleur, J.W. (1990). Calibration of SWMM runoff quality model with expert system. Journal of Water Resources Planning and Management, Vol. 116 No. 2, pp 247-261.

Bailey, H., Kajese, T. and Koro, E. (1996). Report on pollution issues of Lake Chivero and catchment. Unpublished paper, Environment 2000, Harare.

Bakker, N., Dubbeling M., Gundel S., Sabel-Koschella U. and De Zeeuw H. (eds.). (2000). Growing cities, growing food: Urban agriculture on the policy agenda. Feldafing: DSE.

Balbo, M., Visser, C., and Argenti, O. (2001). Food supply and distribution to cities in developing countries. A guide for urban planners and managers. Food into Cities Collection, DT/44-01E. URL: http://www.fao.org/ag/ags/agsm/sada/docs/doc/dt4401e.doc. Accessed July 2002.

Baldwin, P.L., Henty, C.J. (1978). Some suggestions for improving the implementation of fertiliser projects in developing countries No 172.

Balkema, A.J., Preisig, H.A., Otterpohl, R., Lambert, A.J.D. and Weijers, S.R. (2001). Developing a model based decision support tool for the identification of sustainable treatment options for domestic wastewater. War Sci Tech, Vol. 43, No 7, pp 265-269.

Baran, E. (1988). Use of sewage sludge from tertiary treatment as phosphate fertiliser. In: Welte, E. and Szabolcs, I. (eds.). Agricultural waste management and environmental protection. Proceedings of the 4th International Symposium of CIEC, Braunschweig, May 11-14, 1987. FAL, Braunschweig-Volkenrode, pp 403-410.

Barber, B. (1989). Phosphate in Zimbabwe. Mineral resources series No. 24, pp 31. Zimbabwe Geological Survey, Harare.

Barber, S.A. (1980). Soil-plant interactions in the phosphorus nutrition of plants. In: Khasawneh, E., Sample, E.C. and Kamprath, E.J. (eds.). The role of phosphorus in agriculture. Amer. Soc. Agron., Madison, Wisconsin, pp 591-616.

Barber, S.A. (1984). Soil nutrient bioavailability: a mechanistic approach. John Wiley and Sons, New York. 398 pp.

Barber, S.A.; Olson, R.A. (1968). Fertiliser use on corn. In: Nelson, L.B. (ed): Changing patterns in fertiliser use. Soil Sci. Soc. Amer., Madison.

Barbier, E.B. (1990). Economics, natural resource scarcity and development: Conventional and alternative views. Earthscan, London.

Barrow, N.J. (1961). Phosphorus in soil organic matter. Soils and Fertiliser Abstracts, 24, 169-173.

Barrow, N.J. (1983). A mechanistic model for describing the sorption and desorption of phosphate by soil. J. Soil Sci., Vol. 34, pp 733-750.

Bartlett, R.J. (1986). Soil redox behaviour, pp 179-206. In: Sparks, D.L. (ed). Soil physical chemistry. CRC Press, Boca Raton, Florida.

Baumgartner, A. and Reichel, E. (1975). The world water balance. Elsevier Publications, Amsterdam.

Baveye, P. and Valocchi, A. (1989). An evaluation of mathematical models of the transport of biologically reacting solutes in saturated soils and aquifers, Water Resources Research, 25(6), pp 1413-1421.

Beasley, D.B., Monke, E.J., Miller, E.R. and Huggins, L.F. (1985). Using simulation to assess the impacts of conservation tillage on movement of sediment and phosphorus into Lake Erie. J. Soil Water Conservation, 40, pp 233-237.

Beaton, J.D. and Murphy, L.S. (1988). Recent developments and trends in fertilisation techniques with particular reference to fluid fertilisers. Presented at IFA/ANDA Meeting on Fertilisers and Agriculture. Rio De Janeiro, Brazil, March 15-17, 1988.

Beaton, J.D., Roberts, T.L., Halstead, E.D. and Cowell, L.E. (1995). Global transfers of P in fertiliser materials and agricultural commodities pp 7-26. In: Tiessen, H. (ed.). Phosphorus in the global environment: Transfers, cycles and management. Scientific Committee on Problems of the Environment, (SCOPE), Vol. 54, pp 452, John Wiley & Sons, Chichester.

Beck, M.B. (1987). Water quality modelling: A review of the analysis of uncertainty. Water Resources Research 23(8): pp 1393-1442.

Beck, M.B. (1997). Applying systems analysis in managing the water environment: Towards a new Agenda. Watermatex 97, Wat. Sci. Tech. pp 1-17.

Beck, M.B. and Cummings, R.G. (1996). Wastewater infrastructure: Challenges for the sustainable city in the new Millennium. Habitat International, Vol. 20, No. 3, pp 405-420.

Beck, M.B., Chen J., Saul A.J., and Butler D. (1994). Urban drainage in the 21st Century: Assessment of new technology on the basis of global material flows. Wat. Sci. Tech., Vol. 30 No. 2, pp 1-12.

Belevi, H. (2000). Material flow analysis as a strategic planning tool for regional wastewater and solid waste management. Department of Water and Sanitation in Developing Countries (SANDEC), Swiss Federal Institute for Environmental Science and Technology (EAWAG), Duebendorf.

Belyaev, A.V. (1991). Freshwater. World Resources 1990-91. Oxford University Press, pp 161-178.

Bennet, J.G. (1985). A field guide to soil and site description in Zimbabwe. Zimbabwe Agricultural Journal, Technical Handbook No. 6, Government Printers, Harare.

Bennett, D. and Bowden, J.W. (1976). Decide: An aid to efficient use of phosphorus. In: Blair, G.J. (ed.) Prospects for improving efficiency of phosphorus utilisation. Proc. of Symp. at University of New England, Armidale, NSW, Australia. Reviews in Rural Science III, pp. 77-81.

Bijleveld, M. (2003). The possibilities for ecological sanitation in the Netherlands. Master Thesis, Civil Engineering, University of Twente, Twente, pp 111.

Birch, H.F. (1960). Soil drying and soil fertility. Trop. Agric. 37, pp 3 10, Trinidad.

Birley, M.H. and Lock, K. (1999). Health and peri-urban natural resource production. Environment and Urbanisation, Vol. 10, No 1, pp 89-106.

Biswas, A.K. (1978). Water development and the environment. In: Lohani, B.N. and Thanh, N.C. (eds.). Water pollution control in developing countries, Vol. II. Proceedings of the International Conference held in Bangkok, Thailand.

Blair, G.J., Till, A.R., Smith, R.C.G. (1977). The phosphorus cycle: what are the sensitive areas? In: Blair, G.J. (ed.) Prospects for improving the efficiency of phosphorus utilisation. Reviews in Rural Science, 3, pp 9-19.

Blumenthal, U. J., Strauss, M., Mara, D. D. and Cairncross, S. (1989). Generalised model for the effect of different control measures in reducing health risks from waste reuse. Wat. Sci. Tech., Vol. 21, pp. 567-577.

Blumenthal, U.J, Peasey, A., Ruiz-Palacios, G. and Mara, D.D. (2001). Guidelines for wastewater reuse in agriculture and aquaculture: Recommended revisions based on new research evidence. WELL Study, Task No: 68, Part 1. URL: http://www.lboro.ac.uk/well/

Blumenthal, U.J., Mara, D.D., Peasey, A., Ruiz-Palacios, G. and Stott, R. (2000). Guidelines for the microbiological quality of treated wastewater used in agriculture. Recommendations for revising WHO guidelines. Bulletin of the World Health Organisation 78: pp 1104-1116.

Bohn, H. (1976). Estimates of organic carbon in world soils. Soil Sci. Am. J., Vol. 40, pp 468-470.

Bolin, B. (1979). On the role of the atmosphere in biogeochemical cycles. Quarterly Journal Met. Soc., Vol. 105, pp 25-42.

Bolin, B. and Cook, R.B. (eds.). (1983). Interactions of the major biogeochemical cycles: Global change and human impacts. Scientific Committee on Problems of the Environment (SCOPE) No 21, 554 pp, Wiley, Chichester.

Bolin, B., Rosswall, T., Richey, J.E., Freney, J.R., Ivanov, M.V. and Rodhe, H. (1992). C, N, P, and S cycles: Major reservoirs and fluxes. In: Butcher, S.S., Charlson, R.J., Orians, G.H. and Wolfe, G.V. (eds.). Global biogeochemical cycles. Academic Press, London.

Bostrom, B., Persson, G. and Broberg, B. (1988). Bioavailability of different phosphorus forms in freshwater systems. Hydrobiologia, 170, pp 133-155.

Bourque, M. and Canizares, K. (2000). Urban agriculture in Havana, Cuba. Urban agriculture magazine, Vol. 1 No. 1, June 2000, pp 27-29.

Bowyer-Bower T.A.S. and Tengbeh G.T. (1997). Environmental implications of (Illegal) urban agriculture in Harare: A preliminary report of field research (1994/95). In Zinyama, L.M. (ed.). Africa dilemmas of current practice and policy. Geographical Journal of Zimbabwe No. 28, Harare..

Bowyer-Bower T.A.S., Mapaure I., and Drummond R.B. (1996). Ecological degradation in cities: Impact of urban agriculture in Harare, Zimbabwe. Journal of Applied Science in Southern Africa (JASSA), Vol. 2, No 2, pp 53-67.

Bowyer-Bower, T.A.S. (1997). The potential for urban agriculture to contribute to urban development. In Zinyama, L.M. (ed.). Africa dilemmas of current practice and policy. Geographical Journal of Zimbabwe No. 28, Harare.

Bratton, M. and Truscott, K. (1985). Fertiliser packages, maize yields and economic returns: An evaluation in Wedza communal lands. Zimbabwe Agric. J., Vol. 82 (1), pp 1-8.

Bray, R.H. and Kurtz, L.T. (1945). Determination of total, organic and available forms of phosphorus in soils. Soil Sci., Vol. 64, pp 101-109.

Breman, H. (1990). Integrating crops and livestock in southern Mali: Rural development or environmental degradation. In: Rabbinge, R., Goudriaan, J., Van Keulen, H., Penning De Vries, F.W.T. and Van Laar, H.H. (eds.) Theoretical production ecology: Reflections and prospects. Simulation Monographs 34, PUDOC, Wageningen, pp 277-294.

Breslin, E.D. and Dos Santos, F. (2001). Introducing ecological sanitation in rural and peri-urban areas of northern Mozambique. Paper presented during the 1st International Conference on Ecological Sanitation, 5-8 November, 2001, Nanning, China.

Briscoe, J. and De Ferranti D. (1988). Water for rural conmunities; helping people help themselves. The World Bank, Washington DC.

Broberg, O. and Pettersson, K. (1988). Analytical determination of orthophosphate in water. Hydrobiologia, 170, pp 45-59.

Broecker, W. S. (1974). Chemical Oceanography. Harcourt Brace Jovanovich, New York.

Brown, L.R., Flavin C., and Postel S. (1990). Picturing a sustainable society. In: The state of the World 1990. A World-watch Institute report on progress toward a sustainable society.

Brunner, P.H. and Baccini, P. (1992). Regional materials management and environmental protection. Waste Management and Research 10, pp 203-212.

Brunner, P.H., Baccini, P., Deistler, M., Lahner, T., Lohm, U., Obernosterer, R. and Van der Voet, E. (1998). Materials accounting as a tool for decision making in environmental policy (MacTEmPo). Summary Report. 4th European Commission programme for environment and climate, Research Area III, Economic and social aspects of the environment, ENV4-CT-96-0230. Institute for water quality and waste management, Vienna University of Technology.

Bumb, B., and Baanante, C. (1996). The role of fertiliser in sustaining food security and protecting the environment to 2020. 2020 Vision Discussion Paper 17 International Food Policy Research Institute (IFPRI). Washington, DC.

Burwell, R.E., Schuman, G.E., Heinemann, H.G. and Spomer, R.G. (1977). Nitrogen and phosphorus movement from agricultural watersheds. J. Soil and Water Conserv., 32, 226-230.

Butler, D., and Parknson J. (1997). Towards sustainable urban drainage. Wat. Sci. Tech., Vol. 35, No 9, pp 53-63.

Butler, D., Friendler, E. and Gatt, K. (1995). Characterising the quantity and quality of domestic wastewater inflows. Wat. Sci. Tech., Vol. 31, No 7, pp 13-24.

Cairncross, S. (1992). Sanitation and water supply: Practical lessons from the Decade. Water and Sanitation Discussion paper No 9. The World Bank, Washington DC.

Cairncross, S., Blumenthal, U.J., Kolsky, P., Moraes, L. and Tayeh, A. (1995). The public and domestic domains in the transmission of disease. Tropical Medicine and International Health, Vol. 39, pp 173-176.

Cairncross, S., Carruthers, I., Curtis, D. Feachem, R., Bradley, D. and Baldwin, G. (1980). Evaluation for village water supply planning. John Wiley and Sons. Chichester.

Caraco, N.F. (1995). Influence of human populations on P-transfers to aquatic systems: A regional scale study using large rivers pp 235-244. In: Tiessen, H. (ed.). Phosphorus in the global environment: Transfers, cycles and management, Scientific Committee on Problems of the Environment, (SCOPE), Vol. 54, John Wiley & Sons, Chichester.

Carignan, R. and Kalff, J. (1980). Phosphorus sources for aquatic weeds: Water or sediments? Science, 207, pp 987-989.

Carpenter, S., Caraco, N.F., Correll, D.L, Howarth, R.W, Sharpley, A.N. and Smith V.H. (1998). Non-point pollution of surface waters with phosphorus and nitrogen. Issues in Ecology No. 3. Ecological Society of America, Washington. URL: http://esa.sdsc.edu/

Cate, R.B. Jr. and Nelson, L.A. (1971). A simple statistical procedure for partitioning soil test correlation data into two classes. Soil Sci. Soc. Amer. Proc., 35, pp 658-659.

CEEP. (1997). Scope Newsletter No. 21. Centre Européen d'Etudes des Polyphosphates (CEEP), Brussels.

CEEP. (1998). Phosphates a sustainable future in recycling. Publicity leaflet by Centre Europeen d'Etudes des Polyphosphates (CEEP), sector group of the European Chemical Industry Council (CEFIC), Brussels. URL: http://www.nhm.ac.uk/mineralogy/phos/index.htm

Chambers, N., Simmons, C. and Wackernagel, M. (2000). Sharing nature's interest: Ecological footprints as an indicator for sustainability, Earthscan, London.

Chapin, F.S. III, Follett, J.M. and O'Connor, K.F. (1982). Growth, phosphate absorption, and phosphorus chemical fractions in two Chionocloa species. J. Ecol., 70, pp 305-321.

Chapman, D. (ed.) (1992). Water quality assessments. A guide to the use of biota, sediments and water in environmental monitoring. Chapman and Hall, 585 pp.

Characklis G.W. and Wiesner M.R. (1997). Particles, metals, and water quality in runoff from large urban watersheds. J. of Env. Eng. Vol. 123, No 8, pp 753-759.

Chauhan, B. S., Stewart, J. W. B., and Paul, E. A. (1979). Effect of carbon additions on soil labile inorganic, organic and microbially held phosphate. Can. J. Soil Sci., 59, pp 387-396.

Chauhan, B. S., Stewart, J. W. B., and Paul, E. A. (1981). Effect of labile inorganic phosphate status and organic carbon additions on the microbial uptake of phosphorus in soils. Can. J. Soil Sci., 61, pp 373-385.

Chen, J. and Beck, M.B. (1997). Towards designing sustainable urban wastewater infrastructure: A screening analysis. Wat. Sci. Tech., Vol. 35 No 9, pp 99-112.

Chen, Y.D., McCutcheon, S.C., Rasmussen, T.C., Nutter, W.L. and Carsel, R.F. (1993). Integrating water quality modelling with ecological risk assessment for non-point source pollution control: A conceptual framework. Wat. Sci. Tech., Vol. 28, No 3-5, pp 431-440.

Chidavaenzi M. (2001). A rainfall-runoff model for a composite micro-catchment and the potential impact of alternative land uses on runoff from an urban residential area. Unpublished MSc thesis, Water Resources Engineering and Management programme, Department of Civil Engineering University of Zimbabwe.

Chitsiku, I.C. (1991). Nutritive value of foods of Zimbabwe. University of Zimabwe Publications, Harare.

City of Salisbury. (1980). Design of sewerage reticulation systems in African townships: Practice Guideline S1.

Clark, G.A. (2001). Eco-sanitation in Mexico: Strategies for sustainable replication. Paper presented during the 1st International Conference on Ecological Sanitation, 5-8 November, 2001, Nanning, China.

Clarke, R.T. (1998) Stochastic processes for water scientists: developments and applications, John Wiley and Sons, Chichester.

Clemen, R.T. (1996). Making hard decisions: An introduction to decision analysis. Second edition, Duxbury Press Belmont, CA.

CLM. (1988). Nutreint accounting as a farm management tool. Information brochure, Centre for Agriculture and Environment (CLM), Utrecht, The Netherlands.

Club of Rome. (1972). The limits to growth. Universe Books, New York.

Cobb, J.B. Jr. (1995). Is it too late? A theology of ecology. Revised edition, Environmental Ethics Books, Denton, Texas.

Codd, G.A., Edwards, C., Beattle, K.A., Barr, W.M. and Gunn, G.J. (1992). Fatal attraction to cyano-bacteria? Nature, 359, pp 110-111.

COH. (1997). Mabvuku-Tafara sewage treatment works and outfall sewers, Feasibility study report. Volume 3 appendices, Final report by Inter-consult, Zimbabwe in association with GFJ, May 1997. City of Harare.

Cointreau, S.J. (1982). Environmental management of urban solid wastes in developing countries: A project guide. Urban development paper No. 5, World Bank, Washington DC.

Cole, C. V., and Heil, R. D. (1981). Phosphorus effects on terrestrial nitrogen cycling. In: Clark, F. E., and Rosswall, T. (eds.). Terrestrial nitrogen cycles, processes, ecosystem and management impact. Ecological Bulletin, Stockholm, 33, pp 363-374.

Cole, C. V., Elliott, E. T., Hunt, H. W., and Coleman, D.C. (1978). Trophic interactions in soils as they affect energy and nutrient dynamics: Phosphorus transformations. Microbiology and Ecology, 4, pp 381-387.

Cole, C. V., Innis, G. S., and Stewart, J. W. B. (1977). Simulation of phosphorus cycling in semi-arid grassland. Ecology, 58, pp 1-15.

Collenbach, E. (1975). Ecotopia. Bantam, New York.

Colwell, J.D. (1963). The estimation of phosphorus fertiliser requirements of wheat in southern New South Wales by soil analysis. Aust. J. Exp. Agric. Anim. Husb., 3, pp 190-197.

Conway, G. R., and Pretty, J. N. (1991). Unwelcome harvest: Agriculture and pollution. Earthscan Publications Ltd, London.

Cook, D.J., Dickinson, W.T. and Rudra, R.P. (1985). The Guelph Model for Evaluating the Effects of Agricultural Management Systems in Erosion and Sedimentation - GAMES. User's Manual. Tech. Rep. 126-71. School of Engineering, University of Guelph, Guelph, Ontario.

Cook, P.J., Banerjee, D.M. and Southgate, P.N. (1990). The phosphorus resources of Asia and Oceania. In: Phosphorus requirements for sustainable agriculture in Asia and Oceania. International Rice Research Institute, Manila, pp 97-114.

Cooke, G.W. (1986). The intercontinental transfer of plant nutrients. In: Nutrient balances and the need for potassium. Proceedings of the 13th International Potash Institute Congress, Reims, France, International Potash Institute, Basle.

Cooper, G.RC. and Fenner, R.J. (1981). General fertiliser recommendations. Zimbabwe Agric. J., Vol 78 (3), pp 123-128.

Cooper, R.L. (1991). Public health concerns in wastewater reuse. Wat. Sci. Tech. Vol. 24, pp 55-65.

Corbitt R.A. (1989). Standard handbook of environmental engineering. McGraw Hill Inc. New York, NY.

Cordova, A. and Knuth, B.A. (2003). Implementing large-scale urban dry sanitation: An agenda for research and action. Paper presented during the 2nd International Symposium on Ecological Sanitation, 7-11 April, 2003, Lubeck, Baltic Sea, Germany.

Cornforth, I.S. and Sinclair, A.G. (1982). Model for calculating maintenance phosphate requirements for grazed pastures. New Zealand J. Exp. Agric., 10, pp 53-61.

Cosgrove, W.J. and Rijsberman, F.R. (2000). World water vision. Earthscan Publications, London.

Costanza, R., D'Arge, R., De Groot, R., Farber, S., Grasso, M., Hannon, B., Limburg, K., Naeem, S., O'Neill, R,V, Paruelo, J., Raskin, R.G., Sutton, P. and Van der Belt, M. (1997). The value of the world's ecosystem services and natural capital. Nature 387: pp 253-260.

Coster, R. (1991). Alleviating fertiliser supply constraints in West Africa. In: Mokwunye, A. U. (ed.). Alleviating soil fertility constraints to increased crop production in West Africa. Kluwer Academic Publishers, Dordrecht.

Cotton A. and Franceys R., (1991), Services for shelter, infrastructure for urban low-income housing, Liverpool planning manual 3, WEDC, Liverpool University Press in association with Fairstead Press.

Cowan, W.F. and Lee, G.F. (1975). Leaves as a source of phosphorus. Environ. Sci. Technol., 7, No 9, pp 853-854.

Coyle, G. (1996). System dynamics modelling: A practical approach. Chapman and Hall, New York.

Crites, R. and Tchobanoglous, G. (1998). Small and decentralised wastewater management systems. McGraw-Hill, ISBN 0-07-289087-8.

Cropper, M.L., (1988). A note on the extinction of renewable resources. J. Environmental Econ. Management, 15, pp 64-70.

Cruz, M.C. and Medina, R.S. (2003). Agriculture in the City. A key to sustainability in Havana, Cuba. Ian Randle Publishers and IDRC 244 pp, Kington and Ottawa.

CSIR, (1995), Guidelines for the provision of engineering services and amenities in residential township development; *"The Red Book"*, The Department of Housing in Collaboration with The National Housing Board, published by the Centre for Scientific and Industrial Research (CSIR), Division of Building Technology, Pretoria, South Africa.

CSO. (1992). Census 1992: Provincial profile, Harare. Central Statistical Office (CSO). Government Printers, Harare.

CSO. (1998). Poverty in Zimbabwe, pp. 129. Central Statistics Office (CSO). Harare.

CSO. (2000). Quarterly digest of statistics. September 2000. Central Statistical Office (CSO). Harare.

Dahnke, W.C. and Olsen, R.A. (1990). Soil test correlation, calibration and recommendation. In: Westerman, R.L. (ed.). Soil testing and plant analysis. 3rd edition, Book series No 3. Soil Sci. Soc. America Inc., Madison, USA, pp. 45-72.

Dalal, R.C. (1977). Soil organic phosphorus. Advances in Agronomy, 29, pp 83-113.

Dalemo, M. (1999). Environmental systems analysis of organic waste management: The ORWARE model and the sewage plant and anaerobic digestion sub-models. PhD thesis, Swedish University of Agricultural Sciences, Uppsala, ISBN 91-576-5453-0.

Daly, H. (1977). Steady-state economics. W.H. Freeman, San Francisco.

Daly, H. E. (1992) Steady-state economics: Concepts, questions, policies, pp 333-338. GAIA I.

Daly, H.E. (1996). Beyond growth: The economics of sustainable development, pp 254. Beacon Press, Boston.

Daly, H.E. and Cobb, J.B. Jr. (1989). For the common good: Redirecting the economy towards community, the environment, and a sustainable future. Beacon Press, Boston.

De Datta, S.K., Biswas, T.K. and Charenchamratcheep, C. (1990). Phosphorus requirements and management for lowland rice In: Phosphorus requirements for sustainable agriculture in Asia and Oceania. International Rice Research Institute, Manilla, pp. 307-323.

De Groen, M.M. (2002). Modelling interception and transpiration at monthy time steps: Introducing daily variability through Markov chains. PhD thesis, IHE-Delft, Swets and Zeitlinger B.V., Lisse, The Netherlands.

De Groot, R.S. (1992) Functions of nature: Evaluation of nature in environmental planning, management and decision making Wolters-Noordhoff, Groningen, ISBN 9001 35594 3.

De Jong, A.L., De Oude, N.T., Smits, A.H. and Volz, J. (1989). The phosphate load of the river Rhine 1975-1986. Aqua 38, pp 176-188.

De Kruijff, G.J. (1981). Infrastructure design standards in Zimbabwe, Housing Development Services Branch, Ministry of Local government and Housing, Salisbury.

De Oude, N.T., (1989). Anthropogenic sources of phosphorus detergents. In: Tiessen, H. (ed.) Phosphorus cycles in terrestrial and aquatic ecosystems. Regional Workshop 1: Europe. SCOPE/UNEP Proceedings, University of Saskatchewan, Saskatoon, Canada, pp. 214-220.

De Vooys, C.G.N. (1979). Primary production in aquatic environments pp 259-292. In: Bolin, B., Degens, E. T., Kempe, S., and Ketner, P. (eds.). The global carbon cycle. SCOPE Report No 13. Wiley, Chichester.

De Wit, M. (1998). Nutrient emissions and transport from source to river load: A spatial and dynamic analysis applied to the Rhine and the Elbe. Ph.D. thesis, Utrecht University, the Netherlands.

De Zeuw, H., Gundel, S. and Waibel, H. (2000). The integration of agriculture in urban policies. Urban agriculture magazine, Vol. 1 No 1, June 2000, pp 13-15.

Dear, B.S., Helyar, K.R., Muller, W.J. and Loveland, B. (1992). The P fertiliser requirements of subterranean clover, and the soil P status, sorption and buffering capacities from two P analyses. Aust. J. Soil Res., 30, pp 27-44.

Debo, T. N. and Reese, A. J. (1995). Municipal storm water management. CRC Press, Lewis Publishers, Boca Raton, Florida.

Deevey, E. S. (1970). Mineral cycles. Scient. Amer., Vol 223, pp 149-158.

Degens, E.T., Kempe, S. and Richey, J.E. (1991). Biogeochemistry of major world rivers. John Wiley & Sons, New York.

Del Porto, D., and Steinfeld, C. (1999). The composting toilet system book. The Centre for Ecological Pollution Prevention (CEPP), 235 pp. Concord, MA.

Demayo, A. and Steel A. (1996). Data handling and presentation. In: Chapman, D. (ed.). Water quality assessments: A guide to the use of biota, sediments and water in environmental monitoring, 2nd Edition, WHO, UNESCO, UNEP, Chapman and Hall, London.

Devai, I., Felfody, L., Wittner, I. and Plosz, S. (1988). Detection of phosphine: New aspects of the phosphorus cycle in the hydrosphere. Nature 333: pp 343-345.

Dibb, D.W., Fixen, P.E. and Murphy, L.S. (1990). Balanced fertilisation with particular reference to phosphates: interaction of phosphorus with other inputs and management practices. Fertiliser Research, 26, pp 29-52.

DMS. (1981). Climate handbook of Zimbabwe. Department of Meteorological Services (DMS), Salisbury (Harare).

Donigian Jr., A.S., Beyerlein, D.C., Davis Jr., H.H. and Crawford, N.H. (1977). Agricultural runoff management (ARM) model verison II: Refinement and testing. United States Environmental Protection Agency. EPA 600/3-77-098.

Doudoroff, M. and Aelberg, E.A. (1970). The microbial world. 3rd edition. Prentice Hall, Inc.

Dowall, D. and Giles, C. (1996). A framework for reforming urban land policies in developing countries. Urban management programme discussion Paper No 7, 53 pp. The World Bank, Washington DC.

Drakakis-Smith, D.W. (1992). Strategies for meeting basic food needs in Harare. In: Baker, J. and Pederson, P.O. (eds.). The rural-urban interface in Africa. Nordiska Afrikainstitutet, Uppsala, pp 258-283.

Drangert, J.O. (1996). Perception of human excreta and possibilities to reuse urine in peri-urban areas. Stockholm Water Symposium, August 6-9, Stockholm.

Drangert, J.O. (1997). Perceptions, urine blindness and urban agriculture pp 30-38. In: Drangert, J.O., Bew, J. and Winblad, U. (eds.). Ecological alternatives in sanitation. Publications on Water Resources: No 9. Department for Natural Reasources and the Environment, Sida, Stockholm.

Drangert, J.O. (2003). Requirements on sanitation systems: The flush toilet sets the standard for ecosan options. Paper presented during the 2nd International Symposium on Ecological Sanitation, 7-11 April, 2003, Lubeck, Baltic Sea, Germany.

Drangert, J.O., Bew, J. and Winblad, U. (eds.). (1997). Ecological alternatives in sanitation. Publications on Water Resources: No 9. Department for Natural Reasources and the Environment, Sida, Stockholm.

Drescher, A. and Iaquinta, D. (1999). Urban and peri-urban agriculture: A new challenge for the United Nations Food and Agriculture Organisation (FAO). FAO - Internal report. Rome.

Drescher, A.W. (2000). Technical tools for urban land use planning. Paper presented during the electronic conference on 'Urban and peri-urban agriculture on the policy agenda', 21 August to 30 September 2000 hosted by the United Nations Food and Agricultural Organisation (FAO) and Resource Centre for Urban Agriculture and Forestry (RUAF).

Driessen, P.M. and Konijn, N.T. (1992). Land-use systems analysis. Wageningen Agricultural University, Department of Soil Science and Geology: Malang: INRES.

Driscoll, E.D., Shelley P.E., and Strecker E.W. (1990). Pollutant loadings and impacts from storm-water runoff, Vol. III, Analytical investigation and Research Report FHWA-RD-88-008, Federal Highway Administration, Washington DC.

Driver, J. (1998). Welcome and introduction, sustainable development; Why does the phosphate industry care. International Conference on Phosphorus recovery from sewage and animal wastes, 6-7 May 1998, Warwick University UK.

Driver, J., Lijmbach, D. and Steen I. (1999). Why recover phosphorus for recycling, and how? Environmental Technology, Vol. 20 No 7, pp 651-662.

Dubbeling M. (2000). Appropriate methodologies for development of a facilitating framework for planning and policy in urban agriculture. Paper presented during the electronic conference on 'Urban and peri-urban agriculture on the policy agenda', 21 August to 30 September 2000, hosted by the united Nations Food and Agricultural Organisation (FAO) and Resource Centre for Urban Agriculture and Forestry (RUAF).

Duce, R. A. (1983). Biogeochemical cycles and the air-sea exchange of aerosols. In: Bolin, B. and Cook, R.B. (eds.). Interactions of the major biogeochemical cycles: Global change and human impacts. Scientific Committee on Problems of the Environment (SCOPE) No 21, 554 pp, Wiley, Chichester.

Duce, R.A., Liss, P.S., Merrill, J.T., Atlans, E.L., Buat-Menard, P. Hicks, B.B., Miller, J.M., Prospero, J.M., Atimoto, R., Church, T.M., Ellis, W., Galloway, J.N., Hansen, L., Jickells, T.D., Knap, A.H., Reinhardt, K.H., Schneider, B., Soudine, A., Tokos, J.J., Tsunogai, S., Wollast, R. and Zhou, M. (1991). The atmospheric input of trace species to the world ocean. Global Biogeochemical Cycles, 5, pp 193-259.

Edwards, P. (1992). Reuse of human wastes in aquaculture: A technical review. UNDP-World Bank, Washington DC.

Eicher, C.K. and Baker, D.C. (1982). Research on agricultural development in sub-Saharan Africa: A critical survey. Department of Agricultural Economics, Michigan State University Development Paper No. 1, pp 124-128.

Ekins, P. and Cooper, I. (1993). Cities and sustainability. Background to a Research Programme, Clean Technology Unit, UK Science and Engineering Research Council, Swindon.

Eklind, Y. and Kirchmann, H. (2000). Composting and storage of organic household waste with different litter amendments I: Carbon turnover. Bio-resource Technology, 74, pp 115-124.

Ekvall, A. (1995). Metal speciation and toxicity in sewage sludge. PhD thesis, Department of Sanitary Engineering, Chalmers University of Technology, Gothenburg, Sweden.

Ellis, J.B. (1989). The management and control of urban runoff quality. Journal Institute of Water and Environmental Management, April 1989.

Elwell, H.A. (1975). Programme for estimating soil loss and runoff in southern Africa. Research Bulletin No 17, Department of Conservation and Extension, Salisbury.

Elwell, H.A. (1977). A soil loss estimation system for southern Africa. Research Bulletin No. 22, Department of Conservation and Extension, Salisbury.

Elwell, H.A. and Stocking, M.A. (1982). Developing a simple yet practical method of soil loss estimation. Tropical Agriculture, 59: pp 43-48, Trinidad.

Emery, K.O. (1968). Relict sediment on continental shelves of the world. Bull. Am. Ass. Petr. Geol., 52, 445-464.

Emsley, J. (2000). The 13th Element: The Sordid Tale of Murder, Fire, and Phosphorus pp. 327. Wiley, London.

ENDA. (1995). Urban agriculture in Harare: Results and recommendations of a household survey conducted in Harare. Research, Development and Consulting Division (REDEC), and Environment and Development Activities (ENDA), Harare.

Engelman, R. and LeRoy, P. (1993). Sustaining water; population and the future of renewable water supplies. Population and Environment Programme. Population Action International, Washington.

Engle, D.L. and Sarnelle, O. (1990). Algal use of sedimentary phosphorus from an Amazon floodplain lake: Implications for total phosphorus analysis in turbid waters. Limnol. Oceanogr., 35, pp 483-490.

Esrey, S.A. (2000). Rethinking Sanitation: Panacea or pandora's box. In: Chorus, I., Ringelband, U., Schlag, G. and Schmoll, O. (eds.). Water, sanitation and health. IWA Publication, London.

Esrey, S.A., Andersson I., Hillers A., Sawyer R. (2001). Closing the loop: Ecological sanitation for food security. Publications on Water Resources No 18, Swedish International Development Cooperation Agency (Sida), Mexico, ISBN 91-586-8935-4.

Esrey, S.A., Gough J., Rapaport D., Sawyer R., Herbert M.S., Vargas J., and Winblad U. (eds.). (1998). Ecological sanitation. Swedish International Development Co-operation Agency (Sida), Stockholm.

Esser, G. and Kohlmaier, G.H. (1991). Modelling terrestrial sources of nitrogen, phosphorus, sulphur and organic carbon to rivers. In: Degens, E.T., Kempe, S. and Richey, J.E. (eds.). Biogeochemistry of major world rivers, pp 297-322, John Wiley & Sons, Chichester.

Eugster, H.P. and Hardie, L.A. (1978). Saline lakes. In: Lerman, A. (ed.). Lakes, chemistry, geology, physics. Springer-Verlag. pp 237-294.

Evans, P., Pollard, R., Narayan-Parker, D., Boydell, R., Kerwin, M. and McNeil, M. (1990). Rural sanitation in Lesotho: From pilot project to national programme. UNDP-World Bank Water and Sanitation Program Discussion Paper No 3., Washington D.C.

Fair, G.M. and Geyer, J.C. (1954). Water supply and wastewater disposal. John Wiley and Sons, Inc, New York.

Falkenmark, M. (1995). Coping with water scarcity under rapid population growth. Conference of SADC Ministers, Pretoria 23-24 November 1995.

Falkenmark, M., Klohn, W. Lundqvist, J. Postel, S. Rockstrom, J. Seckler, D. Shuval, H. and Wallace, J. (1998). Water scarcity as a key factor behind global food insecurity: Round Table Discussion. Ambio, Vol. 27, No 2.

Fantel, R.J., Peterson, G.R. and Stowasser, W.F. (1985). The worldwide availability of phosphate rock. Natural Resources Forum, 9, pp 5-24.

FAO, IFA and IFDC. (1994). Fertiliser use by crop 2. ESS/MISC/1994/4. Food and Agriculture Organisation of the United Nations. Rome.

FAO. (1965). Soil erosion by water: Some measures for its control on cultivated lands. Food and Agricultural Organisation of the United Nations (FAO), Series No 7 and FAO Agricultural Development paper No 81, Land and Water Development, Rome.

FAO. (1968). Food composition table for use in Africa. Food and Agricultural Organisation of the United Nations (FAO), Food consumption and planning branch, Nutrition division and US Department of Health, Education and Culture, Maryland.

FAO. (1973). Soil map of the World. Revised legend. World soil resources report 60. Rome, 119 pp.

FAO. (1979) Yield response to water Irrigation and drainage paper no. 33, Food and Agriculture Organisation, Rome, Italy

FAO. (1979). Agriculture: toward 2000. Report C79/24 for FAO conference 20th session November 1979. Food and Agriculture Organisation of the United Nations, Rome.

FAO. (1980a). Maximising the efficiency of fertiliser use by grain crops. FAO Fertiliser and Plant Nutrition Bulletin 3. Food and Agriculture Organisation of the United Nations, Rome (Reprinted in 1988), 30 pp.

FAO. (1980b). Organic recycling in Africa. Food and Agricultural Organisation (FAO) of the United Nations, Soils Bulletin 43, Rome.

FAO. (1981). Crop production levels and fertiliser use. FAO Fertiliser and Plant Nutrition Bulletin 2. Food and Agriculture Organisation of the United Nations, Rome, 69 pp.

FAO. (1984). Fertiliser and plant nutrition guide. Food and Agricultural Organisation of the United Nations (FAO) Bulletin No. 9, Fertiliser and Plant Nutrition Service, Land and Water Development Division, Rome.

FAO. (1984). Land, food and people. Based on the FAO/UNFPA/IIASA report 'Potential population supporting capacities of lands in the developing world'. Food and Agriculture Organisation of the United Nations, Rome.

FAO. (1985). Soil conservation and management in developing countries. Food and Agricultural Organisation of the United Nations (FAO), Soils Bulletin No. 33, Soil Resources, Management and Conservation Service, Land and Water Development Division, Rome.

FAO. (1986a). Efficient fertiliser use in acid upland soils of the humid tropics. Food and Agricultural Organisation of the United Nations (FAO) Bulletin No 10. Fertiliser and Plant Nutrition Service, Land and Water Development Division, Rome.

FAO. (1986b). Soil tillage in Africa: needs and challenges. Food and Agricultural Organisation of the United Nations (FAO), Soils Bulletin No 69, Soil Resources, Management and Conservation Service, Land and Water Development Division, Rome.

FAO. (1986c). Yield response to water. FAO Irrigation and Drainage Paper 33. Food and Agriculture Organisation of the United nations, Rome, 193 pp.

FAO. (1987a). Fertiliser strategies. FAO Land Use and Water Development Series, No 10, Rome, 148 pp.

FAO. (1987b). Soil and water conservation in semi arid-areas. Food and Agricultural Organisation of the United Nations (FAO), Soils Bulletin No 57, Soil Resources, Management and Conservation Service, Land and Water Development Division, Rome.

FAO. (1991). Agrostat PC: Computerised Information Series 1/3: Land use. FAO Publications Division. Food and Agriculture Organisation of the United Nations, Rome.

FAO. (1992). Fertiliser yearbook 1991. FAO Statistics Series No 106, Vol. 41, Rome.

FAO. (1993). Agriculture: Towards 2010. FAO Conference report C93/24. Food and Agriculture Organisation of the United Nations, Rome.

FAO. (1993). Field measurement of soil erosion and runoff. Food and Agricultural Organisation of the United Nations (FAO), Soils Bulletin No. 68, Soil Resources, Management and Conservation Service, Land and Water Development Division, Rome.

FAO. (1994). Land degradation in south Asia: its severity, causes and effects upon the people. World Soil Resources Report 78. Food and Agriculture Organisation of the United Nations, Rome.

FAO. (1995). Forest resources assessment 1990. Global synthesis. FAO Forestry Paper 124, Food and Agriculture Organisation of the United Nations, Rome.

FAO. (1998). Crop evpo-transpiration; Guidelines for computing crop water requirements. Food and Agricultural Organisation of the United Nations, Irrigation and Drainage paper No 56 by Allen, R.G., Pereira, L.S., Raes, D., and Smith, M., Rome.

FAO. (1999a). Crop Production, Food and Agriculture Organisation of the United Nations. URL: http://www.fao.org/docrep/w5146e/w5146e08.htm. Accessed in August 2000.

FAO. (1999b). Fertiliser strategies. Revised edition, Food and Agriculture Organisation (FAO) of the United Nations in collaboration with the International Fertiliser Industry Association (IFA), Rome.

FAO. (1999c). Issues in urban agriculture, Food and Agriculture Organisation of the United Nations, Rome.

FAO. (2000a). Crops and drops –Making the best use of water for agriculture, Rome.

FAO. (2000b). Deficit irrigation practices. Water paper No 22. Food and Agriculture Organisation, Rome.

FAO. (2000c). Agriculture:Towards 2015/30. Technical Interim Report. Food and Agriculture Organisation (FAO) of the United Nations, Rome.

FAO. (2001). Food trade and production series. FAOSTAT Agricultural Database. Food and Agricultural Organisation, Rome. URL http://apps.fao.org/, accessed in August 2001.

FAO. (2003). Aquastat country profiles. Food and Agricultural Organisation, Rome.

FAOSTAT. (2000). FAOSTAT Statistics Database. Food and Agriculture Organisation (FAO) of the United Nations, Rome. URL: http://www.fao.org

FAO-UNESCO. (1978). Report on the agro-ecological zones project. Vol. 1. Methodology and results for Africa. World soil resources report 48. Rome, pp 158.

FAO-UNESCO. (1988a). Soil map of the world. Revised legend. World soil resources report 60. Food and Agricultural Organisation (FAO) and United Nations Education, Scientific and Cultural Organisation (UNESCO), Rome, 119 pp.

FAO-UNESCO. (1988b). Soil survey interpretation for engineering purposes. Soils Bulletin 19. Rome, 24 pp.

Fardeau, J.C., Morel, C. and Jahiel, M. (1988). Does long contact with the soil improve the efficiency of rock phosphate? Results of isotopic studies. Fertilier Reserach, 17, pp 3-19.

Faul-Doyle, R.C. (1999). Ecological sanitation: Exploring options for a better future. Workshop Report October 2-6, 1999. Mvuramanzi Trust/UNICEF/WHO/AFRO, Harare.

Fazal, S. (2000). Urban expansion and loss of agricultural land: a GIS-based study of Saharanpur, India. Environment and Urbanization Vol. 12, No 2, pp 133-150.

Feachem, R.G., Bradley J.B., Garelick H. and Mara D.D. (1983). Sanitation and disease: Health aspects of excreta and wastewater management. Published for the World Bank, Washington DC, John Wiley and Sons, Chichester, 501 pp.

Fedra, K. (1985). A modular interactive simulation system for eutrophication and regional development, Water Resources Research, Vol. 12, pp 2-17.

Fernandes, E. and Varley A. (eds.). (1998). Illegal cities: law and urban change in developing countries. Institute of Commonwealth Studies, University of London, Plymbridge Distributors, Estover, Plymouth 256 pp.

Fernandes, T.R.C. (1989). The phosphate industry in Zimbabwe. Industrial Minerals, November 1989.

Fisher, T.R., Melack, J.M., Grobbelaar, J.U. and Howarth, R.W. (1995). Nutrient limitation of phytoplankton and eutrophication of inland, estuarine, and marine waters pp 301-322. In: Tiessen, H. (ed.). Phosphorus in the global environment: Transfers, cycles and management. Scientific Committee on Problems of the Environment, (SCOPE), Vol. 54, John Wiley & Sons, Chichester.

Fittschen, I., and Niemczynowicz, J. (1997). Experiences with dry sanitation and grey water treatment in the ecovillage Toarp, Sweden. Wat. Sci. Tech. Vol. 35, No 9, pp 161-170.

Fixen, P.E. and Grove, J.H. (1990). Testing soils for phosphorus. In: Soil testing and plant analysis, 3rd Edition. Book series No. 3. Soil Sci. Soc. America Inc, Madison, Wisconsin. pp. 141-180.

Flintoff, F. (1976). Management of solid wastes in developing countries. WHO regional publication, South East Asia Series No. 1, WHO New Delhi.

Flynn, K. (1999). An overview of public health and urban agriculture: water, soil and crop contamination and emerging zoonosis. Cities Feeding People Report 30, IDRC, Ottawa.

Ford, A. (1999). Modelling the environment: an introduction to systems dynamics modelling of environmental systems 401 pp. Island Press, Washington, D.C.

Forrester, J. (1961). Industrial dynamics. Pegasus Communications, Walthan, MA.

Fox, R.L. and Kamprath, E.J. (1970). Phosphate sorption isotherms for evaluating the phosphate requirement of soils. Soil Sci. Soc. Am. Proc., 34, pp 902-907.

Fox, R.L. and Kamprath, E.J. (1971). Adsorption and leaching of P in acid organic soils and high organic matter sand. Soil Sci. Soc. Am. Proc., 35, pp 154-156.

Fox, R.L. (1981). External phosphorus requirements of crops. In: Chemistry in the Soil Environment. ASA Special Publication No. 40. Am. Soc. Agron., Madison, Wisconsin. pp 223-240.

Foxon, T.J. (2000). A multi-criteria analysis approach to the assessment of sustainability of urban water systems, proceedings of the 15th European Junior Scientist Workshop, held May 11-14 2000, in Stavoren, The Netherlands, pp 119-128.

Franceys, R., Pickford, J. and Reed, R. (1992). A Guide to the development of on-site sanitation. WHO, Geneva.

Froelich, P.N. (1988). Kinetic control of dissolved phosphate in natural rivers and estuaries: A primer on the phosphate buffer mechanism. Limnol. Oceanogr., 33, pp 649-668.

Froelich, P.N., Bender, M.L., Luedtke, N.A., Heath, G.R. and Devries, T. (1982). The marine phosphorus cycle. Amer. J. Sci., 282, pp 474-511.

Frossrad, E., Brossard, M., Hedley, M.J. and Metherell, A. (1995). Reaction controlling the cycling of phodphorus in soils pp 107-137. In: Tiessen, H. (ed.). Phosphorus in the global environment: Transfers, cycles and management. Scientific Committee on Problems of the Environment, (SCOPE), Vol. 54, John Wiley & Sons, Chichester.

Fullstone, M.J. (1980). A review of the factors affecting the disposal of effluents containing nitrogen and phosphate in Zimbabwe. Zimbabwe Agricultural Journal, Vol. 77, No 5.

Furedy, C. (1992). Garbage: Exploring non-conventional options in Asian Cities. Environment and Urbanisation Vol. 4, No 2, pp 42-61.

Furedy, C. and Chowdhury, T. (1996). Solid waste reuse and urban agriculture - Dilemmas in developing countries: The bad news and the good news. Urban agriculture notes, City Farmer, Vancouver. URL: http://www.cityfarmer.org/Furedy.html

Gallopin, G. (1994). Impoverishment and sustainable development: A System Approach. Report of the IISD, Winnipeg.

Gardner, G. (1998). Recycling organic waste: From urban pollutant to farm resource. Worldwatch Paper No. 135, 24 pp. Worldwatch Institute, Washington DC.

Garrels, R.M., Mackenzie, F.T., and Hunt, C. (1973). Chemical cycles and the global environment. Los Altos, Calif., Kufman Inc., 206 pp.

Gassmann, G. and Glindemann, D. (1993). Phosphane (PH_3) in the biosphere. Angewandte Chemie, International edition in English 32, pp 761-763.

Gayman, M. (2000). A glimpse into London's early sewers, Cleaner Magazine, Cole Publishing Inc., Three Lakes.

GEMS. (2002). Water quality data tables for 82 major river basins. The Global Environmental Monitoring System GEMS-Water Programme UNEP, WHO, UNESCO and WMO.

Gerner, H., and Baanante, C. (1995). Economic aspects of phosphate rock application for sustainable agriculture in West Africa. In: Gerner, H. and Mokwunye, A. U. (eds.). Use of phosphate rock for sustainable agriculture in West Africa. Miscellaneous Fertiliser Studies No. 11. International Fertiliser Development Centre (IFDC) Africa, Lome.

Gerritse, R.G. and Vriesma, R. (1984). Phosphate distribution in animal waste slurries. J. Agric. Sci., 102, pp 159-161.

GESAMP. (1987). Land and sea boundary flux of contaminants: contributions from rivers. Report and studies No 32 of the IMCO/FAO/UNESCO/WHOM/WHO/IAEA/UNEP Joint Group of Experts on the Scientific Aspects of Marine Pollution. UNESCO, Paris.

Gijzen, H.J. and Mulder, A. (2001). The nitrogen cycle out of balance. Water 21, August 2001, pp 38-40.

Gilliland, M.W., and Baxter-Potter, W. (1987). A geographic information system to predict non-point source pollution potential. Water Resources Bulletin, 23 (2), pp 281-291.

Giorgini, A. and Zingales, F. (eds.). (1986). Agricultural non-point source pollution: Model selection and application. Elsevier, Amsterdam.

Girardet, H. (1992). The Gaia atlas of cities. New directions for sustainable urban living. Gaia Books Ltd., London.

GKW Consult, Brian Colquhoun Hugh O'Donnel and Partners (BCHOD) and Hydro Utilities (HU). (1996). Harare water supply study stage II: HWS 1993 intakes, transmission, treatment and supply. First interim report, Vol. II: Existing water supply facilites, City of Harare, Government of the Republic of Zimbabwe.

Gleick, P.H. (1996). Basic water requirements for human activities: Meeting basic needs. Water International 21, pp 83-92.

Gleick, P.H. (2000). The world's water. The biennial report on freshwater resources, 2000-2001. Island Press, Washington DC.

Godden, D.P. and Helyar, K.R. (1980). An alternative method for deriving optimal fertiliser rates. Review of Marketing and Agricultural Economics, 48, pp 83-97.

Golub, A.L. (1997). Decision analysis: an integrated approach. John Wiley, New York.

Gopalakrishna Pillai, K., Krishnamurthy, K. and Ramaprasad, A. (1984). Studies on varietal tolerance of rice varieties to phosphorus deficiency. Fertilisers and Agriculture, 87, pp 17-20.

Gotaas, H.B. (1956). Composting: Sanitary disposal and reclamation of organic wastes. World Health Organisation (WHO) Monograph series No 31, Geneva.

Goubert, J.P. (1989). The conquest of water: The advent of health in the industrial age. Princeton University Press, Princeton.

Graham, W. F., and Duce, R. A. (1979). Atmospheric pathways of the phosphorus cycle, Geochim. Cosmochimica Acta., 43, pp 1195-1208.

Granat, L., Rodhe, H. and Hallberg, R.O. (1976). The global sulphur cycle. In: Svensson, B.H. and

Soderlund, R. (eds.). Nitrogen, phosphorus and sulphur - Global cycles. SCOPE Report 7, Ecological Bulletins No.22, pp 89-134, Stockholm.

Grant, P.M. (1981). The fertilisation of sandy soils in peasant agriculture. Zimbabwe Agric. J., Vol. 78 (5), pp 169-175.

Gray, N.F. (1999). Water technology, an introduction for environmental engineers. Arnold Publishers, London, ISBN 0 340 67645 0.

Greenwood, D.J., Cleaver, T.J., Turner, M.K., Hunt, J., Niendorf, K.B. and Loquens, S.M.H. (1980). Comparison of the effects of phosphate fertiliser on the yield, phosphate content and quality of 22 different vegetable and agricultural crops. J. Agric. Sci. Camb., 95, pp 457-469.

Gregg, P.E.H. and Currie, L.D. (1992). The use of wastes as fertilisers and soil amendments. Occasional Report No. 6. Fertiliser and Lime Research Centre, Massey University, Palmerston North.

Gregory, J.M., Wigley, T.M.L., and Jones, P.D. (1993). Application of Markov models to area-average daily precipitation series and inter-annual variability in seasonal totals. Climate Dynamics 8, 299 pp.

Griffith, E.J., Ponnameperuma, C. and Gabel, N.W. (1977). Phosphorus, A key to life on the primitive Earth. Origins of Life Vol 8, pp 71-85.

Grigg, J.L. (1965). Prediction of plant response to fertiliser by means of soil tests 1. Correlation of yields of potatoes grown on recent and gley recent soils with results of various methods of assessing available soil phosphorus. N.Z. J. Agric. Res., 8, pp 893-904.

Grobbelaar, J.U. and House, W.A. (1995). Phosphorus as a limiting resource in inland waters; Interactions with nitrogen pp 255-273. In: Tiessen, H. (ed.). Phosphorus in the global environment: Transfers, cycles and management. Scientific Committee on Problems of the Environment, (SCOPE), Vol. 54, John Wiley & Sons, Chichester.

Grobler, D.C. (1985). Phosphorus budget models for simulating the fate of phosphorus in South African reservoirs. Water S.A., Vol. 11, pp 219-230.

Grobler, D.C., Roussouw, J.N., Van Eeden, P., and Oliveira, M. (1989). Decision support system for selecting eutrophication control strategies. Wat. Sci. Tech., pp 219-230.

Grottker, M. (1998). The myth of cycles versus sustainable water and material flux management. Proceedings of the 11th European Junior Scientist Workshop, Wildpark Eekholt, Germany, 12-15 February, 1998.

Gruhn, P. Goletti, F. and Yudelman, M. (2000). Integrated nutrient management, soil fertility, and sustainable agriculture: Current issues and future challenges 38 pp. Food, Agriculture, and the Environment Vision 2020 Discussion Paper No. 32. International Food Policy Research Institute (IFPRI), Washington, D.C.

Gumbo, B. (1995). Reuse of wastewater for potable supplies, Bulawayo, Zimbabwe. MSc Thesis, Loughborough University, Loughborough.

Gumbo, B. (1997a). Integrated water quality management in Harare. Proc. of the 23rd WEDC Conference "Water and Sanitation for All: Partnerships and Innovations", 1-5 September, Durban, 4 pp.

Gumbo, B. (1997b). Reuse of wastewater for potable supplies in Bulawayo, Zimbabwe: Is it a resource or a nuisance? Proc. of the 2nd southern African Water and Wastewater Conference, 15-19 September 1997, Harare, 7pp.

Gumbo, B. (1998a). Dual water supply systems: Is it just another pipe dream? Proc. of the International WEDO Conference, on small and medium size domestic water conservation, wastewater treatment and reuse, Bethlehem, Palestine, 21-24 February 1998, 17 pp.

Gumbo, B. (1998b). Rainwater harvesting in the urban environment: Options for water conservation and environmental protection in Harare. National conference on rain water harvesting: "An alternative water supply", 13-16 October 1998, Masvingo, Zimbabwe, 13 pp.

Gumbo, B. (1999a). Establishing phosphorus fluxes through material flow accounting and systems thinking in an urban-shed in Harare, Zimbabwe. Sharing Scarce Resources Zimbabwe (SSRZ) Seminar II, OTD St Lucia Park, Harare, November 4-6 1999, pp 15.

Gumbo, B. (1999b). Short-cutting phosphorus cycles in an urban drainage system. Proc. of the XXIV General Assembly of the European Geophysical Society. April 19-23, The Hague, 12 pp.

Gumbo, B. (2000a). Re-engineering the urban drainage systems for resource recovery and protection of drinking water supplies. In: Chorus, I., Ringelband, U., Schlag, G. and Schmoll, O. (eds.). Water, sanitation and health. IWA Publication, London. pp 15-21.

Gumbo, B. (2000b). Mass balancing as a tool for assessing integrated urban water management. Proc. of the 1st WARFSA/WaterNet Symposium, "Sustainable Use of Water Resources; Advances in Education and Research", 30 October to 3 November 2000, Maputo, Mozambique.

Gumbo, B. (2000c). Urban agriculture in Harare: Integrating the past, present and the future perspectives, IFNFS-UZ/IAC Food and Nutrition Security in Urban Areas Regional Course: Strengthening Food and Nutrition Training in Southern Africa, Harare 4-16 December 2000.

Gumbo, B. and Savenije, H.H.G. (2001). Inventory of phosphorus fluxes and stocks in an urban-shed: Options for local nutrient recycling. 1st International Conference on Ecological Sanitation held on 5-8 November 2001, Nanning, China.   Extended abstract at http//www.wise-china.com/english/e39.htm

Gumbo, B. and Savenije H.H.G. (2002). Ecologising societal metabolism and recycling of phosphorus at household and neighbourhood level. Poster presentation, European Geophysical Society, XXVII General Assembly, Nice, France, 21-26 April 2002.

Gumbo, B., and Van der Zaag, P. (2002). Water losses and the political constraints to demand management: the case of the City of Mutare, Zimbabwe. Physics and Chemistry of the Earth 27: pp. 805-813.

Gumbo, B., Savenije H.H.G., Kelderman, P. (2002a). Ecologising societal metabolism. The case of phosphorus, 3rd International Conference on Environmental Management, Eskom Conference Centre, Johannesburg, 27-30 August 2002.

Gumbo, B., Munyamba N., Sithole G., Savenije H.H.G. (2002b). Coupling of digital elevation model and rainfall-runoff model in storm drainage network design. Physics and Chemistry of the Earth 27: pp. 755-764.

Gumbo, B. Savenije, H.H.G. and Kelderman, P. (2004a). The phosphorus calculator: A planning tool for closing nutrient cycles in urban eco-systems. In: Ecosan-closing the loop, proceedings of the 2nd International Symposium on Ecological Sanitation, incorporating the 1st IWA specialist group conference on sustainable sanitation, April 7-11 2003, Lubeck, Baltic Sea, Germany. GTZ, Eschborn.

Gumbo, B., (2003a). Non-waterborne sanitation and water conservation, Encyclopaedia of Life Support Systems (EOLSS), EOLSS Publishers Co. Ltd, Oxford.

Gumbo, B., Mlilo, S., Broome, J. and Lumbroso, D. (2003b). Industrial water demand management and cleaner production: A case of three industries in Bulawayo, Zimbabwe. Physics and Chemistry of the Earth 28: pp. 797-84.

Gumbo, B., Juizo, D., van der Zaag, P. (2003c). Information is a prerequisite for water demand management: Experiences from four cities in Southern Africa. Physics and Chemistry of the Earth 28: pp. 827-837.

Gumbo, B. (2004). The status of water demand management in selected cities of southern Africa. Physics and Chemistry of the Earth 29: pp. 1225-1231.

Gumbo, B., Van der Zaag, P., Robinson, P., Jonker, L. and Buckle, J. (2004b). Training needs for water demand management. Physics and Chemistry of the Earth 29: pp. 1365-1373.

Guzha, E. (2001). Ecological sanitation pracrice and technological development in Southern Africa and Zimbabwe case study. Paper presented during the 1st International Conference on Ecological Sanitation, 5-8 November, 2001, Nanning, China.

Haarhoff, J., and Van der Merwe, B. (1996). Twenty-five years of wastewater reclamation in Windhoek, Namibia. Wat. Sci. Tech., Vol. 33, No 10-11, pp 25–35.

Haith, D.A. and Shoemaker, L.L. (1987). Generalised watershed loading functions for stream flow nutrients. Water Resources Bulletin 23(3), pp 471-478.

Hallsmith, G. (2003). The key to sustainable cities: Meeting human needs, transforming community systems, 256 pp. New Society Publishers, Gabriola Island.

Halwiel, B. (2001). Organic gold rush. Worldwatch Institute, Washington DC.

Hammond, J.S., Keeney, R.L., and Raiffa, H. (1999). Smart choices: a practical guide to making better decisions. Harvard University Press, Boston, MA.

Hands, M.R., Harrison, A.F. and Bayliss-Smith, T. (1995). Phosphorus dynamics in slash-and-burn and alley cropping sytems of the humid tropics pp155-170. In: Tiessen, H. (ed.). Phosphorus in the global environment: Transfers, cycles and management, Scientific Committee on Problems of the Environment, (SCOPE), Vol. 54, John Wiley & Sons, Chichester.

Hannon, B. and Ruth, M. (1994). Dynamic modelling. Springer Verlag, New York.

Hannon, B. and Ruth, M. (1997). Modelling dynamic biological systems. Springer Verlag, New York.

Harben, P.W. and Kuzvart, M. (1996). A global geology. Industrial Minerals Information Ltd, Surrey.

Hardin, G. (1968). The tragedy of the commons. Science 162.

Hardoy, J.E., Cairncross, S. and Satterthwaite, D. (eds.). (1990). The poor die young: Housing and health in Third World Cities. Earthscan, London.

Hardoy, J.E., Mitlin, D. and Satterthwaite D. (1992). Environmental problems in Third World Cities. Earthscan Publications, London.

Harender, R. and Bhardwaj, M.L. (2001). Earthworms' role in soil biology. Agriculture Tribune, The Tribune, Monday, March 5, 2001, Chandigarh, India online edition.

Harremoes, P. (1996). Dilemmas in ethics: Towards a sustainable society. Ambio, Vol. 25, pp 390-395.

Harremoes, P. (1997). Integrated water and wastewater management. Wat. Sci. Tech. Vol. 35, No 9, pp 11-20.

Harremoes, P. (2000). Advanced water treatment as a tool in water scarcity management. Wat. Sci. Tech. Vol. 42, No 12, pp 73-92.

Harremoes, P., Bundgaard, E. and Henze, M. (1991). Developments in wastewater treatment for nutrient removal. Water Pollution Control, Vol. 1, pp 19-23.

Harrison, A.F., Miles, J. and Howard, D.M. (1988). Phosphorus uptake by birch from various depths in the soil. Forestry, 61, pp 349-358.

Harrison, A.F. (1985). Effects of environment and management on phosphorus cycling in terrestrial ecosystems. Journal of Environ. Manag., 20, pp 163-179.

Harrison, A.F. (1987). Soil organic phosphorus. A review of world literature. CAB International, Oxon, 257 pp.

Haug, R.T. (1993). The practical handbook of compost engineering. Lewis Publishers, CNC Press Florida. ISBN 0873713737.

Hawken, P., Lovins A., and Hunter-Lovins, L. (1999). Natural Capitalism: Creating the next industrial revolution, pp 396. Little Brown and Company, Boston.

Haygarth, P. (1997). Agriculture as a source of phosphorus transfer to water: sources and pathways, SCOPE Newsletter No 21, June 1997.

Haynes, R.J. (1990). Active uptake and maintenance of cation-anion balance: a critical examination of their roles in regulating rhizosphere pH. Plant and Soil, 126, pp 247-264.

Hayward, K. (1997). Separate ways. Water Quality International Magazine. January/February 1997, pp 18-19.

HCMP. (1992). Harare Combination Master Plan, Report, City of Harare Zimbabwe.

HDS and BCHOD. (1982). Design approach to water and sewerage problems relative to urban (predominantly high density and rural communities in Zimbabwe. Housing Development Services (HDS) and Brian Coulquhoun, Hugh O' Donnel and Partners (BCHOD). Harare.

Hedley, M. J., Stewart, J. W. B., and Chauhan, B. S. (1982). Changes in inorganic and organic soil phosphorus fractions induced by cultivation practices. Soil Sci. Soc. Amer. J. 46, pp 970-976.

Hedley, M.J., Hussin, A. and Bolan, N.S. (1990). New approaches to phosphorus fertilisation. In: Phosphorus requirements for sustainable agriculture in Asia and Oceania. International Rice Reserach Institute, Manila, pp 125-142.

Hedley, M.J., Kirk, G.J.D. and Santos, M.B. (1993). Phosphorus efficiency and the forms of soil phosphorus utilised by upland rice cultivars. Plant and Soil, 158, pp 53-62.

Hedley, M.J., Mortvedt, J.J., Bolan, N.S. and Syers, J.K. (1995). Phosphorus fertility management in agrosystems pp 59-92. In: Tiessen, H. (ed.). Phosphorus in the global environment: Transfers, cycles and management, Scientific Committee on Problems of the Environment, (SCOPE), Vol. 54, pp 452, John Wiley & Sons, Chichester.

Heiney, P. W. and Mwangi, W. (1997). Fertiliser use and maize production. In: Gerner, H. and Mokwunye, A. U. (eds.). Use of phosphate rock for sustainable agriculture in West Africa. Miscellaneous Fertiliser Studies No. 11. Lome: International Fertiliser Development Centre (IFDC) Africa.

Hellmann, H. (1994). Load trends of selected parameters of water quality and of trace substances in the river Rhine between 1955 and 1988. Wat. Sci. Tech. 29(3), pp 69-76.

Hellstrom, D. (1998). Nutrient management in sewerage systems: Investigations of components and exergy analysis. PhD thesis, Department of Environmental Engineering, Division of Sanitary Engineering, Lulea University of Technology Sweden.

Hellstrom, D. and Karrman, E. (1996). Nitrogen and phosphorus in fresh and stored urine. Environmental Research Forum, Vol. 5 No. 6, pp 221-226.

Hellstrom, D., Baky, A., Palm, O., Jeppsson, U. and Palmquist, H. (2003). Comparisoon of resource efficiency of systems for management of toilet waste and organic household waste. Paper presented during the 2nd International Symposium on Ecological Sanitation, 7-11 April, 2003, Lubeck, Baltic Sea, Germany.

Helyar, K.R. and Godden, D.P. (1977). Soil phosphate as a capital asset. In: Blair, G.J. (ed.) Prospects for improving the efficiency of phosphorus utilisation. Reviews in Rural Science, 3, 23-30.

Helyar, K.R. and Godden, D.P. (1977). The biology and modelling of fertiliser response. J. Aust. Inst. Agric. Sci., 43, pp 22-30.

Herdendorf, C.E. (1990). Distribution of the worlds large lakes. In: Tilzer, M.M. and Serruya, C. (eds.). Large lakes, ecological structure and function, pp 3-38.

Hermans, L.M. (1999). Design of phosphorus management strategies for the Cannonsville Basin. Master of Science Degree, Systems Engineering, Policy Management, Delft University of Technology, Delft, The Netherlands.

Herrmann, T. and Klaus, U. (1997). Fluxes of nutrients in urban drainage systems: assessment of sources, pathways and treatment techniques. Wat. Sci. Tech., Vol. 36, No 8-9, pp.167-172.

Herschel, C. (1973). The two books on the water supply of the City of Rome of Sextus Julius Frontinus: Water Commissioner of the City of Rome A.D. 97. A translation into English, and explanatory chapters, New England Water Works Association.

Heynike, J.J.C. and Wiechers, H.N.S. (1984). Situation statement on detergent phosphates and their impact on eutrophication in South Africa. Internal Report of the Water Research Commission, Pretoria.

Hill, J.P. (1978). Experience of the design and operation of fertiliser plants in developing countries. No 174.

Hillbricht-Ilkowska, A., Ryszkowski, L. and Sharpley, A.N. (1995). Phosphorus transfers and landscape structure: Riparian sites and diversified land use patterns, pp201-228. In: Tiessen, H. (ed.). Phosphorus in the global environment: Transfers, cycles and management. Scientific Committee on Problems of the Environment, (SCOPE), Vol. 54, pp 452, John Wiley & Sons, Chichester.

Hillel, D. (1980). Introduction to soil physics. Academic Press Inc., Orlando.

Hirji, R., Johnson, P., Maro, P. and Matiza-Chuita, T. (eds.). (2002). Defining and mainstreaming environmental sustainability in water resources management in Southern Africa. SADC, IUCN, SARDC, World Bank: Maseru/Harare/Washington DC.

Hodges, C.N. (1993). Reversing the flow: water and nutrients from seas to the land. Ambio 22, pp 483-490.

Hoekstra, A.Y. (1998). Perspectives on water: A model-based exploration of the future International Books, Utrecht, the Netherlands.

Hoglund, C. (2001). Evaluation of microbial health risks associated with the reuse of source-separated human urine. PhD thesis, Royal Institute of Technology (KTH), Department of Biotechnology, Applied Microbiology & Swedish Institute for Infectious Disease Control (SMI), Department of Water and Environmental Microbiology, Stockholm.

Hoglund, C., Sternstrom, T.A., Jonsson H., and Sundin, A. (1998). Evaluation of faecal contamination and microbial die-off in urine separating sewage systems. Wat. Sci. Tech. Vol 38, No.6, pp.17-25.

Hoko, Z. (1999). Innovative sanitation for Harare, Zimbabwe. MSc thesis, SEE 080, IHE-Delft the Netherlands.

Holden, R. (2003). Factors which have influenced the acceptance of ecosan in South Africa and development of a marketing strategy. Paper presented during the 2nd International Symposium on Ecological Sanitation, 7-11 April, 2003, Lubeck, Baltic Sea, Germany.

Holford, I.C.R., Morgan, J.M., Bradley, J. and Culls, B.R. (1985). Yield responsibilities and response curvature as essential criteria for the evaluation and calibration of soil phosphate tests for wheat. Aust. J. Soil Res. Vol. 23, pp 167-180.

Holmberg, J. (1995). Socio-ecological principles and indicators for sustainability. PhD Dissertation, Institute of Physical Resource Theory, Goteborg, Sweden.

Holmes, J.R. (1984). Managing solid wastes in developing countries. John Wiley and Sons Ltd, London.

Horward, E. (1965). Garden cities of tomorrow. MIT Press, Cambridge, MA.

Hosper, H. (1997). Clearing lakes: An ecosystem approach to the restoration and management of shallow lakes in the Netherlands. Agricultural University of Wageningen, Wageningen.

Housing Development Services (HDS) and Brian Coulquhoun, Hugh O' Donnel and Partners (BCHOD), (1982), Design approach to water and sewerage problems relative to urban (predominantly high density and rural communities in Zimbabwe, Harare.

Howarth, R.W., Jensen, H.S., Marino, R. and Postma H. (1995). Transport to and processing of phosphorus in near-shore and oceanic waters pp323-345. In: Tiessen, H. (ed.). Phosphorus in the global environment: Transfers, cycles and management. Scientific Committee on Problems of the Environment, (SCOPE), Vol. 54, pp 452, John Wiley & Sons, Chichester.

HPS. (1992). Stella II: An introduction to systems thinking. High Performance Systems (HPS) Inc., Hanover.

HPS. (1993). Stella II: Technical documentation. High Performance Systems (HPS) Inc., Hanover.

HPS. (1994). Getting started with Stella II: A hands-on experience. High Performance Systems (HPS) Inc., Hanover.

HPS. (1996). Stella research: Technical Documentation. High Performance Systems (HPS) Inc., Hanover.

Hranova, R., Gumbo, B., Klein, J., and Van der Zaag, P. (2002). Aspects of the water resources management practice with emphasis on nutrients control in the Chivero basin, Zimbabwe. Physics and Chemistry of the Earth 27, pp 875-885.

Hunter, J.V., Balmat, J., Wilber, W., and Sabationo, T. (1981). Hydrocarbons and heavy metals in urban runoff. Urbanisation, storm-water runoff and the aquatic environment. George Mason University, Fairfax, VA.

Hunter, P.P. (1997). Waterborne disease: epidemiology and ecology. John Wily, Chichester.

Hutchinson, M.F. (1990). A point rainfall model based on a three-state continuous Markov occurrence process. Journal of Hydrology 114, pp 218-236.

ICES. (1995). Income, consumption and expenditure survey. Central Statistics Office (CSO). Government of Zimbabwe, Harare.

ICPR. (1987). Rhine action programme international commission for the protection of the Rhine against Pollution. Koblenz, Germany.

IDRC. (2003). Cities feeding people programme. International Development Research Centre (IDRC), Ottawa. URL: http://network.idrc.ca/en/ev-5911-201-1-DO_TOPIC.html.

IFA and UNEP. (1998). The fertiliser industry, world food supplies and the environment. International Fertiliser Industry Association (IFA) and United Nations Environment Programme (UNEP), Paris.

IFA and UNEP. (2002). Fertiliser: Industry as a partner for sustainable development 65 pp. The International Fertiliser Industry Association and United Nations Environment Programme. Beacon Press, UK.

IFA. (1992). World fertiliser use manual. The International Fertiliser Industry Association. Paris. 632 pp.

IFA. (1994). Nitrogen fertiliser statistics 1988/89 to 1992/93. Information and Market Research Service, International Fertiliser Industry Association, Paris.

IFA. (1998). The fertiliser industry, world food supplies and the environment. Revised edition. The International Fertiliser Industry Association and United Nations Environment Programme (IFA and UNEP), Paris.

IFA. (1999). Supplying plant nutrients for sustainable agriculture – life cycle approaches in the fertiliser industry. In: Industry and Environment, United Nations Environment Programme (UNEP). April – September 1999, Paris.

IFDC. (1992). Fertiliser use statistics and crop yields. International Fertiliser Development Centre. Muscle Shoals, AL. 34 pp.

IFDC. (2003). Free fertiliser statistics. International Fertiliser Development Centre (IFDC). URL: http://www.ifdc.org, accessed in December 2002.

IFOAM. (2000). Basic standards for organic production and processing decided by the IFOAM General Assembly in Basel, Switzerland, September 2000. International Federation of Organic Agriculture Movements. URL: http://www.ifoam.org/, accessed in January 2001.

IFPRI. (1997). The world food situation: recent developments, emerging issues and long-term prospects. International Food Policy Research Institute (IFPR), Washington, DC.

IFPRI. (2001). 2020 Global food outlook: trends, alternatives, and choices. International Food Policy Research Institute (IFPR), Washington, DC.

Ignatieff, V., Doyle, J.J. and Couston, J.W. (1964). Future fertiliser requirements of developing countries and crop response to fertiliser in these countries. No 83.

ILWIS 2.2. (1998). The Integrated Land and Water Information System. The International Institute of Aerospace Survey and Earth Sciences (ITC), Enschede, The Netherlands.

IMWI. (2000). Projected water scarcity in 2025. International Water Management Institute. Colombo.

IRC. (1980). Public standpost water supplies - a design and construction manual. International Reference Centre for Community Water Supply and Sanitation. Technical Report No 14. The Hague, Netherlands.

IRRI. (1980). Priorities for alleviating soil-related constraints for food production in the tropics. International Rice Research Institute, Manila.

Isermann, K. (1990). Share of agriculture in nitrogen and phosphorus emissions into the surface waters of Western Europe against the background of eutrophication. Fertiliser Research, Vol. 26, pp 253-269.

Isherwood, K. F. (1996). Fertiliser subsidy policies in regions other than Asia and the Pacific. Agro-Chemical News in Brief, Special issue, September edition.

Isherwood, K.F. (1999). The globalisation of the fertiliser industry pp10. Article for The Fertiliser Industry Annual Review, on behalf of the International Fertiliser Industry Association (IFA), Paris.

Isherwood, K.F. (2000). Mineral fertiliser use and the environment. The International Fertiliser Industry Association and United Nations Environment Programme (IFA and UNEP), Paris.

ISO 14040. (1997). Environmental management - Life cycle assessment - Principles and framework.

ISO 14042. (2000). Environmental management - Life cycle assessment - Life cycle impact assessment.

ISO 14043. (2000). Environmental management - Life cycle assessment – Life cycle interpretation.

ISO14041. (1998) Environmental management - Life cycle assessment - Goal and scope definition and Inventory analysis.

ITC and FAO. (1995). World soil resource Report, 80, International Institute of Aerospace Survey and Earth Sciences, Enschede.

IWA. (2002). Anaerobic digection model No 1. Report No 13. International Water Association Publishing.

Jackson, I.J. (1977). Climate, water and agriculture in the tropics. Longman Harlow. 377 pp.

Jacobi, P., Drescher, A. and Amend, J. (2000). Urban agriculture: Justification and planning guidelines. GTZ, Eschborn.

Jaffe, D.A. (1992). The nitrogen cycle. In: Butcher, S.S. Charlson, R.J. Orians, G.H. and Wolfe, G.V. (eds.). Global biogeochemical cycles. Academic Press, London.

Jahnke, R.A. (1992). The phosphorus cycle. In: Butcher, S.S. Charlson, R.J. Orians, G.H. and Wolfe, G.V. (eds.). Global biogeochemical cycles. Academic Press, London.

James, S.W. and Wells, K.L. (1990). Soil sample collection and handling: Technique based on source and degree of field variability In: Westerman, R.L. (ed.). Soil testing and plant analysis, 3rd Book Series No. 3. Soil Sci. Soc. Am., Madison, Wisconsin. pp 25-44.

Jarlov, L. (2000). Urban agriculture as a concept in urban planning in South Africa: Example from Port Elizabeth. Paper presented during the electronic conference on 'Urban and peri-urban agriculture on the policy agenda', 21 August to 30 September 2000 hosted by the united Nations Food and Agricutural Organisation (FAO) and Resource Centre for Urban Agriculture and Forestry (RUAF).

Jarvis, M.J.F, Mitchell, D.S. and Thornton, J.A. (1982). Aquatic macrophytes and echhormia crassipes. In Thornton, J.A. and Nduku, W.K. (eds.) Lake McIlwaine – The eutrophication and recovery of a tropical man-made Lake. Dr. W. Junk Publishers, The Hague, pp 137-144.

Jeffrey, P., Seaton R., Parsons S., and Stephenson, P.(1997). Evaluation methods for the design of adaptive water supply systems in urban environments. Wat. Sci. Tech. Vol. 35, No. 9, pp 45-51.

Jenkins, J. (1999). The humanure handbook: A guide to composting human manure 2nd edition. Jenkins Publishing, Chelsea Green Publishing Company. ISBN 0-9644258-9-0 USA.

Jenkinson, D.S., Hart, P.B.S., Rayner, J.H. and Parry, L.C. (1987). Modelling the turnover of organic matter in long-term experiments at Rothamsted. Intecol Bulletin 15, pp 1-8.

Jiang, L. (2001). Ecosan development in Guangxi, China. Paper presented during the 1st International Conference on Ecological Sanitation, 5-8 November, 2001, Nanning, China.

JICA. (1996). The study on water pollution control in Upper Manyame River Basin in the Republic of Zimbabwe. MLGRUD, Nippon Jogesuido Sekkei Co. Ltd., Nippon Koei Co. Ltd., July.

Johanson, R.C., Imhoff, J.C., Davis, H.H. (1984). User's manual for hydrological simulation program - FORTRAN (HSP): Release 7.0. U.S. Environ. Prot. Agency, Athens, GA, 745 pp.

Johansson, M., Jonsson, H., Hoglund C., Richert-Stintzing, A., and Rodhe, L. (2000). Urine separation-closing the nutrient cycle. Final report on the Research and Development project "Source-separated human urine: A future source of fertiliser for agriculture in the Stockholm Region. Stockholm Water Company, Sweden.

Johnes, P.J. (1996). Evaluation and management of the impact of land use change on the nitrogen and phosphorus load delivered to surface waters: The export coefficient modelling approach. Journal of Hydrology 183: 323-349.

Johnston, A.E. (1995). The efficient use of plant nutrients in agriculture. The International Fertiliser Industry (IFA), Paris.

Johnston, A.E. (1997a). Fertilisers and agriculture: Fifty years of developments and challenges. Fertiliser Society Proceedings No 396, New York.

Johnston, A.E. (1997b). Phosphorus, its efficient use in agriculture and its loss from agricultural soils, IACR Rothamsted Publication No. 1997-03. Rothamsted Experimental Station, Harpenden, Hertfordshire AL 5 2 JQ, United Kingdom.

Johnston, A.E. and Steen, I. (2000). Understanding phosphorus and its use in agriculture, 38 pp. European Fertiliser Manufacturers Association (EFMA), Brussels.

Jones, C.A., Sharpley, A.N. and Williams, J.R. (1991). Modelling phosphorus dynamics in the soil-plant system. In: Hanks, J. and Ritchie, J.T. (eds.) Modelling Plant and Soil systems. Agronomy No. 31. Amer. Soc. of Agron., Madison, Wisconsin.

Jonsson H, Dalemo M, Sonesson U, Vinneras, B. (1998). Modelling the sewage system - evaluating urine separation as a complementary function to the conventional sewage treatment. Paper presented at "Systems and engineering models for waste management" International workshop in Goteborg, Sweden, 25-26 February 1998.

Jonsson, H. (1997). Asssessment of sanitation systems and resue of urine pp 11-22. In: Drangert, J.O., Bew, J. and Winblad, U. (eds.). Ecological alternatives in sanitation. Publications on Water Resources: No. 9. Department for Natural Reasources and the Environment, Sida, Stockholm.

Jonsson, H., Sternstrom, T.A, Svensson, J., and Sundin, A. (1997). Source separated urine: nutrient and heavy metal content, water saving and faecal contamination. Wat. Sci. Tech. Vol 35, No.9, pp 145-152.

Jonsson, H., Richert Stintzing, A., Vinneras, B., and Salomon, E. (2004). Guidelines on use of urine and faeces in crop production. Report 2004-2, EcoSanres. Stockholm Environment Institute, Stockholm.

Jordan, C.F. (ed.). (1989). An Amazon rain forest: the structure and function of a nutrient stressed ecosystem and the impact of slash-and-burn agriculture. MAB/UNESCO Series 2, Parthenon Press. 176 pp.

Kabell, T.C. (1984). An assessment of the surface water resources of Zimbabwe and guidelines for development planning. Ministry of Water Resources and Development, Harare.

Kalbermaten, J.M. and Middleton, R.N. (1998). A vision for year 2000. Environmental services, a vision water, life and environment, World Water Council.

Kalbermatten, J.M., Julius, D.S., Gunnerson, C.G. and Mara, D.D. (eds.). (1981). Appropriate technology for water supply and sanitation (12 vols.), The World Bank, Washington DC.

Kamprath, E.J. (1991). Appropriate measurements of phosphorus availability in soils of the semi-arid tropics. In: Johansen, C., Lee, K.K. and Saharwat, KL (eds.). Phosphorus nutrition of grain legumines in the semi arid tropics. ICRISAT, India, pp 23-31.

Karrman, E. (2000). Environmental systems analysis of wastewater management. PhD thesis, Department of Water, Environment and Transport, Chalmers University of Technology, Goteborg.

Katznelson, J. (1977). Phosphorus in the soil-plant-animal ecosystem. An introduction to a model. Oecologia, 26, pp 325-334.

Keatinge, J.D.H. (1995). Integrated fertiliser management: The way forward for the Third World? No 369.

Keeney, R. L. (1992). Value-focused thinking: A path to creative decision making. Harvard University Press, Cambridge, MA.

Kelderman, P. (1984). Sediment-water exchange in Lake Grevelingen under different environmental conditions. Netherland J. Sea Research, 18, pp 286-311.

Kellman, M. (1984). Synergistic relations between fire and low fertility in Neotropical savannas: An hypothesis. Biotropica, 16, pp 158-160.

King, F.H. (1973). Farmers of forty centuries: Permanent agriculture in China, Korea and Japan. Rodale Press, Emmaus PA.

Kirchmann, H. and Pettersson, S. (1995). Human urine: Chemical composition and fertiliser use efficiency. Fertiliser Research 40, pp 149-154.

Kirchmann, H. and Widen, P. (1994). Separately collected organic household waste: Chemical composition and composting characteristics. Swedish J. agric. Res. 24, pp 3-12.

Kirkham, M.B. (1982). Agricultural use of P in sewage sludge. Adv Agron, 35, pp 129-163.

Kirkwood, C. (1995). Vensim tutorial. Department of Decision and Information Systems, Arizona State University, Tempe, AZ.

Knisel, Jr., W.G. (ed.). (1980). CREAMS: A field scale model for chemicals, runoff, and erosion from agricultural management systems. Cons. Res. Rept. No. 26. U.S. Dept. Agric., Washington, D.C. 640 pp.

Kofoed, A.D., Williams, J.H. and L'Hermite, P (1986). Efficient land use of sludge and manure. Elsevier Applied Science, London, 245 pp.

Koide, R.T. (1991). Nutrient supply, nutrient demand and plant response to mycorrhizal infection. New Phytol., 117, pp 365-386.

Kupchella, C.E., and Hyland, M.C. (1993). Environmental science: Living within the system of nature, 3rd Edition, Prentice-Hall Int. Inc.

Laegreid, M., Bockman, O.C. and Kaarstad, O. (1999). Agriculture, fertilisers and the environment. CAB International Publishing, Wallingford.

Lajtha, K. and Harrison, A.F. (1995). Strategies for phosphorus acquisition and conservation by plant species and communities pp139-147. In: Tiessen, H. (ed.). Phosphorus in the global environment: Transfers, cycles and management. Scientific Committee on Problems of the Environment, (SCOPE), Vol. 54, pp 452, John Wiley & Sons, Chichester.

Lardinois, I. and Furedy, C. (eds.). (2000). Source separation of household waste materials: Analysis of case studies from Pakistan, the Phillipines, India, Brazil, Argentina and the Netherlands. Urban Waste Series No. 7, Urban Waste Expertise Programme (UWEP), Gouda, The Netherlands.

Larsen, T.A., and Gujer, W. (1996). Separate management of anthropogenic nutrient solutions (Human urine), Wat. Sci. Tech. Vol. 34, No. 3-4, pp 87-94.

Larsen, T.A., and Gujer, W. (1997). The concept of sustainable water management. Wat. Sci. Tech., No. 9, pp 3-10.

Larson, B.A. and Bromley, D.W. (1990). Property rights, externalities, and resource degradation: Locating the tragedy. J. Development Economics, 33, pp 235-262.

Lean, D.R.S. (1973). Phosphorus dynamics in lake water. Science, 179, pp 678-680.

Leigh, R.A. and Johnston, A.E. (1986). An investigation into the usefulness of phosphorus concentrations in tissue water as indicators of the phosphorus status of field-grown spring barley. Journal of Agricultural Science, Cambridge, 107, pp 329-332.

Lens, P., Zeeman G., Lettinga G. (2001). Decentralised sanitation and reuse: Concepts, systems and implementation. Integrated Environmental Technology Series, IWA Publishing London, ISBN 1-900222-47-7.

Lentner, C., Lentner, C. and Wink, A. (eds.). (1981). Units of measurements, body fluids, composition of the body, nutrition. Scientific tables book 1, Ciba-Geigy Limited, Basle, ISBN 0-914168-50-9.

Lerman, A, MacKenzie, F.T. and Garrels, R.M. (1975). Modeling of geochemical cycles: phosphorus as an example. Mem. Geol. Soc. Amer., 142, pp 205-218.

Lewcock, C. P. (1994). Case study of the use of urban waste by near-urban farmers of Kano, Nigeria. Chatham Maritime. Natural Resources Institute, Project No. A0354.

Liebig, J. von. (1840). Chemistry and its application to agriculture and physiology. Taylor and Walton, London, 4th edition, 352 pp.

Lienert, J. and Larsen, T.A. (2003). Introducing urine separation in Switzerland: Novaquatis, an interdisciplinary research project. Paper presented during the 2nd International Symposium on Ecological Sanitation, 7-11 April, 2003, Lubeck, Baltic Sea, Germany.

Lijklema L., and Tyson J.M. (1993). Urban water quality: Interactions between sewers, treatment plants, and receiving waters, Wat. Sci. Tech. Vol. 27, No. 5-6, pp 29-33.

Likens, G.E. (1972). Nutrients and eutrophication: The limiting nutrient controversy. Am. Soc. Limnol. Oceanogr., Sec., Symp., Vol. 1, pp 1-2.

Likens, G.E. (ed.). (1981). Some perspectives of the major biogeochemical cycles. Scientific Committee on Problems of the Environment, (SCOPE), Vol. 17.

Lindsay, W.L., Vlek, P.L.G. and Chien, S.H. (1989). Phosphate minerals. In: Dixon, J.B. and Weed, S.B. (eds.). Minerals in soil environment 2nd edition, SSSA Monograph. Published by SSSA, Madison, Wisconsin. pp. 1089-1130.

Lock, K. and De Zeeuw, H. (2001). Mitigating the health risks associated with urban and peri-urban agriculture. Urban agriculture magazine, Vol. 1 No. 3, March 2001, pp6-8.

Lock, K. and Van Veenhuizen, R. (2001). Balancing the positive and negative health impacts. Urban agriculture magazine, Vol. 1 No. 3, March 2001, pp 1-5.

Lock, R.R., (1994), Water Pollution Control in Zimbabwe and the role of the Water Pollution Advisory Board, Workshop paper on water Resources Protection and Water Economics in Zimbabwe, 17-20 October, Harare.

Loehr R.C. (1974). Characteristics and comparative magnitude of non-point sources. Journal Water Pollution Control Federation, 46(8), pp 1849-1872.

Loetscher, T. (1999). Appropriate sanitation in developing countries: The development of computerised decision aid. PhD thesis, Department of Chemical Engineering, University of Queensland, Brisbane. ISBN 18649909X.

Loftas, T. (ed.). (1995). Dimensions of need: An atlas of food and agriculture. Food and Agriculture Organisation (FAO), Rome

Lorenz, C.M., Van Dijk, G.M. Van Hattum, A.G.M. and Cofino, W.P. (1997). Concepts in river ecology: Implications for indicator development regulated rivers: Research and Management 13, pp 501-516.

Lovelock, J.E. (1979). Gaia. A new look at life on Earth. Oxford University Press, Oxford.

Lowenson, S.W. (1969). Water pollution control: Legislative aspects. Paper presented to the Rhodesian Scientific Association Symposium in Salisbury, 13 August.

Lundin, M., Molander, S. and Morrison, G.M. (1999). A set of indicators for the assessment of temporal variations in sustainability of sanitary systems. Wat. Sci Tech. Vol. 39 No 5, pp 235-242.

Lyle, J.T. (1985). Design of human ecosystems: landscape, land use, and natural resources. Van Nostrand Reinhold, New York.

Mach, D. M., Ramirez, A. and Holland, H.D. (1987). Organic phosphorus and carbon in marine sediments. Am. J. Sci., 278, pp 429-441.

Machena, C. (1997) The pollution and self-purification capacity of the Mukuvisi River. In: Moyo, N.A.G. (ed.). Lake Chivero: A polluted Lake, University of Zimbabwe Publications, Harare.

Mackay, A.D., Syers, J.K., Tillman, R.W. and Gregg, P.E.H. (1986). A simple model to describe the dissolution of phosphate rock in soils. Soil Sci. Am. J., 50, pp 291-296.

Madimutsa, B. (2000). Cadmium flows through Golden Quarry sanitary landfill. MSc dissertation, Department of Civil Engineering, University of Zimbabwe. Harare.

Maene, L.M. (2000). Efficient fertiliser use and its role in increasing food production and protecting the environment 6th Annual Arab Fertiliser Association (AFA) International Annual Conference, January 2000, Cairo.

Magadza, C.H.D. (1997). Water pollution and catchment management in Lake Chivero. In: Moyo, N.A.G. (ed.). Lake Chivero: A polluted Lake, University of Zimbabwe Publications, Harare.

Malmqvist P.A., and Bennerstedt K., (1997), Future storm-water management in Stochholm. Case Study Hamarby Strand, Proc. Stockholm Water Symposium, August 1997, Stockholm, Sweden.

Malmqvist, P.A. (1997). Ecological sanitation in Sweden: Evaluation pp 5-10. In: Drangert, J.O., Bew, J. and Winblad, U. (eds.). Ecological alternatives in sanitation. Publications on Water Resources: No. 9. Department for Natural Reasources and the Environment, Sida, Stockholm.

Mansson, T. (ed.). (1992). Eco-cycles: The basis of sustainable urban development, 108 pp. A report from the Environmental Advisory Council, Sweden. SOU 1992:43, Stockholm.

Mara, D. D. and Cairncross, S. (1989). Guidelines for the safe use of wastewater and excreta in agriculture and aquaculture. World Health Organisation, Geneva.

Mare, A. (1998). Green water and blue water in Zimbabwe: The Mupfure river basin case. MSc thesis DEW 044, IHE Delft.

Marjanovic, P., Miloradov M., Cukic Z., Bogdamovic S., and Sakulski D., (1995). Integrated cadastre (inventory system) for pollution sources in the Danube basin in Yugoslavia. Wat. Sci. Tech. Vol. 32 No. 5-6, pp265-275.

Marshall, B.E. (1997). Lake Chivero after forty years: The impact of eutrophication. In: Moyo, N.A.G. (ed.). Lake Chivero: A polluted Lake, University of Zimbabwe Publications, Harare.

Marshall, B.E., (1982), The fish of Lake McIlwaine, In: Thornton, J. and Nduku, W. (eds.). Lake McIlwaine: The eutrophication and recovery of a tropical African lake, pp 156-188, Dr. W. Junk, The Hague.

Martin, A., Oudwater, N. and Meadows, K. (2000). Urban agriculture and the livelihoods of the poor in Southern Africa, Case studies from Cape Town and Pretoria, South Africa and Harare, Zimbabwe. Paper presented at the International symposium 'Urban Agriculture and Horticulture – the linkage with urban planning, Berlin 7-9 July 2000.

Martin, J.M. and Meybeck, M. (1979). Elemental mass-balance of material carried by world major rivers. Mar. Chem., 7, pp 173-206

Mathieu, M. De la Vega, J. (1978). Constraints to increased fertiliser use in developing countries and means to overcome them No 173.

Mathuthu, A.S., Mwanga, K. and Simoro, A. (1997). Impact assessment of industrial and sewage effluents on water quality of the receiving Marimba River in Harare. In: Moyo, N.A.G. (ed.). Lake Chivero: A polluted Lake, University of Zimbabwe Publications, Harare.

Matshalaga, N. (1997). Gender dimensions of urban poverty: Urban survival strategies, the case of Tafara. Institute of Development Studies (IDS), University of Zimbabwe.

Matsui, S. (1997). Night-soil collection and treatment in Japan pp 65-74. In: Drangert, J.O., Bew, J. and Winblad, U. (eds.). Ecological alternatives in sanitation. Publications on Water Resources: No. 9. Department for Natural Reasources and the Environment, Sida, Stockholm.

Matthews P., (1996). Water business changes – A driver for green clean compact technologies. Wat. Sci. Tech. Vol. 34 No 12, pp 135-140.

Matthews, P.J. (1992), Sewage sludge disposal in the UK: a new challenge for the next twenty years. J. Institution of Water and Environmental Management, 6, 551-559.

Maxwell, D. (1999). The political economy of urban food security in sub-Saharan Africa. World Development, Vol. 27 No 11, pp 1939.

Maxwell, D., Levin, C. and Csete, J. (1998). Does urban agriculture help prevent malnutrition? Evidence from Kampala. IFPRI Discussion Paper No.45, Washington, DC.

Mbiba B., (1995). Urban agriculture in Zimbabwe: Implication for urban management and poverty. Aldershot, Ashgate Publishing, Avebury.

Mbiba, B. (1998). City harvests: Urban agriculture in Harare (Zimbabwe). In: City Harvest - a Reader on Urban Agriculture. GTZ, Eschborn.

McCall, D.G. and Thorrold, B.S. (1991). Fertiliser history as a useful predictor of soil fertility status. Proceedings New Zealand Grassland Association, 53, pp 191-196.

McCauley, E., Downing, J.A. and Watson, S. (1989). Sigmoid relationships between nutrients and chlorophyll among lakes. Can. J. Fish. Aquatic Sci., 46, pp 1171-1175.

Mckendrick, J. (1982). Water supply and sewage treatment in relation to water quality in Lake McIlwaine. In: Thornton, J.A. and Nduku, W.K. (eds.). Lake McIlwaine; The eutrophication and recovery of a tropical African man-made Lake. Dr W Junk Publishers, The Hague, pp 202-217.

MDG. (2000). Millenuim development goals, United Nations Millenium Goals, World Bank. URL: http://www.developmentgoals.org/index.html, accessed in October 2003.

Meade, R. (1988). Movement and storage of sediment in river systems. In: Lerman, A. and Meybeck, M. (eds.). Physical and chemical weathering in geochemical cycles. Kluwer Academic, Dordrecht. pp. 165-179.

Meadows, D., Meadows, D., Randers, J. and Behrens, W. (1972). The limits to growth. Universe Books.

Meadows, D., Meadows, D.and Randers, J. (1992). Beyond the limits. Chelsea Green Publishing.

Meadows, K. (2000). The social and institutional aspects of urban agriculture in the Cape Flats, Western Cape, South Africa. NRI report. Chatham, UK.

Meinardi, C.R., Beusen, A.H.W., Bollen, M.J.S., Klepper, O. and Willems, W.J. (1995). Vulnerability to diffuse pollution and average nitrate contamination of European soils and groundwater. Wat. Sci. Tech. 31(8), pp 159-165.

Melack, J.M. (1995). Transport and transformations of P, fluvial and lacustrine ecosystems pp 245-254. In: Tiessen, H. (ed.). Phosphorus in the global environment: Transfers, cycles and management, Scientific Committee on Problems of the Environment, (SCOPE), Vol. 54, pp 452, John Wiley & Sons, Chichester.

Melillo, J.M. Field, C.B., and Moldan B. (eds.). (2003). Interactions of the major biogeochemical cycles: global change and human impacts. Scientific Committee on Problems of the Environment (SCOPE) No 61, Island Press, USA.

Mels, A.R., Nieuwenhuijzen, A.F., Van der Graaf, J.H.J.M., Klapwijk, B., De Koning, J., Rulkens, W.H. (1999). Sustainability criteria as a tool in the development of new sewage treatment methods. Wat. Sci. Tech. Vol. 39, No 5, pp 243-250.

Menon, R.G., Chien, S.H. and Gadalla, A.N. (1991). Comparison of Olsen and Pi soil tests for evaluating phosphorus bioavailability in a calcareous soil treated with single superphosphate and partially acidulated phosphate rock. Fert. Res., 29, pp 153-158.

Metcalf and Eddy, Inc. (1991). Wastewater engineering: Treatment disposal, reuse, 3rd edition, pp 1334. McGraw-Hill, New York.

MEWRD. (1978). Guidelines for the Disposal of Sewage Effluent During Wet Weather. Ministry of Energy and Water Resources Development (MEWRD), Rhodesia.

Meybeck, M. and Helmer, R. (1989). The quality of rivers: from pristine state to global pollution', In Palaeogeogr. Palaeoclim., Palaeocol. (Global Planet Change Sect) 75, Elsevier, pp 283-309.

Meybeck, M. (1976). Total dissolved transport by world major rivers. Hydrological Sci. Bull., Vol 21, No 2, pp 265-284.

Meybeck, M. (1982). Carbon, nitrogen and phosphorus transport by world rivers. American Journal of Science Vol. 282, pp 401-450.

Meybeck, M. (1988). How to establish and use world budgets of riverine materials. In: Lerman, A. and Meybeck, M. (eds.). Physical and chemical weathering in geochemical cycles. Kluwer Academic Publ., Dordrecht.pp. 247-272.

Meybeck, M., Chapman, D. and Helmer, R. (eds.). (1989). Global freshwater quality: a first assessment. Blackwell Ref. Oxford. 306 pp.

Mhlanga, A.T. and Madziva, T.J. (1990). Pesticide Residues in Lake McIlwane Zimbabwe. Ambio 19 (8): pp 368-372.

Miller, G.T. (1988). Living in the environment. 6th edition, Wadsworth Belmont California.

Milliman, J.D. and Meade, R.H. (1983). World-wide delivery of river sediment to the oceans. Int. J. Geology, Vol. 91, 1, pp 1-21.

Ministry of Agriculture. (1997). The agricultural sector of Zimbabwe. Statistical Bulletin, Gorvernment of the Republic of Zimbabwe, Harare.

Mitcham, C. (1995). The concept of sustainable development: It's origins and ambivalence. Technology in Society, Vol. 17, No. 3, pp 311-326.

Model Building By-Laws (1977), Guided by section 183 of the Urban Councils Act [Chapter 214], Zimbabwe.

Mokwunye, A. U. (1995). Phosphate rock as capital investment. In: Gerner, H. and Mokwunye, A. U. (eds.). Use of phosphate rock for sustainable agriculture in West Africa. Miscellaneous Fertiliser Studies No. 11. Lome: International Fertiliser Development Centre (IFDC) Africa.

Moldakov, O. (2000). The urban farmers of St Petersburg. Urban agriculture magazine, Vol. 1 No. 1, June 2000, pp 24-26.

Moldan, B. and Billharz, S., (eds.). (1997). Sustainability Indicators: report of the project on Indicators for Sustainable Development. Scientific Committee on Problems of the Environment (SCOPE) No. 58, 440pp, Wiley, Chichester.

MoLG and SALA. (1990a). Sanitation manual design procedures Vol 5. Ministry of Local Government, Zimbabwe (MoLG), in collaboration with the Swedish Association of Local Authorities (SALA), project funded by Swedish International Development Agency (Sida).

MoLG and SALA. (1990b). Water reticulation, treatment and storage manual Vol 4. Ministry of Local Government, Zimbabwe (MoLG), in collaboration with the Swedish Association of Local Authorities (SALA), project funded by Swedish International Development Agency (Sida).

MoLG and SALA. (1990c). Township roads and stormwater drainage manual Vol 6. Ministry of Local Government, Zimbabwe (MoLG), in collaboration with the Swedish Association of Local Authorities (SALA), project funded by Swedish International Development Agency (Sida).

MoLGH. (1981). Design approach to roads and storm-water problems relative to high density housing development in Zimbabwe, Ministry of Local Government and Housing (MoLGH), Zimbabwe.

Mopper, K., and Degens, E.T. (1979). Organic carbon in the ocean: Nature and cycling pp 293-316. In: Bolin, B., Degens, E.T., Kempe, S., and Ketner, P. (eds.). The global carbon cycle, SCOPE Report No. 13. Wiley, Chichester.

Morello, J., Buzai, G.D., Baxendale, C.A., Rodriguez, A.F., Matteucci, S.D., Godagnone, R.E., and Casas, R.R. (2000). Urbanisation and the consumption of fertile land and other ecological changes: The case of Buenos Aires. Environment and Urbanisation, Vol. 12, No. 2, pp 119-131.

Morgan, P., (1999). Ecological sanitation in Zimbabwe-A compilation of manuals and experiences, 130 pp. Conlon printers, Harare.

Morris, R.A., Sattell, R.R. and Christensen, N.W. (1992). Phosphorus sorption and uptake from Sri Lankan Alfisols. Soil Sci. Soc. Am. J., 56, pp 1516-1520.

Mortvedt, J.J. and Beaton, J.D. (1995). Heavy metal and radionuclide contaminants in phosphate fertilisers pp93-105. In: Tiessen, H. (ed.). Phosphorus in the global environment: Transfers, cycles and management. Scientific Committee on Problems of the Environment, (SCOPE), Vol. 54, pp 452, John Wiley & Sons, Chichester.

Mougeot, L.J.A. (2000). Urban agriculture: Concept and definition. Urban agriculture magazine, Vol. 1 No 1, June 2000, pp 5-7.

Moyo, N.A.G. (1997b). Causes of massive fish deaths in Lake Chivero. In: Moyo, N.A.G. (ed.). Lake Chivero: A polluted Lake, University of Zimbabwe Publications, Harare.

Moyo, N.A.G. (ed.). (1997). Lake Chivero: A polluted Lake, University of Zimbabwe Publications, Harare.

Moyo, N.A.G. and Mtetwa, S. (1999). Water pollution control. Paper presented at a Seminar on Water Resources Management in Southern Africa: Enhancing Environmental Sustainability, Harare, October 1999.

Muhleck, R., Grangler, A. and Jekel, M. (2003). Ecological assessment of ecosan concepts and convetional wastewater systems. Paper presented during the 2nd International Symposium on Ecological Sanitation, 7-11 April, 2003, Lubeck, Baltic Sea, Germany.

Mukonoweshuro, F. (2000). Inventory of phosphorus stocks and fluxes in Harare. BSc Civil Engineering final year project. Department of Civil Engineering, University of Zimbabwe. Harare.

Munyirwa, K. and Onganga, O. (1999). Kenya: ecological sanitation as an option where sewerage systems malfunction (the Kisumu case). In Ecological sanitation: Exploring options for a better future, Faul-Doyle, R.C. Workshop Report October 2-6, 1999. Mvuramanzi Trust/UNICEF/WHO/AFRO, Harare.

Mushamba, S. (2002). Different kinds of investment in urban agriculture: Kintyre Lake County and Musikavanhu projects. Urban agriculture magazine, Vol. 7 No. 7, August 2002, pp 26-28.

Nakazato, T. (1997). Improving the water environment with reclaimed wastewater. Water Quality Inetrnational, November/December, pp 14-17.

NASA. (1994). Designing for human presence in space: An introduction to environmental control of life support systems. National Aeronotics and Sapce administration (NASA), Alabama.

Nasr, J. (2000). Agriculture as a sustainable use of urban land. Paper presented during the electronic conference on 'Urban and peri-urban agriculture on the policy agenda', 21 August to 30 September 2000 hosted by the united Nations Food and Agricutural Organisation (FAO) and Resource Centre for Urban Agriculture and Forestry (RUAF).

Nelson, L.A. and Anderson, R.L. (1977). Partitioning of soil test-crop response probability. In: Peck, T.R., Cape Jr., J.T. and Whitney, D.A. (eds.). Soil testing: Correlation and interpreting the analytical results. ASA Special Pub 29. Amer Soc of Agron, Madison, Wisconsin. pp 19-38.

Nelson, M. (1997). Nutrient recycling in Biosphere 2. Life Support and Biosphere Science, Vol. 4, No. 3&4, pp 145-153.

Nhapi, I., Hoko, Z., Siebel, M.A., and Gijzen, H.J. (2002). Assessment of the major water and nutrient flows in the Chivero catchment area, Zimbabwe. Physics and Chemistry of the Earth Vol 27, No. 11-22, pp 783-792.

Nhapi, I. (2004). Options for wastewater management in Harare, Zimbabwe. PhD dissertation, UNESCO-IHE Delft and Wagenningen University. Taylor Francis Group plc. London. ISBN 9058096971.

Nielsen, N.E. and Barber, S.A. (1978). Differences between genotypes of corn in the kinetics of phosphorus uptake. Agron. J., 70, pp 695-698.

Nielsen, N.E. and Schjorring, J.K. (1983). Efficiency and kinetics of phosphorus uptake from soil by various barley genotypes. Plant and Soil, 72, pp 225-230.

Niemczynowicz, J. (1993). New aspects of sewerage and water technology. Ambio, Vol. 22 No. 7.

Niemczynowicz, J. (1997a). Recent trends in urban water management towards development of sustainable solutions, Workshop, UNESCO Centre for Humid Tropics, Hydrology, Kuala Lumpur, Malaysia 12-14 November 1997.

Niemczynowicz, J. (1997b). The water profession and Agenda 21, WQI magazine, March/April 1997, pp 9-11.

Niemczynowicz, J. (1997c). State of the art in urban storm-water design and research, current thinking. Urban hydrology study, Lund University, Sweden.

Niemczynowicz, J. (1998). Comments on Development Report 1997, from Water and Sanitation Perspective. February 1998.

Niemczynowicz, J. (ed.) (1996). Integrated water management in urban areas: Searching for new, realistic approaches with respect to the developing world. Proceedings of the UNESCO-IHP International Symposium, Environmental Research Forum Vol 3-4, 479 pp, Transtec Publications Ltd., Switzerland.

NRC. (1989). Alternative agriculture. National Research Council (NRC). National Academy Press, Washington, DC.

Nugent, R. (2000). The Impact of Urban Agriculture on the Household and Local Economies pp 67-97. In Bakker, N., Dubbeling M., Guendel S., Sabel-Koschella U. and De Zeeuw H. (eds.). Growing Cities, Growing Food, Feldafing: DSE.

Nugent, R.A. (1997). The Significance of urban agriculture. Published by City Farmer, Canada's Office of Urban Agriculture. URL: http://www.cityfarmer.org/racheldraft.html#racheldraft Acessed in July 2000.

NWO. (1992). Sustainability and environmental quality. Priority Research Theme, Nederlandse Organisatie voor Wetenscappelijk Ooderzoek (NWO), The Hague.

Nyamangara, J. and Mzezewa, J. (1999). The effect of long-term sewage sludge application on Zn, Cu,Ni and Pb levels in a clay loam soil under pasture grass in Zimbabwe. Agriculture, Ecosystems and Environment 73, pp 199-204.

Nyamangara, J. and Mzezwa J. (1996). Maize growth and nutrient uptake in Zimbabwean red clay soil amended with anaerobically digested sewage sludge. Journal of Applied Science in Southern Africa (JASSA), The Journal of the University of Zimbabwe, Vol. 2, No. 2, pp83-89.

Nyamapfene, K. (1991). The Soils of Zimbabwe. Nehanda Publishers, Harare, 179 pp.

Nye, P H. and Tinker, P.B. (1977). Solute movement in the soil-root system. Studies in Ecology Volume 4. Blackwell Scientific Publications, 342 pp.

Nye, P.H. and Greenland, D.J. (1964). Changes in the soil after clearing a tropical forest. Plant and Soil, 21, pp 101-112.

Nye, P.H. and Kirk, G.J.D. (1987). The mechanism of rock phosphate solubilisation in the rhizosphere. Plant and Soil, 100, pp 127-134.

Nyiraneza, D. (2001). Kisoro town ecological sanitation: Operation and maintenance experience in South Wsetrn Uganda as contribution to water resource protection. Paper presented during the 1st International Conference on Ecological Sanitation, 5-8 November, 2001, Nanning, China.

O'Dwyer, N. and Partners. (1997). Master plan for water distribution: Vol 4, Future water supply area. City of Harare, Department of Works.

O'Halloran, I.P., Stewart, J.W.B. and De Jong, E. (1987). Changes in P forms and availability as influenced by management practices. Plant and Soil, 100, pp 113-126.

Obeng, L. and Wright, F. (1987). The co-composting of domestic solid and human wastes. World Bank Technical Paper No. 57. Washington, D.C.

Obernosterer, R., Brunner, P.H., Daxbeck, H., Gagan, T., Glenck, E. Hendriks, C., Morf, L., Puamann, R. and Reiner, I. (1998). Urban maetabolism: The City of Vienna. Materials accounting as a tool for decision making in environmental policy (MacTEmPo). Case Study Report 1, European Commission programme for environment and climate, 1994-1998, ENV4-CT-96-0230. Department of waste management, Insittute for water quality and waste management, Vienna Univerity of Technology.

Odum, H.T. (1983). Systems ecology: An introduction. John Wiley & Sons, New York.

Ojeda-Benitez, S., De Vega, C.A. and Ramirez-Barreto, M.E. (2000). The potential for recycling household waste: A case study for Mexicali, Mexico. Environment and Urbanisation, Vol. 12, No. 2, pp 163-173.

Oldeman, L. R. (1992). Global extent of soil degradation. In International Soil Reference and Information Centre (ISRIC) Bi-Annual Report 1991-1992. Wageningen, The Netherlands.

Oldeman, L. R., Makkeling, R. T. A and Sombroek, W. G. (1992). World map of the status of human-induced soil degradation: An explanatory note, 2nd edition, International Soil Reference and Information Centre (ISRIC). Wageningen, The Netherlands.

Olkowski, H. Olkowski, W. and Javits, T. (1979). The integral urban house: Self reliant living in the city, pp 494. Sierra Club Books, San Francisco.

Olsen, R.A. (1975). Rate of dissolution of phosphate from minerals and soils. Soil Sci. Soc. Amer. Proc., 39, pp 634-639.

Olsen, S.R., Cole, C.V., Watanabe, F.S. and Dean, L.A. (1954). Estimation of available phosphorus in soils by extraction with sodium bicarbonate. US Department of Agriculture Cicular 939. US Government Printing Office. Washington DC, USA. 19 pp.

Olsson, A., (1995). Source separated human urine; occurrence and survival of faecal micro-organisms and chemical composition. Report 208, Department of Agricultural Engineering Research, Swedish University of Agricultural Sciences, Uppsala.

Otterpohl, R. (2000). Design of highly efficient source control sanitation and practical experiences. In: Lens, P. and Lettinga, G. (eds.). Decentralised sanitation and reuse, IWA Publications, London.

Otterpohl, R. (2001). Black, brown, yellow, grey – the new colours of sanitation. Water 21, October 2001, pp 37-41. IWA, London.

Otterpohl, R., Abold, A. and Grottker, M. (1996). Integration of sanitation into natural cycles: A new concept for Cities. Environmental Research Forum, Vol. 5-6, pp227-232.

Otterpohl, R., Albold, A. and Oldenburg, M. (1999). Source control in urban sanitation and waste management: Ten systems with reuse of resource. Wat. Sci. Tech, Vol 39, No 5, pp 153-160.

Otterpohl, R., Grottker, M. and Lange, J. (1997). Sustainable water and waste management in urban areas, Wat. Sci. and Tech., Vol. 35, No 9, pp.121-134.

Pallett, J. (1997). Sharing water in Southern Africa. Desert Research Foundation of Namibia, Windhoek.

Park, M. (2001). The Fertiliser industry. Woodhead Publishing Ltd, Cambridge.

Parton, W.J., Stewart, J.W.B. and Cole, C.V. (1988). Dynamics of C, N, P and S in grassland soils: a model. Biogeochemistry, 5, pp 109-131.

PASS. (1995). Poverty assessment study survey report (PASS). Ministry of Public Service Labour and Social Welfare, Government of the Republic of Zimbabwe, Harare.

Pearce, D.W. and Turner, R.K. (1990). Economics of natural resources and the environment Harvester Wheatsheaf, New York.

Pegram G.C., Gorgens A.H.M. and Ottermann A.B. (1997). A framework for addressing the information needs of catchment water quality management. Journal Water SA Vol 23 No. 1 January 1997, Water Research Commission, Pretoria.

Pegram, G.C., and Bath A.J. (1995). Role of non-point sources in development of water quality management plan for the Mgeni river catchment. Wat, Sci. Tech. Vol. 32 No. 5-6, pp 175-182.

Pepper W., Leggett J., Swart R.J., Wasson J., Edmonds J. and Mintzer I. (1992). Emission scenarios for the IPCC, an update. Assumptions, methodology and results. Prepared for the Intergovernmental Panel on Climate Change, Working Group I.

Perrings, C., (1989). An optimal path to extinction? Poverty and resource degradation in the open agrarian economy. J. Development Econ., 30, pp 1-24.

Peterson, S. (1994). Software for model building and simulation: An illustration of design philosophy. In: Morecroft, J and Sterman, J. (eds.). Modelling for learning organisations, Pegasus communications, Walthan, MA.

Pettersson, K., Bostrom, B. and Jacobsen, O.S. (1988). Phosphorus in sediments-speciation and analysis. Hydrobiologia, 170, pp 91-101.

Petts, G.E. (1984). Impounded rivers. John Wiley & Sons, New York.

Phillips, L. D., and Phillips, M. C. (1993). Facilitated work groups: theory and practice. Journal of the Operational Research Society, 44(6), pp 533-49.

Pickford, J. (1995). Low-cost sanitation: A survey of practical experience, IT Publications, London.

Pierrou, U. (1976). The global phosphorus cycle' In: Svensson, B.H. and Soderlund, R. (eds.). Nitrogen, phosphorus and sulphur - Global cycles, SCOPE Report 7, Ecological Bulletins No.22, pp 75-88, Stockholm, Sweden.

Pinkham, R., (1996). 21st century water systems: Scenarios, visions, and drivers, 20 pp, Rocky Mountain Institute USA.

Pitman, W. V. (1973). A Mathematical model for generating monthly river flows from meterological data in South Africa. Report no. 2173, Hydrological Research Unit, University of the Witwatersrand, Johannesburg.

Poels, R.L.H. (1987). Soils, water and nutrients in a forest ecosystem in Suriname. PhD thesis, Agricultural University Wageningen, The Netherlands, pp 253.

Poerbo, H. (1991). Urban solid waste management in Bandung: Towards and integrated resources recovery system. Environment and Urbanisation Vol 3 No 1 pp 60-69, London.

Polprasert, C. (1996). Organic waste recycling; Technology and management. John Wiley and Sons, Chichester.

Polprasert, C., Lohani, B.N. and Chan, C.B. (1981). Human faeces, urine and their itilisation. ENSIC Traslation Committee.

Ponting, C. (1991). A green history of the World. Penguin Books.

Pothukuchi, K. and Kaufman, J.L. (2000). The food system: A stranger to the planning field. Journal of the American Planning Association, Vol. 66, No. 2.

Powersim Corporation. (1996). Powersim 2.5 refernce manual. Powersim Corporation Inc., Herndon.

Pratt, C.J., Mantle, J.D.G. and Schofield, P.A., (1988). Urban stormwater reduction and quality improvement through the use of permeable pavements, Wat. Sci. and Tech., pp 769-778.

Pretorius, W.A. (1983). Should the phosphate concentration in sewage effluents be restricted? IMIESA 8, pp 23-29.

Quon, S. (1999). Planning for urban agriculture: A Review of tools and strategies for urban planners. Cities Feeding People (CFP) Report No. 28. International Development Research Centre (IDRC), Ottawa.

Redefining Progress. (2002). Ecological Footprint Accounts: Moving Sustainability from Concept to Measurable Goal, Redefining Progress (RP), Oakland. URL: http://www.redefiningprogress.org/programs/sustainability/ef/projects, Acessed in August 2002.

Redfield, A.C. (1958). The biological control of chemical factors in the environment. Am. Sci., 46, pp 205-221.

Rees, W.E. (1992). Ecological footprints and appropriated carrying capacity: What urban economics leaves out. Environment and Urbanisation Vol. 4, No. 2, pp 121-130.

Rhode, H. (1989). Acidification in a global perspective. In Ambio, 18, 3, pp 155-159.

Richardson, G. and Pugh, A. (1981). Introduction to system dynamics modelling with Dynamo. Pegasus Communications, Walthan, MA.

Richey, J.E. (1982). Interactions of C, N, P, and S in river systems: A biogeochemical model In: Bolin, B. and Cook, R.B. (eds.). The major biogeochemical cycles and their interactions, John Wiley & Sons, Chichester.

Richey, J.E. (1983). The phosphorus cycle. In Interactions of the major biogeochemical cycles: Global change and human impacts. In: Bolin, B. and Cook, R.B. (eds). Scientific Committee on Problems of the Environment (SCOPE) No. 21, 554 pp, Wiley, Chichester.

Robarts, R.D. and Southall, G.C. (1977). Nutrient limitation of phytoplankton growth in seven tropical man-made Lakes, with specific reference to Lake McIlwaine, Rhodesia. Arch. Hydrobiol. 79-1, pp 1-35

Rockstrom, J. (1997). On-farm agrohrodrological analysis of the sahelian yield crisis: Rainfall partiniong, soil nutrients and water use efficiency of pearl millet. PhD thesis, Natural Resources Management, Department of Sytems Ecology, Stockholm, Sweden.

Rodda, J.C., and Matalas N.C. (1987). Water for the future, Hydrology in perspective, Proc. of the Rome Symp., IAHS, Publ. No. 164.

Roelofs, J. (1996). Greening cities: Building just and sustainable communities. A Toes book, The Bootstrap Press, New York ISBN 0-942850-35-1.

Rose, G. D. (1999), Community-based technologies for domestic wastewater treatment and reuse: options for urban agriculture, Cities feeding people CFP Report Series, Report 27, Spring 1999. URL: http://www.idrc.ca/cfp/index_e.html

Rosenberg, L. and Furedy, C. (eds.). (1996). International source book on environmentally sound technologies for municipal solid waste management. United Nations Environment Programme (UNEP), International Environmental Technology Centre, Osaka.

Ross H., Poungsomlee, A., Punpuing, S. and Archavanitkul, K. (2000). Integrative analysis of city systems: Bangkok "Man and the Biosphere" programme study. Environment and Urbanisation, Vol. 12, No. 2, pp 151-161.

Rosswall, T. (1976). The internal nitrogen cycle between micro-organisms, vegetation and soil. In: Svensson, B.H. and Söderlund, R. (eds.). Nitrogen, phosphorus and sulphur - Global cycles, SCOPE Report 7, Ecological Bulletins No.22, pp 157-167, Stockholm, Sweden.

Rosswall, T. (1981). The biogeochemical nitrogen cycle. In Likens, G.E. (ed.). Some perspectives of the major biogeochemical cycles, SCOPE 17. New York, John Wiley and Sons, pp 25-49.

Rothamsted Experimental Station. (1991). Guide to the classical field experiments. Institute of Arable Crops Research (AFRC) Harpenden, UK.

RUAF. (2003). Resorce Centre on Urban Agricuture and Forestry (RUAF). Leusden, The Netherlands. URL: http://www.ruaf.org/.

Ruel, M., Haddad, L. and Garrett, J. (1999). Some urban facts of life: Implications for research and policy. World Development, Vol 27 No.11.

Runge-Metzger, A. (1995). Closing the cycle: Obstacles to efficient P management for improved global food security pp 27-42. In: Tiessen, H. (ed.). Phosphorus in the global environment: Transfers, cycles and management. Scientific Committee on Problems of the Environment, (SCOPE), Vol. 54, pp 452, John Wiley & Sons, Chichester.

Ruttenberg, K.C. (1990). Diagenesis and burial of phosphorus in marine sediments: Implications for the marine phosphorus budget. PhD dissertation, Yale Univ., New Haven, Connecticut.

Ruttenberg, K.C. (1993). Reassessment of the oceanic residence time of phosphorus. Chemical Geology Vol 107, pp 405-409.

SADC. (2000). SADC statistics: facts and figures 2000, SADC secretariat: Gaberone.

Saggar, S., Dev, G. and Sharma, K.R. (1974). Efficiency of five high yielding varieties for absorption of fertiliser phosphorus. J. Res., (PAU) 11, pp 392-396.

Saggar, S., Hedley, M.J., White, R.E. (1990a). A simplified resin membrane technique for extracting phosphorus from soils. Fert. Res., 24, pp 173-180.

Saggar, S., Mackay, A.D., Hedley, M.J., Lambert, M.G. and Clark, D.A. (1990b). A nutrient-transfer model to explain the fate of phosphorus and sulphur in a grazed hill-country pasture. Agric. Eco. and Environ., 30, pp 295-315.

Salcedo, I.H. and Medeiros, C. (1995). Phosphorus transfer from tropical terrestrial to aquatic systems-mangroves pp347-36. In: Tiessen, H. (ed.). Phosphorus in the global environment: Transfers, cycles and management. Scientific Committee on Problems of the Environment, (SCOPE), Vol. 54, pp 452, John Wiley & Sons, Chichester.

Sanchez, P., Izac, A.M. Buresh, R. Shepherd, K. Soule, M. Mokwunye, U. Palm, C. Woomer, P. and Nderitu, C. (1997). Soil fertility replenishment in Africa as an investment in natural resource capital. In: Buresh, R. and Sanchez, P. (eds). Replenishing soil fertility in Africa. Soil Science Society of America (SSSA) Special Publication No. 51. Madison.

Sanchez, P.A. (1976). Properties and management of soil in the tropics. John Wiley, New York. 618 pp.

Sanchez, P.A. and Salinas, J.G. (1981). Low-input technology for managing Oxisols and Ultisols in tropical America. Adv. in Agron., 34, 280-406.

Sanchez, P.A. and Uehara, G. (1980). Management consideration for acid soils with high phosphorus fixation capacity. In: Khaswana, F.E., Sample, E.C. and Kamprath, E.J. (eds.). The role of phosphorus in agriculture. Am. Soc. Agron., Madison, Wisconsin. pp 471-514.

SANDEC and WSSCC. (2000) Environmental Sanitation in the 21st Century; Summary Report of Bellagio Expert Consultation 1-4 February; SANDEC, Duebendorf.

Satterthwaite, D. (ed.). (1999). The Earthscan reader in sustainable cities. Earthscan Publications, London, pp 478.

Sauchelli, V. (1965). Phosphates in agriculture. Reinhold Publishing Corporation, New York, pp 277.

Savenije, H. H.G. (2002). Why water is not an ordinary economic good, or why the girl is special. Physics and Chemistry of the Earth, Vol 27 pp 741–744.

Savenije, H.H.G. (1995). New definitions for moisture recycling and the relationship with land use changes in the Sahel. Journal of Hydrology 167: 57-78.

Savenije, H.H.G. (1996). The runoff coefficient as the key to moisture recycling. Journal of Hydrology 176: 219-225.

Savenije, H.H.G. (1997). Determination of evaporation from a catchment water balance at a monthly timescale. Hydrology and Earth System Sciences 1, pp 93-100.

Savenije, H.H.G. (1998). How do we feed a growing world population in a situation of water scarcity? 8th Stockholm Water Symposium, SIWI Report 3, Water: the key to socio-economic development and quality of life. Stockholm, pp 49-58.

Savenije, H.H.G. (1999). The role of green water in food production in sub-Saharan Africa. FAO internet-email Conference on "Water for Food in sub-Saharan Africa", 1999, Rome. URL: http://www.fao.org/waicent/FaoInfo/Agricult/AGL/AGLW/Africvis.htm.

Savenije, H.H.G. (2000). Water scarcity indicators; the deception of numbers. Phys. Chem. Earth, Vol. 25, No. 3, pp 199-204.

Savenije, H.H.G. (2003). Water resources. In: 'Principles of integrated water resources management,' Chapter 4. Unpublished lecture notes, WaterNet module IWRM0.1, University of Zimbabwe, Harare.

Savenije, H.H.G. (2004). The importance of interception and why we should delete the term evapotranspiration from our vocabulary. Hydrological Processes 18(8), pp 1507-1511.

Savenije, H.H.G. and Van der Zaag, P. (2002). Water as an economic good and water demand management: Paradigms with pitfalls. Water International, Vol. 27, No. 1, pp 98-104.

Savoury, A. (1988). Holistic Resource Management. Island Press, Covelo.

Schattauer, H., Tushabe, A.A. and Nalubega, M. (2001). Experiences with ecological sanitation in South Western Uganda. Paper presented during the 1st International Conference on Ecological Sanitation, 5-8 November, 2001, Nanning, China.

Schertenleib, R. (2001). Principles and implications of household centred approach in environmental sanitation, extended abstract, The first International Ecological Sanitation Conference, Nanning, China, URL: http://www.ias.unu.edu/proceedings/icibs/ecosan/schertenlieb.html

Schertenleib, R., Gujer W. (2000) On the path to new strategies in urban water management; EAWAG News 48e.

Schlesinger, W.H. and Hartley, A.E. (1992). A global budget for atmospheric NH_3. Biogeochemistry 15, pp 191-211.

Schlesinger, W.H. and Melack, J.M. (1981). Transport of organic carbon in the world's rivers. Tellus, 33, pp 172-187.

Schlesinger, W.H., (1997). Biogeochemistry: An analysis of global change. 2nd Edition, pp. 588. Academic Press Inc, San Diego California.

Schoemaker, P.J.H. (1991). When and how to use scenario planning: a heuristic approach with illustration. Journal of Forecasting, Vol. 10, pp 549-564.

Schonning, C., and Stenstrom, T.A. (2004). Guidelines for the safe use of urine and faeces in ecological sanitation systems. Report 2004-1, EcoSanres. Stockholm Environment Institute, Stockholm.

Schroeder, W. and Strongman, J. (1974). Adapting urban dynamics to Lowell. In: Mass, N. (ed.). Readings in urban dynamics, Vol. 1, Pegasus Communications, Walthan, MA.

Schulze R. E., Schmidt E. J. and Smithers J. C. (1992). SCS-SA User Manual PC Based SCS Design Flood Estimates for Small Catchments in Southern Africa, Report No. 40, Department of Agricultural Engineering University of Natal, Pietermaritzburg, South Africa.

Schulze, R.E. (1987). Hydrological science and hydrological practice. In: Hughes, D.A. and Stone, A.W. (eds.). Reflections as we approach the 1990s,. Proceedings of the Hydrological Sciences Symposium, Department of Geography, Rhodes University, Grahamstown, pp 1-19.

Schulze-Rettmer R. (1991). The simultaneous chemical precipitation of ammonium and phosphate in the form of magnesium-ammonium-phosphate. Wat. Sci. Tech., 23(4-6), pp 461-469.

Schumacher, E.F. (1973). Small is beautiful-A study of economics as if people mattered, Abacus, London.

Scobie, G.M. and St. Pierre, N.R. (1987). Economics of phosphorus fertiliser use on pastures. 2. Incorporating the residual effect. New Zealand Journal of Experimental Agriculture, 15, pp 445-451.

Senge, P. (1990). The fifth discipline. Dobleday Currency, New York.

Senge, P., Kleiner, A., Roberts, C., Ross, R. and Smith, B. (1994). The fifth discipline fieldbook. Dobleday Currency, New York.

Sharpley, A.N. (1983). Effect of soil properties on the kinetics of phosphorus desorption. Soil Sci. Soc. Am. J., 47, pp 805-809.

Sharpley, A.N. (1985a). Depth of surface soil-runoff interaction as affected by rainfall, soil slope and management. Soil Sci. Soc. Am. J., 49, pp 1010-1015.

Sharpley, A.N. (1985b). The selective erosion of plant nutrients in runoff. Soil Sci. Soc. Am. J., 49, 1527-1534.

Sharpley, A.N. (1991). Soil phosphorus extracted by iron-aluminium-oxide-impregnated filter paper. Soil Sci. Soc. Am. J., 55, pp 1038-1041.

Sharpley, A.N. and Smith, S.J. (1983). Distribution of phosphorus forms in virgin and cultivated soils and potential erosion losses. Soil Sci. Soc. Am. J., 47, pp 581-586.

Sharpley, A.N. and Syers, J.K. (1979). Phosphorus inputs into a stream draining an agricultural watershed: II. Amounts and relative significance of runoff types. Water, Air and Soil Pollut., 11, pp 417-428.

Sharpley, A.N. and Williams, J.R. (eds.). (1990). EPIC-Erosion/Productivity Impact Calculator. 1. Model documentation. USDA Technical Bull. 1768. U.S. Govt. Print. Office, Washington, D.C. 235 pp.

Sharpley, A.N., Hedley, M.J., Sibbesen, E., Hillbricht-Ilkowska, A., House, W.A. and Ryszkowski, L. (1995). Phosphorus transfers from terrestrial to aquatic ecosystems pp 171-199. In: Tiessen, H. (ed.). Phosphorus in the global environment: Transfers, cycles and management. Scientific Committee on Problems of the Environment, (SCOPE), Vol. 54, pp 452, John Wiley & Sons, Chichester.

Sharpley, A.N., Jones, C.A., Grey, C. and Cole, C.V. (1984). A simplified soil and plant phosphorus model II: Predication of labile, organic and sorbed phosphorus. Soil Sci. Soc. Am. J., 48, pp 805-809.

Sheldon, R.P. (1987). Industrial minerals with emphasis on phosphate rock. In: McLaren, H. and Skinner, D. (eds). Resources and World Development. J. Wiley and Son, New York.

Sheldrick, W.F. (1976). The role of the World Bank in helping to meet the fertiliser requirements of developing countries No 160.

Shiming, L. (2001). The utilisation of human excreta in Chinese agriculture and the challenge faced. Paper presented during the 1st International Conference on Ecological Sanitation, 5-8 November 2001, Nanning, China

Shuttleworth, W.J. (1997) Evaporation. In *Handbook of Hydrology* ed. Maidment, David R., McGraw-Hill, New York, USA. pp. 4.1-4.53.

Shuval, H. I., Gunnerson, C. G. and Julius, D. S. (1981). Night-soil composting, Vol. 10. In: Kalbermatten, J.M., Julius, D.S., Gunnerson, C.G. and Mara, D.D. (eds.). Appropriate technology for water supply and sanitation (12 Vols.), The World Bank, Washington DC.

Shuval, H.I., Adin, A., Fattal, B., Rawitz, E. and Yekutiel, P. (1986). Wastewater irrigation in developing countries: Health effects and echnical solutions. World Bank Technical Paper No. 51. The World Bank, Washington.

SI 274/2000. (2000). Water (Waste and effluent disposal) regulations of Zimbabwe. Satutorty Instrument 274 of 2000 Chapter 20:24. Government Printers, Harare.

Sibbesen, E. (1983). Phosphate soil tests and levels and their suitability to assess the phosphate status of soil. J. Sci. Food Agric., 34, pp 1368-1374.

Sibbesen, E. and Runge-Metzger, A. (1995). Phosphorus balance in European Agriculture: Status and policy options pp 43-57. In: Tiessen, H. (ed.). Phosphorus in the global environment: Transfers, cycles and management. Scientific Committee on Problems of the Environment, (SCOPE), Vol. 54, pp 452, John Wiley & Sons, Chichester.

Silverstone, S. (1993). Eating in: From the field to the kitchen in Biosphere 2. Synergetic Press, Santa Fe, New Mexico, pp 112. ISBN 1 88 248 04 8.

Simpson-Hebert, M. (2001). Ecological sanitation and urban sustainability. Paper presented during the 1st International Conference on Ecological Sanitation, 5-8 November, 2001, Nanning, China.

Sinclair, A.G. and Cornforth, I.S. (1984). A modification of the 'superchoice' phosphate maintenance model. New Zealand J. Exp. Agric., 12, pp 141-144.

Sissingh, H.A. (1971). Analytical technique of the PW method used for the assessment of the phosphate status of arable soils in the Netherlands. Plant and Soil, 34, pp 483-486.

Smaling, E. (1993). An agro-ecological framework for integrated nutrient management, with special reference to Kenya. Doctoral thesis. Wageningen Agricultural University. Wageningen. 250 pp.

Smaling, E. M. A., and Braun, A. R. (1996). Soil fertility research in Sub-Saharan Africa: New dimensions, new challenges. Communications in Soil Science and Plant Analysis Vol. 27 Nos. 3 and 4, pp 365-386.

Smaling, E.M.A., Stoorvogel, J.J., and Windmeijer, P.N. (1993). Calculating soil nutrient balances in Africa at different scales. II. District scale. Fertiliser Research, 35: pp 237-250.

Smeck, N.E. (1985). Phosphorus dynamics in soils and landscapes. Geoderma, 36, pp 185-199.

Smil, V. (1990). Nitrogen and phosphorus. In: Turner, B.L. (ed.). The Earth as transformed by human action. Cambridge University Press, pp 423-426.

Smil, V. (1991). Population growth and nitrogen: An exploration of a critical existential link. Population and Development Review 17(4): pp 569-601.

Smil, V. (1997). Global population and the nitrogen cycle. Scientific American, July 1997.

Smil,V. (1999). Long-range perspectives on inorganic fertilisers in global agriculture. Travis P. Hignett Lecture, International Centre for Soil Fertility and Agricultural Development (IFDC), Muscle Shoals, Alabama.

Smil,V. (2001). Enriching the Earth: Fritz Haber, Carl Bosch and the transformation of world food production. MIT Press, Cambridge, Massachusetts.

Smit, J. (1996). What would the world be Like in the 21st Century if Cities were nutritionally self-reliant? The prospect for urban agriculture, paper presented at a Global Roundtable in Marmaris, Turkey (UNDP & UNCHS) on April 20, 1996. URL: http://www.cityfarmer.org/ Acessed in July 2000.

Smit, J. (2000). Urban agriculture and biodiversity: Urbanisaion and diminishing biodiversity. Urban agriculture magazine, Vol. 1 No. 1, June 2000, pp 11-12.

Smit, J., and Nasr, J. (1992). Urban agriculture for sustainable cities: Using wastes and idle land and water bodies as resources. Environment and Urbanisation, Vol. 4, No. 2 pp 141-152, London.

Smith, D.W. and Tevera, D.S. (1997). Socio-economic context for the householder of urban agriculture in Harare, Zimbabwe, Geographical Journal of Zimbabwe No. 28, L.M. Zinyama (ed.), Geographical Association of Zimbabwe, Harare.

Smith, S.R. (1996). Agricultural recycling of sewage sludge and the environment. CAB International, Wallingford, Oxon.

Soderlund, R. and Svensson, B.H. (1976). The global nitrogen cycle. In: Svensson, B.H. and Soderlund, R. (eds.). Nitrogen, phosphorus and sulphur - Global cycles. SCOPE Report 7, Ecological Bulletins No.22, pp 23-73, Stockholm, Sweden.

Soil Map. (1979). Provisional soil map of Zimbabwe-Rhodesia, Edition 2, 1:1000000, compiled by Pedology and Soil Survey Section, Chemistry and Soil Research and Specialist Services, Published by the Surveyor-General, Salisbury, Zimbabwe-Rhodesia

Soltzberg, L. (1996). The dynamic environment: Computer models to accompany: consider a spherical cow. University Science Books, Sausalito, CA.

Somlyody, L. (1992). Water Quality 2000, Yearbook 1992-93, International Association on Water Quality, London, UK, pp 3-10.

Somlyody, L. (1995). Water quality management: Can we improve integration to face future problems? Wat. Sci. Tech., Vol. 31 No. 8, pp 249-259.

Sonesson, U. (1998). Systems analysis of waste management, the ORWARE model, transport and compost sub-models, Doctoral Thesis, Swedish University of Agricultural Sciences, Uppsala 1998, ISBN 91-576-5470-0.

Sowman, M. and Urquhart, P. (1998). A place called home: Environmental issues and low-cost housing. University of Cape Town Press, Cape Town.

Sparling, G.P. (1985). The soil biomass. In: Vaughan, D. and Malcolm, R.E. (eds.). Soil organic matter and biological activity. Developments in Plant and Soil Science, 18, pp. pp 223-263.

Spurway, J.K.R (1995). World soil resource Report, 80, pp 123-128 in ITC and FAO International Institute of aerospace survey and earth sciences.

Stangel, P.J. and von Uexkull, H.R. (1990). Regional food security; demographic and geographic implications In: Phosphorus requirements for sustainable agriculture in Asia and Oceania, IRRI, Manilla, Philippines. pp 21-43.

Staudenmann J., Schonborn A., Etnier C., (eds.). (1996). Recycling the resource: Ecological engineering for wastewater treatment. Proceedings of the Second International Conference on Ecological Engineering for Wastewater Treatment, Environmental Research Forum Vol 5-6, 479 pp, Transtec Publications Ltd., Switzerland.

Steen, P. (1998). Phosphorus recovery in the 21st century: Management of a non-renewable resource. Phosphorus and Potassium Journal, Issue No. 217.

STOA. (2001). Ecological foot-printing. Directorate General for Research, Division Industry, Research, Energy, Environment, and Scientific and Technological Options Assessment (STOA), commissioned by the European Parliament, URL: http://www.europarl.eu.int/stoa/publi/pdf/00-09-03_en.pdf, accessed in August 2002.

Stocking, M. (1986). The cost of soil erosion in Zimbabwe in terms of the loss of three major nutrients. Consultants' Working Paper No. 3, Soil Conservation Programme, Land and Water Development Division, FAO, Rome.

Stoorvogel, J. and Smaling. E.M.A. (1990). Assessment of soil nutrient depletion in Sub-Saharan Africa. Report 28. Wageningen: Winand Staring Centre.

Stoorvogel, J. J., Smaling, E. M. A. and Janssen, B. H. (1993). Calculating soil nutrient balances in Africa at different scales. Fertiliser Research. No. 35: pp 227-335.

Stowasser, W. (1986). Phosphate rock: Mineral facts and problems. United States Bureau of Mines, Bulletin 675, p585, Washington, DC.

Stowasser, W. and Fantel, R. (1985). The outlook for the United States phosphate rock industry and its place in the world. Paper presented at the SME-AIME Annual Meeting, New York, February. 24-28, 1985, Society of Mining Engineers Reprint No. 85-116.

Stumm, W. (1973). The acceleration of the hydro-geochemical cycling of phosphorus. Water Resources, Vol. 17, pp 131-144.

Stumm, W. (1980). Water: an endangered ecosystem. Ambio 15, pp 201-207.

Stumm, W. and Morgan, J.J. (1981). Aquatic Chemistry. 2nd edtion. John Wiley and Sons New York, 780 pp.

Sturm, A., Wackernagel, M. and Muller, K. (2000). The winners and losers in global competition: Why Eco-efficiency Reinforces Competitiveness: A Study of 44 Nations, Ruegger, Chur Zurich.

Suess, M.J. (ed.). (1985). Solid waste management: Selected topics. WHO regional office for Europe, Copenhagen.

Suess, M.J. (ed.). (1995). Examination of water of pollution control. Vol. 2, p 310. ISBN 0-08-025255-9

Sweeney, R.E., Liu, K.K., and Kaplan, I.R. (1977). Oceanic nitrogen isotopes and their uses in determining the sources of sedimentary nitrogen. In Robinson, B. W. (ed.). Stable isotopes in the earth sciences, DSIR Bulletin, 220, 9-26, Wellington, New Zealand Dept. of Scientific and Industrial Research.

Syers, J.K. and Curtin, D. (1989). Inorganic reactions controlling phosphorus cycling. In: Tiessen, H. (ed.). Phosphorus cycles in terrestrial and aquatic ecosystems. Regional workshop 1: Europe. SCOPE/UNEP Proceedings, University of Saskatchewan, Saskatoon, Canada. pp 17-29.

Syers, J.K. Williams, J.D.H., Cambell, A.S. and Walker, T.W. (1967). The significance of apatite inclusions in soil phosphorus studies. Soil Sci. Soc. Am. Proc., 31, pp 752-756.

Takahashi, T., Broecker, W. S., and Bainbridge, A. E. (1981). Supplement to the alkalinity and total carbon dioxide concentration in the world oceans pp159-200. In: Bolin, B. (ed.). Carbon cycle modelling, SCOPE Report No. 16. Wiley, Chichester.

Tandon, H. L. S. (1992). Fertilisers, organic manures, recyclable wastes and biofertilisers: Components of integrated plant nutrition. Fertiliser Development and Consultation Organisation, New Delhi.

Tandon, H. L. S. (1998). Use of external inputs and the state of efficiency of plant nutrient supplies in irrigated cropping systems in Uttar Pradesh, India. In Proceedings of the IFPRI/FAO workshop on soil fertility, plant nutrient management, and sustainable agriculture: The future through 2020. Gruhn, P., Goletti, F., and Roy, R. N (eds.). International Food Policy Research

Institute (IFPRI) and Food and Agriculture Organisation of the United Nations (FAO). Washington, DC and Rome.

Tandon, H.L.S. (1987). Phosphorus research and agricultural production in India. Fertiliser Development and Consultation Organisation, New Delhi. 160 pp.

Tanner, P.D. (1984). Effects of the method of placement of granular compound fertiliser on growth, potassium and phosphorus uptake, and yield of maize. Zimbabwe Agric. J., Vol. 81 (1), pp. 3-8.

Taras, M.J. (ed.). (1981). The quest for pure water. 2nd edition, Vol II. American Water Works Association.

Taylor, P. and Mudege, N.R. (1997). Urban Sanitation in Zimbabwe and the relation to environmental pollution. Institute of Water and Sanitation Development, Harare.

Taylor, S. R. (1964). Abundance of chemical elements in the continental crust: a new table, Geochim. Cosmochim. Acta, 28, pp 1273-1285.

Tchobanoglous, G., Theisen, H. and Vigil S.A. (1993). Integrated solid waste management: Engineering priciples and management issues. McGraw Hill International Editions, New York.

Teboh, J. F. (1995). Phosphate rock as a soil amendment: Who should bear the cost? In Use of phosphate rock for sustainable agriculture in West Africa, Gerner, H. and Mokwunye, A. U. (eds). Miscellaneous Fertiliser Studies No. 11. Lome: International Fertiliser Development Centre (IFDC) Africa.

Terpstra P.M.J. (1998). Sustainable water usage systems; Models for the sustainable utilisation of domestic water in urban areas. In International WIMEK Congress on Options for Closed Water Systems, Sustainable Water Management, March 11-13, Wageningen, The Netherlands.

Terrefe, A. and Edstrom, G. (1999). Ethiopia: Economical ecological sanitation (Ecosan). In Ecological sanitation: Exploring optionbs for a better future, Faul-Doyle, R.C. Workshop Report October 2-6, 1999. Mvuramanzi Trust/UNICEF/WHO/AFRO, Harare.

Terrefe, A. and Edstrom, G. (2001). Valuable use of urine, aeces, household waste and some grey water or Ecosan – ecology, economy, sanitation. Proceedings of the International Symposium, 30-31 October 2000, Bonn. GTZ, Eschborn.

Tevera, D.S. (1989). Waste-recycling as a livelihood in the informal sector: The case of Harare's Teviotdale dump scavengers. Paper presented at the Geographical Association of Zimbabwe Conference, Harare.

Tevera, D.S. (1991). Solid waste disposal in Harare and its effects on the environment: Some preliminary observations. The Zimbabwe Science News, Vol 25 Nos 1-3, pp 9-13.

The Catalyst. (1993). Newsletter of the Halifax Project, No. 1, April, 1993.

The Ecologist. (1993). Whose common future? New Society Publishers, Philadelphia pp 3-20.

Thompson, L. M., and Troeh, F. R. (1973). Soils and soil fertility. 3rd edition McGraw-Hill, New York.

Thompson, J.G. (1972). What is a vlei? Rhodesia Agricultural Journal Technical Bulletin, No. 15, pp 153-154.

Thornthwaite, C.W., & Mather, J.R. (1955). The water balance. Publications in Climatology, Drexel Institute of Technology, Laboratory of Climatoloty, New Jersey 8, pp 1-104.

Thornthwaite, C.W., & Mather, J.R. (1957) Instructions and tables for computing potential evapotranspiration and the water balance. Publications in Climatology, Drexel Institute of Technology, Laboratory of Climatology, New Jersey 10: 185-195.

Thornton J.A., and Nduku W.K. (eds.). (1982). Lake McIlwane: The Eutrophication and Recovery of a Tropical Lake, Dr W. Junk, The Hague.

Thornton, J A. (1980). A comparison of the summer phosphorous loading to three Zimbabwean water-supply reservoirs of varying trophic states. Water SA Vol. 64, pp 163-170.

Thornton, K.W., Kimmel, B.L. and Payne, F.E. (1990). Reservoir limnology: ecological perspectives. John Wiley & Sons, New York.

Tiessen, H. (1995). Introduction and synthesis pp 1-6. In: Tiessen, H. (ed.). Phosphorus in the global environment: Transfers, cycles and management. Scientific Committee on Problems of the Environment, (SCOPE), Vol. 54, pp 452, John Wiley & Sons, Chichester.

Tiessen, H., Stewart, J.W.B. and Cole, C.V. (1984). Pathways of phosphorus transformations in soil of differing pedogenesis. Soil Sci. Soc. Am. J., 48, pp 853-858.

Tiessen, H., Stewart, J.W.B. and Moir, J.O. (1983). Changes in organic and inorganic phosphorus fractions of two soils during 60 to 90 years of cultivation. Journal of Soil Science, 34, pp 815-823.

Tilth People. (1982). The future is abundant: A guide to sustainable agriculture, Arlington, WA.

Tisdale, S. L., and Nelson, W. L. (1975). Soil fertility and fertilisers. 3rd edition Macmillan New York.

Tjallingii, S.P. (1995). Ecopolis: Strategies for sound urban development. Backhuys Publishers, Leiden.

Todd, N.J. and Todd, J. (1994). From eco-cities to living machines- Principles of ecological design. North Atlantic Books. Berkeley, California.

Toerien, D.F. (1977). A review of eutrophication and guidelines for its control in South Africa. Special Report WAT. 48, National Institute for Water Research, CSIR, Pretoria.

Toerien, D.F., Hyman, K.L. and Bruwer, M.J. (1975). A preliminary trophic status classification of some South African impoundments. Water SA 1, pp 15-23.

Tomlinson, R.W. and Wurzel, P. (1977). Aspects of site and situation. In: Kay, G. and Smout, M., (eds.). Salisbury: A Geographical survey of the capital of Rhodesia. Hodder and Stoughton, London.

Troug, E. (1930). The determination of readily available phosphorus in soils. J. Am. Soc. Agron., 22, 874-882.

Tunney, H., Carton, O.T., Brookes, P. C. and Johnston, A.E. (eds.). (1997). Phosphorus loss from soil to water 467 pp. CAB International, Wallingford, UK.

Uhl, C. (1987). Factors controlling succession following slash-and-burn agriculture in Amazonia. J. Ecol., 75, pp 377-407.

UNAIDS. (2000). Aids epidemic update. December 2000. Joint United Nations Programme for HIV/AIDS Geneva.

UNCHS. (1989). Solid waste management in low-income housing projects: The scope for community participation. United Nations Centre for Human Settlements (UNCHS). Nairobi. ISBN 92-1-131015-6.

UNDP. (1996). Urban Agriculture: Food, jobs and sustainable cities, United Nations Development Programme, Publication Series for Habitat II, Volume 1, 302 pp, New York.

UNDP. (2001). Human development report 2001; making new technologies work for human development. United Nations Development Programme. Oxford University Press, New York.

UNDP. (2003). Indicators for monitoring the millennium development goals: Definitions, rationale, concepts and sources. Draft, July 2003, United Nations, New York.

UNDSD. (1996). Indicators of sustainable development: Guidelines and methodologies. United Nations Division for Sustainable Development. (UDSD). United Nations, New York. URL: www.un.org/esa/sustdev/natlinfo/indicators/isd.htm, accessed in October 2003.

UNEP, WHO, HABITAT, and WSSCC. (2004). Guidelines on municipal wastewater management. The United Nations Environmennt Programme (UNEP) in collaboration with World Health Organisation of the United Nations (WHO), the United Nations Human Settlements Programme (UN-HABITAT) and the Water Supply and Sanitation Collaborative Council (WSSCC). URL: http://www.sanicon.net

UNEP. (1994). The pollution of lakes and reservoirs. UNEP Environment Series No. 12. Nairobi, Kenya, pp 35.

UNEP. (2002). Melbourne principles for sustainable cities. Integrative Management series No 1, United nations environmental Programme, Division of Technology, Industry and Economics, International Environmental Technology Centre (UNEP-DTIE-IETC). Osaka.

UNIDO. (1998). Fertiliser Manual. United Nations Industrial Development Organisation (UNIDO) and the International Fertiliser Development Center (IFDC). Kluwer Academic Publishers, The Netherlands.

United Nations. (2002). Report of the World Summit on Sustainable Development (WSSD). Johannesburg, South Africa, August-September 2002. United Nations, New York.

Urban Councils Act. (1996). Urban Councils Act of Zimbabwe Chapter 29:15. Government Printers, Harare.

USDA. (1989). The second RCA appraisal: Soil, water and related resource on non-federal land in the United States, analysis of conditions and trends. United States Department of Agriculture (USDA).Washington, DC.

USDoA. (1986). United States Department of Agriculture. Soil Conservation Service: Urban Hydrology for Small Watersheds. Technical Release 55. National Technical Information Service, Springfield, VA.

USEPA. (1983). Phosphorus, all forms. Method 365.1 (Colorimetric, Automated, Ascorbic Acid), pp 365-1.1 365-1.7. In: Methods for Chemical Analysis of Water and Wastes, EPA-600/ 4-79-020. United Sates Environmental Protection Agency (USEPA), Cincinnati, Ohio.

USEPA. (1984). Environmental regulation and technology: Use and disposal of municipal wastewater sludge. United States Environmental Protection Agency (USEPA), Washington, DC.

Van Bennekom, A.J., and Salomons, W. (1981). Pathways of organic nutrients and organic matter from land to ocean through rivers. In Martin, J.M., Burton, J.D., Eisma, D., (eds.). River input to the ocean system: UNESCO-UNEP, SCOR Workshop, Rome, 1979, pp 33-51.

Van Boheemen, P.J.M. (1987). Extent, effects and tackling of a regional manure surplus, a case-study for Dutch region. In: Van der Meer, H.G. (eds.). Animal manure on grassland and fodder crops. Martinus Nijhoff, Dordrecht. pp 175-193.

Van der Pol, F. (1992). Soil mining: An unseen contributor to farm income in southern Mali. Bulletin 325. KIT, Amsterdam.

Van der Ryn, S. (1995). The toilet papers. Recycling waste and conserving water. Ecological Design Press, Sausalito, Carlifornia.

Van der Ryn, S. and Calthorpe. (1991). Sustainable communities: a new design synthesis for cities, suburbs, and towns. Sierra Club Books, San Francisco.

Van der Ryn, S. and Cowan, S. (1996). Ecological design, pp 203. Island Press, Washington D.C.

Van der Vleuten-Balkema, A.J. (2003). Sustainable wastewater treatment: developing a methodology and selecting promising systems. PhD Thesis, Technical University Eindhoven, The Netherlands. Eindhoven, University Press, ISBN 90-386-1805-0.

Van Noordwijk, M., De Willigen, P., Ehlert, P.A.L. and Chardon, W.J. (1990). A simple model of P uptake by crops as a possible basis for P fertiliser recommendations. Netherlands J. Agric. Sci., 38, pp 317-332.

Van Raij, B., Quaggio, J.A. and da Silva, N.M. (1986). Extraction of phosphorus, potassium, calcium and magnesium from soils by an ion-exchange resin procedure. Commun. Soil Sci. Plant Anal., 17, pp 547-566.

Van Ruiten, L. (1998). Phosphate recovery from animal manure, the possibilities. A report commissioned A report by Centre Europeen d' Etudes des Polyphophates (CEEP). URL: http://www.nhm.ac.uk/mineralogy/phos/manure.htm

Van Starkenburg, W. and Rijs, G.B.J., 1989. Phosphate in Sewage and Sewage Treatment. In: Tiessen, H. (ed.). Phosphorus cycles in terrestrial and aquatic ecosystems. Regional Workshop 1: Europe. SCOPE/UNEP Proceedings, University of Saskatchewan, Saskatoon, Canada, pp. 221-231.

Van Wazer, J.R. (1958). Phosphorus and its compounds. 1: Chemistry. Interscience, New York. 954 pp.

Van Wazer, R. (1973). The compounds of phosphorus pp 169-178. In: Griffith, E., Beeton, A., Spencer, J., and Mitchell, D. (eds.). Environmental phosphorus Handbook. Wiley, New York.

Vark, W. and Van der Lee, H. (1989). Soil and plant analysis: plant analysis procedures. Department of Soil Science and Plant Nutrition, Wageningen Agriculture University, Wageningen.

Vaughan, D. and Ord, B.G. (1985). Soil organic matter - a perspective on its nature, extraction, turnover and role in soil fertility. In: Vaughan, D. and Malcolm, R.E. (eds.). Soil organic matter and biological activity. Developments in Plant and Soil Science, 18, pp 1-35.

Ventana Systems. (1995). Vensim user's guide, Version 1.62. Ventana Systems, Inc., Belmont, MA.

Ventana Systems. (1996). Vensim personal learning edition: User's guide, Version 1.62. Ventana Systems, Inc., Belmont, MA.

Ventura, S.J. and Kim K., (1993), Modelling urban non-point source pollution with a geographic information system, Water Resources Bulletin, 29(2), pp 189-198.

Viessman, W. and Lewis, G. L. (1996). Introduction to hydrology. 4th Harper Collins College Publishers, New York.

Vinneras, B. (2002). Possibilities for sustainable nutrient recycling by faecal separation combined with urine diversion. Doctoral thesis Agraria 353, Department of Agricultural Engineering, Swedish University of Agricultural Sciences. Uppsala.

Vinneras, B. and Jonsson, H. (2003). Separation of faeces combined with urine diversion: function and efficiency. Paper presented during the 2nd International Symposium on Ecological Sanitation, 7-11 April, 2003, Lubeck, Baltic Sea, Germany.

Vitousek, P.M., Aber, J., Howarth, R.W., Likens, G.E., Matson, P.A., Schindler, D.W., Schlesinger, W.H. and Tilman, G.D. (1998). Human alteration of the global nitrogen cycle: Causes and consequences. Issues in Ecology, No. 1, Ecological Society of America.

Vollenweider, R.A. (1968). Scientific fundamentals of eutrophication of lakes and flowing waters, with particular reference to nitrogen and phosphorus as factors in eutrophication. OECD Report DAS/CSI/68.27.

Vollenweider, R.A. (1981). Eutrophication - A global problem. Water Qual. Bull., 6, pp 59-89.

Von Bertalanffy, L. (1975). Perspectives on general systems theory, Braziller.

Von Peter, A. (1980). Fertiliser requirements in developing countries No 188.

Von Uexkull, H.R. and Mutert, E. (1995). Rehabilitation of anthropic savanna pp 149-154. In: Tiessen, H. (ed.). Phosphorus in the global environment: Transfers, cycles and management. Scientific Committee on Problems of the Environment, (SCOPE), Vol. 54, pp 452, John Wiley & Sons, Chichester.

Von Weizsacker, E. U., Lovins, A., and Hunter-Lovins, L. (1997). Factor four: Doubling wealth - Halving resource use. A report to the Club of Rome. Earthscan, London.

Wackernagel, M., Rees W., (1996). Our ecological footprint: Reducing human impact on the earth. The new Catalyst, Bioregional series, 160 pp, New Society Publishers, Canada.

Wackernagel, M. (2001). Advancing sustainable resource management: Using ecological footprint analysis for problem formulation. Policy Development, and Communication, Prepared for DGXI, European Commission, Redefining Progress, Oakland.

Wackernagel, M., Monfreda, C. and Deumling, D. (2002b). Ecological footprint of nations November 2002 update: How much do they use? How much nature do they have? Sustainability issue brief, November, Redefining Proress (RP), Oakland, CA.

Wackernagel, M., Schulz, N.B., Deumling, D., Linares, A.C., Jenkins, M., Kapos, V., Monfreda, C., Loh, J., Myers, N., Norgaard, R. and Randers, J. (2002a). Tracking the ecological overshoot of the human economy, Proc. Natl. Acad. Sci. USA, Vol. 99, Issue 14, pp 9266-9271.

Walbridge, M.R. (1991). Phosphorus availability in acid organic soils of the lower North Carolina coastal plain. Ecology, 72, pp 2083-2100.

Walker, W.W. (1983), Significance of eutrophication in water supply reservoirs. J. American Water works Assoc., Vol. 75, pp 38-42.

Walker, J.C.G. (1991). Numerical adventures with geochemical cycles. Oxford University Press,Oxford, pp 192.

Walker, T.W. and Syers, J.K. (1976). The fate of phosphorus during pedogenesis. Geoderma, 15, pp 1-19.

Walling, D.E. and Webb, B.W. (1983). Patterns of sediment yield. In: Gregory K.J. (ed.). Background to Palaeohydrology, John Wiley, pp 69-100.

Wang, K.M., Strecker E.W., and Stenstrom M.F. (1997). GIS to estimate storm-water pollutant mass loadings, Journal of Environmental Eng ASCE, Vol. 123 No. 8, pp 737-745.

Ward, J.C., O'Connor, K.F. and Wei-Bin, G. (1990). Phosphorus losses through transfer, runoff, and soil erosion. In: IRRI (ed.), Phosphorus Requirements for Sustainable Agriculture in Asia and Oceania. Proceedings of a symposium, March 6-10, 1989. IRRI, Los Banos, Philippines. pp 167-182.

Water Act. (1998). Water Act of Zimbabwe Chapter 20:24. Government Printers, Harare.

WECD. (1987). Our common future. World Commission on Environment and Development, Oxford University Press, Oxford, UK.

Weibel S.R., Anderson R.J. and Woodward R.L. (1964). Urban land runoff as a factor in stream pollution, J. Water Pollution Control Fed., 36(7), pp 914-924.

Welch, E.B. and Lindell, T. (1980). Ecological effects of waste water. Cambridge University Press, London.

Wendland, C. and Oldenburg, M. (2003). Operation experiences with a source separating project. Paper presented during the 2nd International Symposium on Ecological Sanitation, 7-11 April, 2003, Lubeck, Baltic Sea, Germany.

Wenhua, L. and Rusong, W. (2001). System consideration of eco-sanitation in China. Paper presented during the 1st International Conference on Ecological Sanitation, 5-8 November, 2001, Nanning, China.

Werner, C. (2001). Ecosan: A holistic approach to material flow management in sanitation. Ecosan - closing the loop in wastewater management and sanitation. Proceedings of the International Symposium, 30-31 October 2000, Bonn. GTZ, Eschborn.

Werner, C., Mang, H.P. and Kessen, V. (2003). Key activities, services and current pilot projects of the international ecosan programme of GTZ. Paper presented during the 2nd International Symposium on Ecological Sanitation, 7-11 April, 2003, Lubeck, Baltic Sea, Germany.

Wetzel, R.G. (1983). Limnology. Saunders, New York, London, 767 pp.

Whelans, Maunsell, and Palmer. (1993). Water sensitive urban (residential) design guidelines for the Perth Metropolitan Region. Prepared for the Department of Planning and Water Authority of Western Australia, Perth.

White, R.E. (1979). Introduction to the principles and practice of soil science, Blackwell Scientific Publications, London.

Whitlow, R. (1988). Land degradation in Zimbabwe. Department of Natural Resources. Harare.

Whittaker, R. H., and Likens, G. E. (1975). The biosphere and man pp 305-328. In: Lieth, H. (ed.). Primary productivity of the biosphere, 14. Springer-Verlag, Berlin.

WHO. (1989). Health guidelines for the use of wastewater in agriculture and aquaculture. Technical Report Series No 778 of a World Health Organisation (WHO) Scientific Group, Geneva.

WHO. (1991). Urban solid waste management. IRIS, for World Health Organisation (WHO), Firenze.

WHO. (1992). Our planet, our health. Report of the WHO Commission on health and the environment. World Health Organisation, Geneva.

WHO. (1996a). Creating healthy cities in the 21st Century. World Helath Organisation, Geneva.

WHO. (1996b). The problem of sanitation. Report of WHO Collaborative Council Working Group on Promotion of Sanitation, fifth draft, Geneva, December 1996.

WHO. (1997). Health and environment in sustainable development. Five years after the Earth Summit. World Health Organisation. Geneva.

Wild, A. (ed.). (1988). Russells soil conditions and Plant growth. 11th Edition. Longman Scientific and Technical, London.

Wild, H. and Barbosa, L.A., (1967). Flora zambesiaca supplement; vegetation map of the Flora Zambesiaca area, M.O. Collins, Harare.

Wildung, R.E., Schmidt, R.L. and Gahler, A.R. (1974). The phosphorus status of eutrophic lake sediments as related to changes in limnological conditions - total, inorganic, and organic phosphorus. J. Environ. Qual., 3, pp 133-138.

Winblad, U. (1996). Towards ecological approach to sanitation, Proc. Of Internat, Toilet Symp., Toyama, October 9, pp 5-6.

Winblad, U. (1997). Ecologiocal sanitation: A global overview pp 3-4. In: Drangert, J.O., Bew, J. and Winblad, U. (eds). Ecological alternatives in sanitation, Publications on Water Resources: No. 9. Department for Natural Reasources and the Environment, Sida, Stockholm.

Winblad, U. (2000). Sanitation without pollution. In: Chorus, I., Ringelband, U., Schlag, G. and Schmoll, O. (eds). Water, sanitation and health, IWA Publication, London.

Winblad, U. and Kilama, W. (1985). Sanitation without water. Revised and enlarged edition, 161 pp, Macmillan Press, London.

Winograd, M. (1997). Vertical and horizontal linkages in the context of indicators of sustainable developement. In: Moldan, B. and Billharz, S. (eds). SCOPE 58, Sustainability Indicators: report of the project on Indicators for Sustainable Development 440pp, Wiley, Chichester.

Withers, P.J.A. (1999). Phosphate and potash fertiliser recommendations for cereals: Current issues and future needs. The Home-Grown Cereals Authority (HGCA), Report Review No. 40. Winchester, Hampshire, UK.

Withers, P.J.A., Edwards, A.C. and Foy, R.H. (2001). Phosphorus cycling in UK agriculture and implications for phosphorus loss from soil. Soil Use and Management, Vol 17 No. 3, pp 139-149.

WMO. (1997). The world's water: Is there enough? World Meteorological Office (WMO)-No. 857, Paris. ISBN 92-63-10857-9.

Wollast, R. (1983). Interactions in estuaries and coastal waters. In Bolin, B. and Cook, R.B. (eds.). Major biogeochemical cycles and their interactions, Scientific Committee on Problems of the Environment, (SCOPE), Vol. 21. John Wiley & Sons, Chichester.

Wollast, R. (1992). Interactions in estuaries and coastal waters. In: Butcher, S.S., Charlson, R.J. Orians, G.H. and Wolfe, G.V. (eds.). Global biogeochemical cycles. Academic Press, London, UK.

World Bank and FAO. (1996). Recapitalisation of soil productivity in sub-Saharan Africa. Washington, D.C.: World Bank.

World Bank. (1996). Natural resource degradation in sub-Saharan Africa. Restoration of soil fertility. A concept note and action plan. Washington, D.C.: Africa Region, World Bank.

WRI, UNEP and UNDP. (1992). World resources 1992-93: A guide to global environments: Toward sustainable development. World Resources Institute (WRI), United Nations Environment Programme (UNEP), and United Nations Development Programme (UNDP). Oxford University Press, New York.

WSSCC. (2000). Vision 21: A shared vision for hygiene, sanitation and water supply and a framework for action, 62 pp. Water Supply and Sanitation Collaborative Council (WSSCC).

WWF. (2002). Living Planet Report 2002. World-Wide Fund (WWF) for Nature International, UNEP World Conservation Monitoring Centre, Redefining Progress, Centre for Sustainability Studies, Gland, Switzerland.

Yeung, Y. M. (1993). Urban agriculture research in East and Southeast Asia: Record, capacities and opportunities. Cities Feeding People Report 6. International Development Research Centre (IDRC), Ottawa.

Yi-Zhang, C. (1999). Case study: urban agriculture in Shanghai. GATE Technology and Development 2 (April-June): pp 18-19.

Young, D.D. (1985). Reuse via rivers for water supply. In: Resue of sewage effluent, proceedings of the International Symposium organised by the Institution of Civil Engineers, 30-31 October, 1984, London. Thomas Telford, London.

Young, R.A., Onstad, C.A., Bosch, D.D. and Anderson, W.P. (1989). AGNPS: A non-point-source pollution model for evaluating agricultural watersheds. J. Soil Water Conserv., 44, pp 168-173.

Young, W.J., Farley, T.F. and Davis, J.R. (1995). Nutrient management at the catchment scale using a decision support system. Wat. Sci. Tech. 32(5-6), pp 277-282.

Zanamwe, L. (1997). Population growth and land use in the Manyame catchment area. In: Moyo, N.A.G. (ed.). Lake Chivero: A polluted Lake, University of Zimbabwe Publications, Harare.

Zaranyika, M.F. (1997). Sources and levels of pollution along Mukuvisi River: A review. In: Moyo, N.A.G. (ed.). Lake Chivero: A polluted Lake, University of Zimbabwe Publications, Harare.

Zata, L. (1996). Solid waste management in Harare. Unpublished lecture notes, Water and public health engineering, deopartment of Civil Engineering, University of Zimbabwe, Harare.

Zeeman, G. (1991). Mesophilic and psychrophilic digestion of liquid manure. PhD thesis, Wageningen University, Wageningen, The Netherlands.

Zeeman, G. and Lettinga, G. (1998). The role of anerobic digestion of domestic sewage in closing the water and nutrient cycle at community level. Wat. Sci. Tech. Vol. 39, No. 5 pp 187-194.

Zimbabwe National Water Authority (ZINWA) Act. (1998). Chapter 20:25, Printed by the Government Printer, Harare.

Appendix A

Regional food, fibre and P-fertiliser balances

Table A-1 Regional population estimates for 2000

Region	Total		Rural		Urban	
	'000'	As %of total	'000'	%	'000'	%
Africa	793626	13	498400	63	295227	37
Asia	2389901	39	1478025	62	911882	38
China	1282437	21	818798	64	463640	36
Europe	727304	12	193242	27	534062	73
North America	314113	5	71115	23	242999	77
Oceania	30520	1	7911	26	22608	74
South America	345739	6	70476	20	275263	80
Central America & Caribbean	173070	3	56989	33	116077	67
Total or average	**6056710**	**100**	**3194956**	**53**	**2861758**	**47**

Data source: FAO (2001)

Table A-2 Regional land-use categories in year 2000

	Total Area	Land Area	Permanent Crops	%^{a}	Permanent Pasture	%	Agricultural Area	%	Arable land	%
					Area is in Gm^2					
C. America & Caribbean	2715	2648	63	2.3	989	37.3	1417	53.5	365	13.8
China	9598	9327	115	1.2	4000	42.9	5487	58.8	1371	14.7
Africa	30312	29633	261	0.9	8997	30.4	11061	37.3	1803	6.1
Asia	22271	21655	490	2.2	7100	32.8	11234	51.9	3644	16.8
Europe	22976	22602	170	0.7	1828	8.1	4895	21.7	2896	12.8
North America	19600	18380	22	0.1	2630	14.3	4876	26.5	2224	12.1
Oceania	8564	8491	32	0.4	4195	49.4	4751	56.0	524	6.2
South America	17866	17529	140	0.8	5140	29.3	6394	36.5	1115	6.4
Global total	133902	130266	1293	1.0	34879	26.8	50115	38.5	13942	10.7

a)   All percentages units are % of land area i.e. column number three.
Data source: FAO (2001)

Notes:
Definitions used by reporting countries vary considerably and items classified under the same category often relate to greatly differing kinds of land.   FAO (2001) definitions of land use (land cover) categories are as follows:
Total Area: the total area of the country, including area under inland water bodies.   Data in this category are obtained mainly from the United Nations Statistical Division, New York.   Possible variations in the data may be due to updating and revisions of the country data and not necessarily to any change of area.
Land Area: total area excluding area under inland water bodies.   The definition of inland water bodies generally includes major rivers and lakes.
Agricultural Area: the sum of area under arable land and permanent crops and permanent pastures.

Arable and Permanent Crops: Arable land and land under permanent crops shows the sum of area under arable land and permanent crops.

Arable Land: land under temporary crops (double-cropped areas are counted only once), temporary meadows for mowing or pasture, land under market and kitchen gardens and land temporarily fallow (less than five years). The abandoned land resulting from shifting cultivation is not included in this category. Data for "Arable land" are not meant to indicate the amount of land that is potentially cultivable.

Permanent Crops: land cultivated with crops that occupy the land for long periods and need not be replanted after each harvest, such as cocoa, coffee and rubber; this category includes land under flowering shrubs, fruit trees, nut trees and vines, but excludes land under trees grown for wood or timber.

Permanent Pasture: land used permanently (five years or more) for herbaceous forage crops, either cultivated or growing wild (wild prairie or grazing land). The dividing line between this category and the category "Forests and woodland"; is rather indefinite, especially in the case of shrubs, savannah, etc., which may have been reported under either of these two categories.

Forests and Woodland: land under natural or planted stands of trees, whether productive or not. This category includes land from which forests have been cleared but that will be reforested in the foreseeable future, but it excludes woodland or forest used only for recreation purposes. The question of shrub land, savannah, etc. raises the same problem as in the category "Permanent meadows and pastures".

Non arable and non-permanent crops: from 1995 this element includes any other land not specifically listed under Arable land and Permanent crops i.e.: permanent pastures, forests and woodland, built on areas, roads, barren lands, etc.

For: *Greenland:* "Total area" refers to area free from ice. *Russian Federation:* Data on total area exclude the portion of land under the White Sea and the Azov Sea.

Table A-3 Regions and countries used in inter-regional P-transfer analysis

Africa	Africa	Africa	Asia	Asia	C. America & Caribbean	Europe	Europe	Oceania	Oceania	South America
Algeria	Kenya	Tanzania	Jordan	United Arab Em	Mexico	Albania	Latvia	Amer Samoa	Tokelau	Argentina
Angola	Lesotho	Togo	Kazakhstan	Uzbekistan	Martinique	Andorra	Liechtensten	Australia	Tonga	Bolivia
Benin	Liberia	Tunisia	Korea D P Rp	Viet Nam	Montserrat	Austria	Lithuania	Canton Is	Tuvalu	Brazil
Botswana	Libya	Uganda	Korea Rep	Yemen	Nethantilles	Bel-lux	Luxembourg	Christmas Is	Us Minor Is	Chile
Br Ind Oc Tr	Madagascar	W. Sahara	Kuwait	**C. America & Caribbean**	Nicaragua	Belarus	Macedonia	Cocos Is	Vanuatu	Colombia
Burkina Faso	Malawi	Zambia	Kyrgyzstan	Anguilla	Panama	Belgium	Malta	Cook Is	Wake Is	Ecuador
Burundi	Mali	Zimbabwe	Laos	Antigua Barb	Puerto Rico	Bosnia	Moldova Rep	Fiji Islands	Wallis Fut I	Falkland Is
Cameroon	Mauritania	**Asia**	Lebanon	Aruba	St Kitts Nev	Bulgaria	Monaco	Fr Polynesia		Fr Guiana
Cape Verde	Mauritius	Afghanistan	Malaysia	Bahamas	St Lucia	Channel Is	Netherlands	Guam		Guyana
Cent Afr Rep	Mayotte	Armenia	Maldives	Barbados	St Vincent	Croatia	Norway	Johnston Is		Paraguay
Chad	Morocco	Azerbaijan	Mongolia	Belize	Trinidad Tobago	Czech Rep	Poland	Kiribati		Peru
Comoros	Mozambique	Bahrain	Myanmar	Br Virgin Is	Turks Caicos	Czechoslovak	Portugal	Marshall Is		S. Georgia
Congo, D R	Namibia	Bangladesh	Nepal	Cayman Is	US Virgin Is	Denmark	Romania	Micronesia		Suriname
Congo, Rep	Niger	Bhutan	Oman	Costa Rica	**North America**	Estonia	Russian Fed	Midway Is		Uruguay
Cote de voire	Nigeria	Brunei	Pakistan	Cuba	Canada	Faeroe Is	San Marino	N Marianas		Venezuela
Djibouti	Reunion	Cambodia	Philippines	Dominica	USA	Finland	Serbia-monte	Nauru		
Egypt	Rwanda	Cyprus	Qatar	Dominican Rp	Bermuda	France	Slovakia	New Zealand		
Eq Guinea	Sao Tome Prn	Gaza Strip	Saudi Arabia	El Salvador	Greenland	Germany	Slovenia	Newcaledonia		
Eritrea	Senegal	Georgia	Singapore	Grenada	St Pier Mq	Gibraltar	Spain	Niue		
Ethiopia	Seychelles	India	Sri Lanka	Guadeloupe	**China**	Greece	Svalbard Is	Norfolk Is		
Ethiopia Pdr	Sierra Leone	Indonesia	Syria	Guatemala	Mainland China	Holy See	Sweden	Pacific Is		
Gabon	Somalia	Iran	Tajikistan	Haiti	China, Macao	Hungary	Switzerland	Palau		
Gambia	South Africa	Iraq	Thailand	Honduras	China, H. Kong	Iceland	UK	Papua N G.		
Ghana	St Helena	Israel	Timor Leste	Jamaica		Ireland	Ukraine	Pitcairn Is		
Guinea	Sudan	Japan	Turkey			Isle Of Man	Yugoslav SFR	Samoa		
Guinea Bissau	Swaziland		Turkmenistan			Italy		Solomon Is		

Table A-4 Food, fibre and fertiliser trade balance for Africa in year 2000

Commodity	Imports	Exports	P-conc.	P-imports	P-exports	P-Net	Total P-transfer
	000 Mg/a [a]		mg/g	000 Mg/a		000 Mg/a	
Vegetal products							
Cereals	45918	2430	3.3	151.53	8.02	143.51	159.55
Starchy roots	872	499	0.5	0.44	0.25	0.19	0.69
Sugar crops & sweeteners	5523	3735	0.1	0.55	0.37	0.18	0.93
Pulses	957	138	4.2	4.02	0.58	3.44	4.60
Oil-crops & vegetable oils	5130	2435	6	30.78	14.61	16.17	45.39
Vegetables	1172	1130	0.4	0.47	0.45	0.02	0.92
Fruits	683	3901	0.2	0.14	0.78	-0.64	0.92
Stimulants	508	3366	3.9	1.98	13.13	-11.15	15.11
Spices	57	78	0.2	0.01	0.02	0.00	0.03
Beverages	383	602	0.3	0.11	0.18	-0.07	0.30
Fibre crops	181	1092	7.2	1.30	7.86	-6.56	9.16
Tobacco	197	377	0.3	0.06	0.11	-0.05	0.17
Total vegetal products	**61581**	**19782**		**191.39**	**46.36**	**145.03**	**237.76**
Animal products							
Meat	993	135	1.8	1.79	0.24	1.55	2.03
Offal	95	1	1.2	0.11	0.00	0.11	0.12
Fats	453	53	0.3	0.14	0.02	0.12	0.15
Milk	4766	392	0.9	4.29	0.35	3.94	4.64
Eggs	31	7	2.1	0.06	0.02	0.05	0.08
Seafood	2416	1561	1.6	3.87	2.50	1.37	6.36
Total animal products	**8753**	**2148**		**10.26**	**3.12**	**7.13**	**13.38**
P-fertiliser products							
Phosphatic fertiliser	**428**	**2040**	**437**	**186.84**	**891.70**	**-704.86**	**1078.53**
Grand total						-552.70	1329.67

a)   Mg = million grammes is equivalent to one metric tonne

Notes:

P-Net is P-import - P-export

Total P-transfer is P-import + P-export

Stimulant crops include: coffee, green, raw coffee in all forms *Coffea spp.* (*arabica, robusta, liberica*); cocoa beans, the seeds contained in the fruit of the cacao-tree, including whole or broken, raw or roasted. (*Theobroma cacao*) and tea, green tea (unfermented), black tea (fermented), and partially fermented tea. (*Camellia sinensis; Thea sinensis; Thea assaamica*). Excludes green tea eaten as a vegetable but includes dried leaves of certain shrubs prepared in a way similar to tea.

Cotton lint (*Gossypium spp*) consists of fibres from ginning seed cotton that have not been carded or combed. Trade data also include fibres that have been cleaned, bleached, dyed or rendered absorbent.

Wool is a natural fibre taken from sheep or lambs, includes fleece-washed, shorn and pulled wool (from slaughtered animals), but does not include carded or combed wool is not included in the analysis for fibre products.

Table A-5 Food, fibre and fertiliser trade balance for Asia in year 2000

Commodity	Imports	Exports	P-conc.	P- imports	P- exports	P-Net	Total P- transfer
	000 Mg/a		mg/g	000 Mg/a		000 Mg/a	
Vegetal products							
Cereals	120884	40178	3.3	398.92	132.59	266.33	531.50
Starchy roots	10067	15338	0.5	5.03	7.67	-2.64	12.70
Sugar crops & sweeteners	16330	7998	0.1	1.63	0.80	0.83	2.43
Pulses	2259	2015	4.2	9.49	8.46	1.03	17.95
Oil-crops & vegetable oils	51368	24665	6	308.21	147.99	160.22	456.19
Vegetables	7401	8944	0.4	2.96	3.58	-0.62	6.54
Fruits	12167	12665	0.2	2.43	2.53	-0.10	4.97
Stimulants	1905	3110	3.9	7.43	12.13	-4.70	19.56
Spices	580	841	0.2	0.12	0.17	-0.05	0.28
Beverages	1849	984	0.3	0.55	0.30	0.26	0.85
Fibre crops	3543	2292	7.2	25.51	16.51	9.00	42.01
Tobacco	883	722	0.3	0.26	0.22	0.05	0.48
Total vegetal products	**229237**	**119752**		**762.54**	**332.93**	**429.61**	**1095.48**
Animal products							
Meat	7605	2566	1.8	13.69	4.62	9.07	18.31
Offal	724	243	1.2	0.87	0.29	0.58	1.16
Fats	2060	276	0.3	0.62	0.08	0.54	0.70
Milk	15180	1987	0.9	13.66	1.79	11.87	15.45
Eggs	252	224	2.1	0.53	0.47	0.06	1.00
Seafood	16449	8313	1.6	26.32	13.30	13.02	39.62
Total animal products	**42271**	**13609**		**55.69**	**20.55**	**35.13**	**76.24**
P-fertiliser products							
Phosphatic fertiliser	**3174**	**876**	**437**	**1386.83**	**382.65**	**1004.18**	**1769.48**
Grand total						1468.92	1941.19

Table A-6 Food, fibre and fertiliser trade balance for Central America and the Caribbean in year 2000

Commodity	Imports	Exports	P-conc.	P-imports	P-exports	P-Net	Total P-transfer
	000 Mg/a		mg/g	000 Mg/a		000 Mg/a	
Vegetal products							
Cereals	18239	1073	3.3	60.19	3.54	56.65	63.73
Starchy roots	457	318	0.5	0.23	0.16	0.07	0.39
Sugar crops & sweeteners	516	2625	0.1	0.05	0.26	-0.21	0.31
Pulses	199	213	4.2	0.83	0.89	-0.06	1.73
Oil-crops & vegetable oils	6855	525	6	41.13	3.15	37.98	44.28
Vegetables	649	4307	0.4	0.26	1.72	-1.46	1.98
Fruits	1326	7190	0.2	0.27	1.44	-1.17	1.70
Stimulants	110	1089	3.9	0.43	4.25	-3.82	4.67
Spices	26	40	0.2	0.01	0.01	0.00	0.01
Beverages	281	1339	0.3	0.08	0.40	-0.32	0.49
Fibre crops	496	30	7.2	3.57	0.21	3.36	3.78
Tobacco	31	69	0.3	0.01	0.02	-0.01	0.03
Total vegetal products	**29184**	**18819**		**107.05**	**16.06**	**90.99**	**123.11**
Animal products							
Meat	1227	160	1.8	2.21	0.29	1.92	2.50
Offal	228	6	1.2	0.27	0.01	0.27	0.28
Fats	778	21	0.3	0.23	0.01	0.23	0.24
Milk	3382	334	0.9	3.04	0.30	2.74	3.34
Eggs	28	13	2.1	0.06	0.03	0.03	0.09
Seafood	291	417	1.6	0.47	0.67	-0.20	1.13
Total animal products	**5934**	**950**		**6.28**	**1.30**	**4.99**	**7.58**
P-fertiliser products							
Phosphatic fertiliser	**388**	**249**	**437**	**165.50**	**108.64**	**60.86**	**278.14**
Grand total						156.85	408.83

Table A-7 Food, fibre and fertiliser trade balance for China in year 2000

Commodity	Imports	Exports	P-conc.	P-imports	P-exports	P-Net	Total P-transfer
	000 Mg/a		mg/g	000 Mg/a		000 Mg/a	
Vegetal products							
Cereals	10565	14416	3.3	34.86	47.57	-12.71	82.44
Starchy roots	3779	567	0.5	1.89	0.28	1.61	2.17
Sugar crops & sweeteners	1428	796	0.1	0.14	0.08	0.06	0.22
Pulses	301	583	4.2	1.26	2.45	-1.18	3.71
Oil-crops & vegetable oils	19450	1653	6	116.70	9.92	106.78	126.62
Vegetables	1113	4153	0.4	0.45	1.66	-1.22	2.11
Fruits	2708	2136	0.2	0.54	0.43	0.11	0.97
Stimulants	139	279	3.9	0.54	1.09	-0.54	1.63
Spices	33	275	0.2	0.01	0.05	-0.05	0.06
Beverages	435	324	0.3	0.13	0.10	0.03	0.23
Fibre crops	376	312	7.2	2.70	2.25	0.46	4.95
Tobacco	73	136	0.3	0.02	0.04	-0.02	0.06
Total vegetal products	**40401**	**25630**		**159.25**	**65.92**	**93.33**	**225.17**
Animal products							
Meat	2713	1673	1.8	4.88	3.01	1.87	7.90
Offal	471	234	1.2	0.57	0.28	0.28	0.85
Fats	1098	226	0.3	0.33	0.07	0.26	0.40
Milk	2110	546	0.9	1.90	0.49	1.41	2.39
Eggs	89	69	2.1	0.19	0.14	0.04	0.33
Seafood	6046	3174	1.6	9.67	5.08	4.60	14.75
Total animal products	**12527**	**5921**		**17.54**	**9.07**	**8.46**	**26.61**
P-fertiliser products							
Phosphatic fertiliser	**2212**	**289**	**437**	**966.43**	**126.29**	**840.13**	**1092.72**
Grand total						941.93	1344.50

Table A-8 Food, fibre and fertiliser trade balance for Europe in year 2000

Commodity	Imports	Exports	P-conc.	P-imports	P-exports	P-Net	Total P-transfer
	000 Mg/a		mg/g	000 Mg/a		000 Mg/a	
Vegetal products							
Cereals	63770	86685	3.3	210.44	286.06	-75.62	496.50
Starchy roots	24891	12828	0.5	12.45	6.41	6.03	18.86
Sugar crops & sweeteners	15605	14516	0.1	1.56	1.45	0.11	3.01
Pulses	3006	1538	4.2	12.63	6.46	6.17	19.09
Oil-crops & vegetable oils	41988	22319	6	251.93	133.91	118.01	385.84
Vegetables	20994	19562	0.4	8.40	7.82	0.57	16.22
Fruits	43974	28608	0.2	8.79	5.72	3.07	14.52
Stimulants	7512	3234	3.9	29.30	12.61	16.68	41.91
Spices	316	159	0.2	0.06	0.03	0.03	0.09
Beverages	9007	11915	0.3	2.70	3.57	-0.87	6.28
Fibre crops	1712	586	7.2	12.33	4.22	8.11	16.55
Tobacco	1801	1182	0.3	0.54	0.35	0.19	0.89
Total vegetal products	**234577**	**203131**		**551.12**	**468.64**	**82.48**	**1019.76**
Animal products							
Meat	11048	11413	1.8	19.89	20.54	-0.66	40.43
Offal	675	939	1.2	0.81	1.13	-0.32	1.94
Fats	3502	3430	0.3	1.05	1.03	0.02	2.08
Milk	40066	55637	0.9	36.06	50.07	-14.01	86.13
Eggs	775	871	2.1	1.63	1.83	-0.20	3.46
Seafood	19548	16362	1.6	31.28	26.18	5.10	57.46
Total animal products	**75614**	**88652**		**90.71**	**100.78**	**-10.07**	**191.49**
P-fertiliser products							
Phophatic fertiliser	**2749**	**4103**	**437**	**1201.26**	**1792.93**	**-591.67**	**2994.19**
Grand P-Net						-519.25	4205.44

Table A-9 Food, fibre and fertiliser trade balance for North America in year 2000

Commodity	Imports	Exports	P-conc.	P-imports	P-exports	P-Net	Total P-transfer
	000 Mg/a		mg/g	000 Mg/a		000 Mg/a	
Vegetal products							
Cereals	9473	113384	3.3	31.26	374.17	-342.91	405.43
Starchy roots	2919	3531	0.5	1.46	1.77	-0.31	3.23
Sugar crops & sweeteners	3399	1524	0.1	0.34	0.15	0.19	0.49
Pulses	283	3269	4.2	1.19	13.73	-12.54	14.92
Oil-crops & vegetable oils	5177	37470	6	31.06	224.82	-193.75	255.88
Vegetables	6912	4543	0.4	2.76	1.82	0.95	4.58
Fruits	19740	8453	0.2	3.95	1.69	2.26	5.64
Stimulants	2924	554	3.9	11.40	2.16	9.24	13.56
Spices	214	31	0.2	0.04	0.01	0.04	0.05
Beverages	4419	1626	0.3	1.33	0.49	0.84	1.81
Fibre crops	85	669	7.2	0.61	4.82	-4.20	5.43
Tobacco	230	434	0.3	0.07	0.13	-0.06	0.20
Total vegetal products	**55776**	**175487**		**85.48**	**625.74**	**-540.27**	**711.22**
Animal products							
Meat	2533	6122	1.8	4.56	11.02	-6.46	15.58
Offal	79	690	1.2	0.09	0.83	-0.73	0.92
Fats	274	1909	0.3	0.08	0.57	-0.49	0.65
Milk	5806	2763	0.9	5.22	2.49	2.74	7.71
Eggs	38	119	2.1	0.08	0.25	-0.17	0.33
Seafood	4278	2523	1.6	6.85	4.04	2.81	10.88
Total animal products	**13008**	**14127**		**16.89**	**19.20**	**-2.31**	**36.08**
P-fertiliser products							
Phosphatic fertiliser	**639**	**4405**	**437**	**279.17**	**1924.81**	**-1645.64**	**2203.98**
Grand total						-2188.21	2951.29

Table A-10 Food, fibre and fertiliser trade balance for Oceania in year 2000

Commodity	Imports	Exports	P-conc.	P-imports	P-exports	P-Net	Total P-transfer
	000 Mg/a		mg/g	000 Mg/a		000 Mg/a	
Vegetal products							
Cereals	1308	19760	3.3	4.32	65.21	-60.89	69.52
Starchy roots	188	156	0.5	0.09	0.08	0.02	0.17
Sugar crops & sweeteners	377	3945	0.1	0.04	0.39	-0.36	0.43
Pulses	28	2646	4.2	0.12	11.11	-11.00	11.23
Oil-crops & vegetable oils	624	2997	6	3.74	17.98	-14.24	21.72
Vegetables	370	673	0.4	0.15	0.27	-0.12	0.42
Fruits	614	1081	0.2	0.12	0.22	-0.09	0.34
Stimulants	148	135	3.9	0.58	0.53	0.05	1.11
Spices	11	8	0.2	0.00	0.00	0.00	0.00
Beverages	216	529	0.3	0.06	0.16	-0.09	0.22
Fibre crops	135	714	7.2	0.97	5.14	-4.17	6.11
Tobacco	21	4	0.3	0.01	0.00	0.01	0.01
Total vegetal products	**4041**	**32646**		**10.20**	**101.09**	**-90.89**	**111.29**
Animal products							
Meat	176	2583	1.8	0.32	4.65	-4.33	4.96
Offal	4	148	1.2	0.01	0.18	-0.17	0.18
Fats	51	1073	0.3	0.02	0.32	-0.31	0.34
Milk	622	16318	0.9	0.56	14.69	-14.13	15.25
Eggs	6	2	2.1	0.01	0.00	0.01	0.02
Seafood	481	777	1.6	0.77	1.24	-0.47	2.01
Total animal products	**1340**	**20900**		**1.68**	**21.08**	**-19.40**	**22.76**
P-fertiliser products							
Phosphatic fertiliser	**816**	**130**	**437**	**356.57**	**18.72**	**299.76**	**413.38**
Grand total						189.47	547.43

Table A-11 Food, fibre and fertiliser trade balance for South America in year 2000

Commodity	Imports	Exports	P-conc.	P-imports	P-exports	P-Net	Total P-transfer
	000 Mg/a		mg/g	000 Mg/a		000 Mg/a	
Vegetal products							
Cereals	23948	26426	3.3	79.03	87.20	-8.18	166.23
Starchy roots	725	313	0.5	0.36	0.16	0.21	0.52
Sugar crops & sweeteners	972	8811	0.1	0.10	0.88	-0.78	0.98
Pulses	524	357	4.2	2.20	1.50	0.70	3.70
Oil-crops & vegetable oils	3562	25397	6	21.37	152.38	-131.01	173.75
Vegetables	912	1145	0.4	0.36	0.46	-0.09	0.82
Fruits	2008	17417	0.2	0.40	3.48	-3.08	3.89
Stimulants	303	2196	3.9	1.18	8.57	-7.39	9.75
Spices	23	44	0.2	0.00	0.01	0.00	0.01
Beverages	409	955	0.3	0.12	0.29	-0.16	0.41
Fibre crops	469	196	7.2	3.37	1.41	1.96	4.79
Tobacco	85	458	0.3	0.03	0.14	-0.11	0.16
Total vegetal products	**33938**	**83716**		**108.53**	**256.48**	**-147.95**	**365.01**
Animal products							
Meat	423	2555	1.8	0.76	4.60	-3.84	5.36
Offal	49	89	1.2	0.06	0.11	-0.05	0.17
Fats	306	605	0.3	0.09	0.18	-0.09	0.27
Milk	3122	2129	0.9	2.81	1.92	0.89	4.73
Eggs	17	19	2.1	0.04	0.04	0.00	0.08
Seafood	809	11796	1.6	1.29	18.87	-17.58	20.17
Total animal products	**4726**	**17193**		**5.05**	**25.72**	**-20.66**	**30.77**
P-fertiliser products							
Phosphatic fertiliser	**1949**	**95**	**437**	**851.73**	**41.58**	**810.15**	**893.31**
Grand total						641.54	1289.09

Table A-12 Food, fibre and fertiliser trade balance for Zimbabwe in year 2000

Commodity	Imports	Exports	P-conc.	P-imports	P-exports	P-Net	Total P-transfer
	000 Mg/a		mg/g	Mg/a		Mg/a	
Vegetal products							
Cereals	123.0	191.6	3.3	405.77	632.12	-226.35	1037.88
Starchy roots	1.0	0.7	0.5	0.52	0.34	0.18	0.86
Sugar crops & sweeteners	1.6	282.1	0.1	0.16	28.21	-28.05	28.37
Pulses	1.8	5.0	4.2	7.48	21.17	-13.69	28.64
Oil-crops & vegetable oils	67.7	98.7	6	406.14	592.38	-186.24	998.52
Vegetables	3.4	12.3	0.4	1.37	4.90	-3.53	6.27
Fruits	9.8	108.3	0.2	1.96	21.66	-19.70	23.61
Stimulants	1.5	23.0	3.9	5.77	89.51	-83.73	95.28
Spices	0.4	10.2	0.2	0.09	2.04	-1.95	2.13
Beverages	2.4	8.4	0.3	0.71	2.51	-1.81	3.22
Fibre crops	49.0	139.6	7.2	352.80	1005.12	-652.32	1357.92
Tobacco	5.6	183.3	0.3	1.68	54.98	-53.30	56.65
Total vegetal products	**267.1**	**1063.0**		**1184.4**	**2454.9**	**-1270.5**	**3639.4**
Animal products							
Meat	0.3	20.0	1.8	0.59	36.04	-35.44	36.63
Offal	0.2	0.0	1.2	0.20	0.04	0.17	0.24
Fats	12.8	0.1	0.3	3.85	0.04	3.81	3.89
Milk	6.7	76.5	0.9	5.99	68.82	-62.84	74.81
Eggs	0.0	2.8	2.1	0.00	5.78	-5.78	5.78
Seafood	20.9	1.7	1.6	33.44	2.66	30.78	36.10
Total animal products	**40.9**	**101.1**		**44.1**	**113.4**	**-69.3**	**157.4**
P-fertiliser products							
Phosphatic fertiliser	**4.5**	**1.0**	**437**	**1966.50**	**437.00**	**1529.50**	**2403.50**
Grand total						189.71	6200.29

Table A-13 Food balance sheet and major non-edible vegetal products for Zimbabwe in year 2000

Product	Production	Imports	Stock changes	Exports	Total supply 000 Mg/a	Feed	Seed	Processing	Waste	Other uses	Food	Supply (kg/p.a)
Vegetal Products												
Cereals - Excluding beer	2537.3	123.0	104.9	191.6	2573.6	404.6	38.7	69.7	128.7	-	1931.9	153
Starchy Roots	207.6	1.0	-	0.7	208.0	-	2.1	0.0	10.5	0.5	194.9	15.43
Sugar crops	4227.5	0.0	-	0.0	4227.5	-	-	4227.5	-	-	0.0	0
Sugar & Sweeteners	590.4	1.6	76.1	282.1	386.0	-	-	0.7	-	30.0	355.5	28.15
Pulses	52.4	1.8	1.6	5.0	50.6	-	6.1	-	2.6	-	41.9	3.32
Tree-nuts	0.9	0.1	0.0	0.1	0.9	-	-	-	-	-	0.9	0.07
Oil-crops	492.5	15.8	25.0	93.7	439.7	4.0	21.0	282.9	10.6	22.9	106.2	8.41
Vegetable Oils	79.7	51.8	-1.9	4.9	124.7	-	-	0.0	-	12.4	111.0	8.79
Vegetables	146.0	3.4	0.0	12.3	137.2	-	-	-	13.2	-	124.4	9.85
Fruits - Excluding Wine	231.6	9.8	12.5	108.3	145.6	-	-	2.7	19.0	0.0	124.2	9.83
Stimulants	31.1	1.5	1.0	23.0	10.6	-	-	-	-	2.9	8.9	0.7
Spices	17.2	0.4	0.5	10.2	8.0	-	-	-	-	-	8.0	0.63
Alcoholic Beverages	366.6	2.4	0.6	8.4	361.2	-	-	-	-	0.4	360.8	28.57
Animal Products												
Meat	186.5	0.3	-	20.0	166.8	-	-	0.0	-	-	167.3	13.25
Offals	17.2	0.2	-	0.0	17.4	-	-	-	-	-	17.4	1.38
Animal Fats	10.9	12.8	10.0	0.1	33.6	-	-	-	-	9.8	23.9	1.89
Milk - Excluding Butter	310.0	6.7	0.0	76.5	240.2	17.3	-	-	11.7	-	211.3	16.73
Eggs	22.0	0.0	-	2.8	19.3	-	1.6	0.0	1.1	-	16.6	1.31
Fish, Seafood	13.3	20.9	0.0	1.7	-	7.2	-	0.0	-	-	25.4	2.01
Non-edible products												
Cotton lint	327.0	49.0	-	139.6	-	-	-	-	-	-	-	-
Tobacco	227.7	5.6	-	183.3	-	-	-	-	-	-	-	-

Data Source: FAO (2001)

Appendix B

Phosphorus content in various goods

Table B-1 Amount of P removed in some harvested products [a]

Crop	Optimum yield (kg/m^2)	Form and state	P removed (g/m^2)	P concentration (σ) in harvest (g/kg)
Barley	0.4	Grain and straw	1.2	3.0
Potatoes	5.0	Bulbs	2.5	0.5
Grass	1.0	Dry matter	3.0	3.0
Maize	1.3	Grain and stover	5.0	3.8
Kale	5.0	Fresh weight	2.5	0.5

a) Detailed information on analytical methods used, variety, maturity, time of harvest, length of storage or exposure in the market, part of sample analysed, part considered inedible, or other factors which would influence P-content are usually not fully provided by the analysts.

Source: White (1979)

Table B-2 Phosphorus concentration of harvested portion of crop commodities

Crop Commodity	P concentration (σ) g/kg
Fruit	0.2
Vegetables	0.4
Oilseed crop seed	6.0
Oilseed cake	10.1
Fibre crops	7.2
Coconut products	3.8
Cocoa products	3.6
Coffee and tea	3.9
Cereals	3.3
Pulses	4.2
Bran	12.4
Sugar	0.1
Malt	10.0
Others	0.1 - 0.5

a) Based on market weight
Source: Beaton *et al.*, (1995)

Table B-3 Phosphorus concentration of livestock commodities

Livestock product	P concentration (σ) g/kg
Live cattle [a]	8.6
Live pigs	4.6
Live sheep	6.5
Fresh meat	2.2
Meat products	1.8
Milk	0.9
Cheese and whey	5.0
Eggs	2.1
Others	0.1 - 0.4

a) Dressed carcass weight (kg/head) for live cattle, pigs and sheep varies from region to region with average weights of 200 kg/head, 70 kg/head, 16 kg/head respectively. Dressing percent for live cattle, pigs (hide attached) and sheep can be assumed to be 58%, 75%, and 50%, respectively (FAO, 1992).

Source: Beaton *et al.*, (1995)

Table B-4 Average P concentration of rock phosphate

Region	P content (σ) g/kg
Africa	147
Asia	136
Europe	156
North & Central America	135
Oceania	125
South America	140
World Average [a]	144

a) Weighted according to 1990 production in each region
Source: Beaton *et al.*, (1995)

Table B-5 P-content and some characteristic of common food consumed in Zimbabwe

Food and description	Food energy (calories)	Moisture Content (%) [a]	Ash (g)	P-content (σ) (g/kg)
Bread	252	39.1	1.1	1.21
White bread	261	36.9	1.7	0.95
Maize corn white variety	357	11.6	1.2	2.20
Mealie meal native ground or unsifted	353	12.2	1.3	2.18
Sifted	368	12.2	1.0	1.78
Perlenta (Ngwerewere)	369	12.0	1.1	1.55
Samp (*coarse hominy*)	366	11.0	0.3	0.40
Semolina from white variety	365	12.3	1.4	1.62
Malt (sorghum product)	366	11.6	1.8	2.09
Malt (barley product)	336	9.1	3.0	3.10
Macaroni (wheat product)	379	8.4	0.7	1.69
Dried and cooked	154	60.0	3.5	0.67
Amadumbe bulbs (*colocasia esculenta*) (raw and cooked)	102	73.1	1.2	0.88
	87	80.0	1.7	0.61
Sweet potato (pale, yelow and deep yellow varieties)	121	68.8	0.9	0.38
Potatoes (*solanum tuberosum*) raw and cooked	82	77.7	1.6	0.51
	84	77.5	1.0	0.56
Banana	88	77.0	0.8	0.21
Peanut; groundnut (*arachis hypogaea*) shelled dried, roasted and boiled	549	6.5	2.5	4.09
	595	1.8	2.4	3.54
	235	44.6	4.0	2.60
Peas; garden and field whole seeds dried (*pissum spp*)	339	11.0	3.6	3.82
Soybean (*glycine max; G. hispida; G. soja*) whole mature seeds, dried	405	9.5	5.0	5.41
Hibiscus, kenaf; kando (*Hibiscus cannabinus*) seed dried	427	8.1	4.6	6.06
Bean leaves, raw	36	86.8	2.6	0.75
Cabbage, common (*brassica oleracea var. capitata*) raw	26	91.4	0.8	0.40
Cowpeas, catjang (*Vigna unguiculata subs. Catjang*) leaves raw	44	85.0	1.7	0.63
Okra; ladies finger (*Hibiscus esculentus*) raw and dried	36	88.6	0.9	0.90
	283	10.2	8.1	3.97
Onion, garden, common (*allium cepa*) mature bulbs, raw and immature bulbs and leaves raw	41	88.5	0.6	0.45
	22	92.9	0.8	0.24
Peanut; groundnut (*arachis hypogaea*) leaves, raw	69	78.5	1.6	0.82
Pumpkin; vegetable marrow (*cucurbita pepo*) leaves raw and dried	27	89.2	2.2	1.36
	220	19.0	11.7	5.33
Rape (*brassica napus*) fresh leaves	100	76.6	2.8	1.19
Spinach (*spinacia oleracea*) raw	26	90.6	1.8	0.46

Table B-5 P-content and some characteristic of common food consumed in Zimbabwe (conti)

Food and description	Food energy (calories)	Moisture Content (%) [a]	Ash (g)	P-content (σ) (g/kg)
Tomato (*solanum lycopersicum; lycopersicon esculentum*) raw, ripe, whole	21	93.5	0.5	0.24
Guava, common (*psidium guajava*), fruit, whole, raw	64	82.2	0.6	0.31
Lemon (*citrus limon*)	29	89.8	0.3	0.18
Mango, common (*mangifera indica*) fruit ripe	60	82.9	0.5	0.22
Mulberry (*morus spp.*) fruit		88.4		0.33
Orange, sweet (*citrus sinensis*)	43	88.0	0.5	0.17
Pawpaw (*carica papaya*)	32	90.8	0.4	0.15
Peach (*prunus persica*)	50	85.8	0.6	0.19
Watermelon (*citrullus lanatus*)	22	93.6	0.3	0.09
Honey, jams	311	19.2	0.3	0.04
Beef (*bos taurus*) medium fat and dried and salted	237	63.1	1.0	1.94
	250	29.4	13.7	9.10
Caterpillars (*bombycomorpha*) raw and dried	86	81.1	1.4	1.39
	430	9.1	5.7	6.17
Chicken (*gallus gallus*) raw total edible	146	72.0	1.0	2.06
Heart, beef	109	77.0	1.3	1.72
Liver, beef	143	70.0	1.3	3.60
Mopani worm (*conimbrasia belina*), dried	444	6.1	6.9	5.76
Mutton, lamb (*ovies aries*)	265	60.7	1.0	1.48
Pork, (*sus scrofa*) medium fat	418	46.1	1.0	1.74
Hen egg, whole, raw	140	77.0	1.0	2.00
Catfish (*synodontis spp.*) raw and dried	90	79.6	1.2	1.97
	362	20.9	4.8	6.00
Fish, average of all kinds, raw and dried	103	73.7	5.0	1.13
	269	13.8	31.5	7.49
Sour milk (*aybe*)	122	73.6	0.9	1.80
Milk, cow, fluid and whole	79	85.2	0.8	0.95
Beer made from maize and sorgum (2% alcohol by weight)		94.5		0.13
Sorghum, kaffir beer strained (2.1% alcohol by weight)	31	93.6		0.07
Unclarified (2.8% alcohol by weight)	117	72.7	0.7	
Coffee (*coffea spp.*), infusion	4	97.5	0.4	0.18
Rooibos tea (*aspalathus contaminatus*), dried	313	10.1	2.5	2.20

a) The moisture content of the foods may fluctuate greatly with season, length of storage etc. The need for determining moisture content whenever possible is important. The moisture content is usually determined by keeping the food in an oven at a temperature of from 100 to 105 °C until constant weight is reached. The food is weighed before and after, and the loss in weight is expressed as a percentage of the original weight.

Source: Various sources; mainly, FAO, (1968); Baccini & Brunner, (1991); Chitsiku, (1991), Specific laboratory tests conducted during the research period for this dissertation (1998 to 2000).

Appendix C

STELLA Model Equations

Rainfall Water Balance mm/month

○ `B = 1-Gamma_zero+Gamma_zero*exp(-1/Gamma_zero)`

DOCUMENT: Transpiration factor (B), which is the slope of the linear part of the relationship between effective rainfall and monthly transpiration.

○ `Beta = R*(1-p11+p01)/(30*p01)`

DOCUMENT: Rainfall scale parameter. Number of days in a month taken as 30.

○ `C = 0.24`

DOCUMENT: Surface runoff coefficient (C) dependent on land use and amount of effective rainfall. Determined for the Marimba catchment using linear regression model and the SCS-SA method. The value obtained compared well with the one calculated from the sum of gauged runoff (Q_s) divided by the sum of effective rainfall (R_{eff}).

○ `D = 3`

DOCUMENT: Daily interception threshold (D), taken as 3 mm/day. Typical value range is 3-5 mm/day (De Groen, 2002; Savenije, 2004).

○ `Gamma_zero = Sb/(Tpot*30)`

DOCUMENT: Dimensionless transpiration time scale parameter (Gamma-zero).

○ `I = R*(1-exp(-D/Beta))`

DOCUMENT: Interception calculated using the Markov process monthly model (De Groen, 2002).

○ `p01 = 0.020*(R)^0.55`

DOCUMENT: The transition probabilities p01 (rain-dry) is a logistic or power functions of the monthly rainfall used in the spells model described by De Groen (2002). The calibrated model parameters to calculate p_{01} are those for Harare i.e. q=0.020, r=0.55.

○ `p11 = 0.2*(R)^0.24`

DOCUMENT: The transition probabilities p11 (rain-rain) is a logistic or power functions of the monthly rainfall used in the spells model described by De Groen (2002). The calibrated model parameters to calculate p11 are those for Harare i.e. u=0.20, v=0.24.

○ `Qg = Reff-Tr-Qs`

DOCUMENT: Groundwater flow or seepage (Q_g), calculated from the model.

○ `Qs = C*Reff`

DOCUMENT: Surface run-on or run-off (Q_s), computed from the model.

○ `Reff = R-I`

DOCUMENT: Monthly effective rainfall $R_{eff}=R-I$.

○ `Sb = 60`

DOCUMENT: Limiting soil moisture S_b, from De Groen (2002), the ratio of S_b/S_{max} lies between 0.5 and 0.8. Assuming a value of 0.5 for $S_{max}=120$ mm, the S_b value of 60 mm is adopted based on soil characteristics in Harare and in the micro-study catchment.

○ `Ti = Wi`

DOCUMENT: Transpiration (T_i) arising from garden irrigation using municipal water supplies.

○ `Tpot = 6`

DOCUMENT: From De Groen (2002), the value for the potential transpiration per day (T_{pot}) for arable crops varies between 8 and 4 mm/day. The median of 6 mm/day is assumed. Number of days in a month is taken as 30 to convert potential daily transpiration to potential monthly transpiration.

○ `T = B*Reff`

DOCUMENT: Transpiration from rainfall (T).

○ `Tt = T+Ti`

○ `R = GRAPH(time)`

Rainfall Water Balance mm/month

```
(1.00, 218), (2.00, 135), (3.00, 105), (4.00, 38.7), (5.00,
12.4), (6.00, 0.333), (7.00, 0.182), (8.00, 2.85), (9.00,
4.10), (10.0, 34.3), (11.0, 73.5), (12.0, 163)
```
DOCUMENT: Year 2000 average monthly rainfall for three stations (Belvedere, Warren Hills and Aspindale). The use of the average rainfall is meant to accommodate areal fluctuations of point rainfall amounts recorded.

Municipal Water Balance mm/month

○ `A = 6500000`

DOCUMENT: A is the total area of the micro-study catchment in square metres, determined from aerial photographs.

○ `Age_group_factor = 0.46*0.50+0.54`

DOCUMENT: The age structure in the study area was taken as 0-14 years, 43%; 15-54, 54% and 65 and over, 3%. It is assumed that the 15-64 age group (adults) excretion 100% of the estimated rate of urine and faeces per month whilst the other age groups excrete 50% of these values.

○ `aW = a_fraction*W`

○ `a_fraction = 0.2`

DOCUMENT: Water use fraction (a), which determines the volume of water used for on-plot garden irrigation. This figure was established through a survey of 45 households (Hoko, 1999; Gumbo, 1998, 2000). Other water use fractions were also determined during this survey i.e. b, c, d, e, f, and h.

○ `b_fraction = 0.47`

DOCUMENT: Determines the grey water volume generated from the kitchen $b_1 = 0.08$; bathroom $b_2 = 0.25$ and laundry $b_3 = 0.15$ ($b = b_1 + b_2 + b_3$).

○ `c_fraction = 0.3`

DOCUMENT: This fraction (c) determines the volume of black water arising from toilet flushing. The average toilet cistern size is 10 litres. The water contained in faecal matter is assumed to be negligible when compared to the flush water volume.

○ `d_fraction = 0.03`

DOCUMENT: Based on assumption that an average adult person in the study area consumes about 2 litres of water per day either directly or indirectly.

○ `Ec = Wc-Wy`

DOCUMENT: E_c is the water flux excreted by the human population mainly through respiration and perspiration (W_c - W_y).

○ `Eg = f_fraction*Wg`

DOCUMENT: E_g is the evaporation due to drying-out of laundry material and kitchen utensils.

○ `eWg = e_fraction*Wg`

○ `e_fraction = 0.15`

DOCUMENT: It is normal practice in the high density suburbs that part of the greywater is applied on agricultural land especially during the dry season and when there are water use restrictions. The diversion of greywater for irrigation is estimated at about 15% of the greywater generated i.e. e = 0.15.

○ `f_fraction = 0.10`

DOCUMENT: Fraction of greywater which evaporates from laundry fabrics, kitchen utensils and human body after washing and bathing (f).

○ `H = 100000`

DOCUMENT: Estimated population in the micro-study catchment in year 2000 projected from the 1992 national census report.

○ `hQ = h_fraction*(Qg+Qs)`

○ `h_fraction = 0.12`

DOCUMENT: This factor is calculated from the proportion of surface and ground water flow which infiltrates the foul sewer system during wet weather (MEWRD, 1978; Hoko, 1999; Gumbo, 2000). During wet weather up to 12% of the storm runoff and ground water flow is channelled through the foul sewer system leading to increased flows at the sewage

Municipal Water Balance mm/month

treatment plants.

○ `Wb = c_fraction*W`

DOCUMENT: Brown water generated from household activity related to toilet flushing after defaecation (W_h).

○ `Wc = d_fraction*W`

DOCUMENT: W_c is the municipal water flux consumed by the population either directly or contained in ingested food products.

○ `Wg = b_fraction*W`

DOCUMENT: Greywater (W_g) generated from activities related to nourishing and cleaning (kitchen, bathroom and laundry).

○ `Wi = aW+eWg`

DOCUMENT: Total municipal water used for garden irrigation $W_i = aW + eW_g$.

○ `Wms = Ws+hQ`

DOCUMENT: W_{ms} is the municipal sewage water flux, which is a combination of yellow, black, and a proportion of grey and storm water conveyed through a pipe to a sewage treatment plant ($W_{ms} = (1 - e - f)W_g + W_b + W_y + hQ$).

○ `Ws = (1-e_fraction-f_fraction)*Wg+Wb+Wy`

DOCUMENT: W_s is the foul sewage or 'black water' flux which is a combination of yellow, brown, and a proportion of greywater ($W_s = W_{ms} - hQ$). This corresponds to dry weather flow.

○ `Wy = 1000*(Age_group_factor*H*y)/A`

DOCUMENT: W_y is the volume of yellow water excreted by an equivalent adult population per month normalised to the micro-catchment area (A).

○ `y = 0.012`

DOCUMENT: The urinary volumetric rate of excretion in l/p.d in winter or summer for an adult person in the study area. The daily excretion rate for summer and winter was determined from observation to be 0.36 l/p.d and 0.45 l/p.d respectively. For the purposes of this model the average value of 0.40 l/p.d is used which translates to 12 l/p.month or 0.012 m3/p.month.

○ `W = GRAPH(time)`
`(1.00, 25.6), (2.00, 23.3), (3.00, 22.9), (4.00, 24.9), (5.00, 26.8), (6.00, 27.2), (7.00, 25.2), (8.00, 26.7), (9.00, 29.7), (10.0, 31.4), (11.0, 26.4), (12.0, 30.3)`

DOCUMENT: Municipal water supply normalised to the catchment area (W) mm/month for year 2000 as recorded by the Mufakose Housing Office.

Household P Balance kg/month

○ `ePg = e_fraction*Pg`

DOCUMENT: Proportion of grey water P-flux applied on land as a result of garden irrigation (= P_{gi} in the agricultural subsystem).

○ `Lamda = 1.35`

DOCUMENT: Is the quantity of dry organic or biodegradable solid waste fraction generated per person per month =1.35 kg/p month.

○ `Pb = Pms-(1-e_fraction)*Pg-Py`

DOCUMENT: P_b is the brown water P-flux emanating from toilet flushing of human faecal material. This was determined indirectly from sewage flow characterisation studies done in Mabvuku and Tafara suburbs.

○ `Pfb = Age_group_factor*H*Total_fb`

DOCUMENT: P_{fb} is the food and beverage P-flux reaching the household subsystem i.e. imported from outside micro-catchment and some produced within.

○ `Pg = Psd`

DOCUMENT: P_g is the grey water P-flux emanating from activities related to nourishing and cleaning.

○ `Pms = A*Sigma_ms*Wms/1000`

○ `Psd = H*Total_sd`

Household P Balance kg/month

DOCUMENT: P_{sd} *is the soap and detergent P-flux reaching the household subsystem.*

○   `Psw = Lamda*H*Sigma_sw`

○   `Py = A*Sigma_y*Wy/1000`

○   `q' = 0.10`

DOCUMENT: From the household survey it was established that about 10% of the organic solid waste by weight was either deliberately composted on-site or due to collection problems dumped into open land where it decomposed and got incorporated into the soil matrix.

○   `q'Psw = q'*Psw`

DOCUMENT: P-flux of organic solid waste which is either deliberately composted or is uncollected and end up being manure on agricultural land (= P_{csw} in the agricultural subsystem).

○   `Sigma_ms = 0.0211`

DOCUMENT: Determined indirectly from sewage flow characterisation studies done in Mabvuku and Tafara suburbs. Average population equivalent of 37.28 g/p month or 0.03728 kg/month was calculated. An average concentration of 0.0211 g/l as P (0.0211 kg/m^3) was determined for an average daily flow rate of 3 500 m^3/day (105 500 m^3/month).

○   `Sigma_sw = 0.0016`

DOCUMENT: Taken as the calculated weighted average P-content of the main components of the food flux =1.6 g/kg or 0.0016 kg/kg.

○   `Sigma_y = 1.89`

DOCUMENT: The P concentration of fresh and stale urine was determined in the laboratory. The median value for fresh urine of 1.89 kg/m^3 for an adult was used in combination with the volumetric excretion rate of 0.012 m^3/p.month.

○   `Total_fb = 0.0416`

DOCUMENT: Sum total of P contained in the food and beverages consumed by the population. Calculated from the P-content (kg/kg) of the individual food and beverage items and quantity consumed per person per month (kg). The main food components consist of eight groups, namely, cereals, meat and fish, milk and eggs, vegetables, nuts, starchy roots, pulses and beverages.

○   `Total_sd = 0.00148`

DOCUMENT: The quantities of soap and detergent used per person per month (kg) were determined from a household survey. These were multiplied by the average P-content of the identified soap and detergents (kg/kg).

Agriculture P Balance kg/month

▢   `Pbs(t) = Pbs(t - dt) + (Pbu_Flow - Pbr_Flow - Phb_Flow) * dt`
      `INIT Pbs = 626.4`

DOCUMENT: The initial biomass stock (P_{bs}) is assumed to be equal to 626.4 kg as of January of each year which corresponds to P_{bu} for October to December of the previous year. This is related to Y_{hb} of 0.20 g/m2. When Y_{hb} changes the initial value of P_{bs} has to be reset. Note that in the month of October each year P_{bs} becomes zero as it is incorporated into the soil matrix after land preparation.

      `INFLOWS:`

⇒   `Pbu_Flow = Pbu`

      `OUTFLOWS:`

⇒   `Pbr_Flow = Pbr`

⇒   `Phb_Flow = Phb`

▢   `P_ss(t) = P_ss(t - dt) + (Pmf_Flow + Pbr_Flow + Pma_Flow + Pgi_Flow - Pbu_Flow - Psl_Flow - Psr_Flow - Ple_Flow) * dt`
      `INIT P_ss = Initial_Pss`

      `INFLOWS:`

⇒   `Pmf_Flow = Pmf`

Agriculture P Balance kg/month

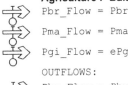

Pbr_Flow = Pbr

Pma_Flow = Pma

Pgi_Flow = ePg

OUTFLOWS:

Pbu_Flow = Pbu

Psl_Flow = Psl

DOCUMENT: The expression is based on research conducted in Zimbabwe for FAO (Stocking, 1986). The value obtained is based on the relationship between phosphorus losses, soil losses and surface runoff.

Psr_Flow = Psr

Ple_Flow = Ple

Ac = 2900000

DOCUMENT: A_c is the total area under cultivation in the micro-study catchment (= 2 900 000 m^2) for both on-plot and off-plot cultivation. Area determined from aerial photographs.

Initial_Pss = Rho_b*V_factor*V*Sigma_ss

DOCUMENT: The initial value of bio-available Pss is calculated from field test results conducted after the harvest period in August and September 2000. Although the model runs from January to December of each year this value is assumed to be indicative of the actual storage value. Adjustments could be made to this value to include the P-fluxes of the three months (October to December) of the previous year. The growing season/hydrological year is different from the calendar year.

Mim = 101000

DOCUMENT: Several households use significant amount of imported organic matter as manure. An estimate of at least 10 kg/household.a of imported manure was made for the 10 100 stands in Mufakose and Marimba suburbs.

Mmf = 0.042

DOCUMENT: Mass of mineral fertiliser applied per annum or growing season (kg/a) as Compound D. The recommended application rate for basal fertiliser in Zimbabwe is about 420 kg/ha or 42 g/m^2 as what is termed 'maizefert'. The composition of Compound D or maizefert is $8:14:7;N:P_2O_5:K_2O$.

Pbr = DF_Pbr*(Pbs-Phb)

DOCUMENT: The residual biomass is what remains within the agricultural subsystem after the harvest period. It is the difference between the biomass P-uptake (P_{bu}) and the harvested P-flux (P_{hb}).

Pbu = DF_Pbu*(0.0018*Ac*Yhb)

DOCUMENT: Based on minimum P-biomass uptake requirements for the water limited potential production for maize (Driessen & Konijn, 1992). The values are distributed as monthly uptakes, assuming the planting period is mid-October each year, and that the uptake follows the relation given in literature (Barber & Olson, 1968; Leigh & Johnston, 1986).

Pcsw = 12*q'Psw

DOCUMENT: From field observations and survey it is noted that 10-20 of the biodegradable solid waste fraction is composted either intentionally or by default as solid waste is not always collected in time by the Municipality resulting in dumping in open spaces which usually are also under cultivation during the rainy season.

Phb = DF_Phb*(0.0011*Ac*Yhb)

DOCUMENT: The growing season i.e. from planting to maturity is about 5 months and harvesting occurs over a period of 3 months, from March to May each year.

Pim = Mim*Sigma_im

Ple = Ac*Qg*Sigma_sr/(1000000*2.2919)

DOCUMENT: P_{le} is the leaching P-flux due to percolation and groundwater flow.

Pma = DF_Pma*(Pcsw+Pim)

Agriculture P Balance kg/month

○ `Pmf = DF_Pmf*(Theta*Ac*Mmf*Sigma_mf)`

○ `Psl = 0.0005*Qs*Ac/1000`

DOCUMENT: Equation based on regression equations relating overland flow (Q_s) and soil loss (S_l) and hence P-loss due to erosion (P_{sl}) established by Stocking (1986). Soil Loss Estimation Model for Southern Africa (SLEMSA).

○ `Psr = Ac*Qs*Sigma_sr/(1000000*2.2919)`

DOCUMENT: P_{sr} is the surface runoff P-flux dissolved in storm water. P_{sr} represents the dissolved soluble P (DP) losses from the agricultural subsystem.

○ `Rho_b = 1400`

DOCUMENT: The average bulk density determined during the field soil sampling was 1 400 kg/m³. This value is within the expected range for soils within the Harare area (Thompson & Purves, 1981; Nyamapfene, 1991).

○ `Sigma_im = 0.0006`

DOCUMENT: Sawdust and tobacco residue has a very low P-content for most samples it was tested P was undetectable.

○ `Sigma_mf = 0.14/2.2919`

DOCUMENT: Compound D fertiliser contains 14% P as P_2O_5 by mass. To convert the P-content from P_2O_5 to as P a factor of 2.2919 is used.

○ `Sigma_ss = 0.000005672`

DOCUMENT: This is the average available or labile P-content of the soil samples taken at eight sites after the harvesting period in year 2000 (September) in the micro-study area. An average value of 13 mg/kg as P_2O_5 or 5.67 mg/kg as P (0.000005672 kg/kg of soil as P) is used.

○ `Theta = 0.15`

DOCUMENT: Is the proportion of the recommended commercial fertiliser application rate applied by urban farmers as Compound D. The value ranges from 0.1 to 0.2.

○ `V = 0.2*Ac`

DOCUMENT: Volume of soil in the plough layer. In Zimbabwe the plough layer is taken as the top 0.2 m of the soil.

○ `V_factor = 0.88`

DOCUMENT: Is the effective volume factor i.e. volume of soil, less the greater than 2 mm soil fraction, which is deemed not to possess P-supply capacity. The value was determined from mechanical soil grading tests.

○ `Yhb = 0.20`

DOCUMENT: Estimated average maize yield from the surveys conducted in the year 1999 and 2000 indicated an average value of 0.20 kg/m².

⊘
```
DF_Pbr = GRAPH(time)
(1.00, 0.00), (2.00, 0.00), (3.00, 0.4), (4.00, 0.3), (5.00,
0.3), (6.00, 0.00), (7.00, 0.00), (8.00, 0.00), (9.00,
0.00), (10.0, 0.00), (11.0, 0.00), (12.0, 0.00)
```

⊘
```
DF_Pbu = GRAPH(time)
(1.00, 0.2), (2.00, 0.1), (3.00, 0.1), (4.00, 0.00), (5.00,
0.00), (6.00, 0.00), (7.00, 0.00), (8.00, 0.00), (9.00,
0.00), (10.0, 0.1), (11.0, 0.2), (12.0, 0.3)
```

⊘
```
DF_Phb = GRAPH(time)
(1.00, 0.00), (2.00, 0.00), (3.00, 0.4), (4.00, 0.3), (5.00,
0.3), (6.00, 0.00), (7.00, 0.00), (8.00, 0.00), (9.00,
0.00), (10.0, 0.00), (11.0, 0.00), (12.0, 0.00)
```

⊘
```
DF_Pma = GRAPH(time)
(1.00, 0.00), (2.00, 0.1), (3.00, 0.2), (4.00, 0.3), (5.00,
0.2), (6.00, 0.00), (7.00, 0.00), (8.00, 0.00), (9.00,
0.00), (10.0, 0.1), (11.0, 0.1), (12.0, 0.00)
```

⊘
```
DF_Pmf = GRAPH(time)
(1.00, 0.00), (2.00, 0.00), (3.00, 0.00), (4.00, 0.00),
(5.00, 0.00), (6.00, 0.00), (7.00, 0.00), (8.00, 0.00),
(9.00, 0.00), (10.0, 0.5), (11.0, 0.3), (12.0, 0.2)
```

Agriculture P Balance kg/month

DOCUMENT: Disaggregating Factors used to convert annual or growing season totals of P_{mf}, P_{ma}, P_{bu}, P_{hh} and P_{br} to monthly P-flux values. The DF's are based on observation of the urban agricultural dynamics in the micro-study area.

```
Sigma_sr = GRAPH(time)
(1.00, 1.45), (2.00, 0.763), (3.00, 0.958), (4.00, 0.61),
(5.00, 0.315), (6.00, 1.54), (7.00, 0.508), (8.00, 1.25),
(9.00, 2.94), (10.0, 2.19), (11.0, 1.41), (12.0, 0.405)
```

DOCUMENT: Based on 4-year water quality data for station CR46 located on the Marimba River on the periphery of the micro-study catchment. City of Harare recorded the monthly values of P concentrations as ortho-phosphate. These values are divided by the conversion factor of 2.2919 to give P-concentration as P.

Appendix D

Sensitivity and Reliability of Model Parameters

This appendix provides a framework for checking the reliability and sensitivity of the model parameters used in STELLA. One of STELLA's most useful features is the ease of conducting sensitivity analysis. This is a collection of simulations that reveals the importance of one of the model inputs.

By calling on the Sensi Specs dialogue box STELLA enables the establishment of the characteristics for a set of sensitivity runs. STELLA allows running the model (sub-models) with the sensitivity analysis turned on. For each input parameter a range of run values was defined using the incremental variation option. Three runs were specified by entering in addition to model value, the start and end values. For example Figure D.1 shows the Sensi Spec dialogue box in STELLA defining the sensitivity runs for the population of the micro-study area (*H*).

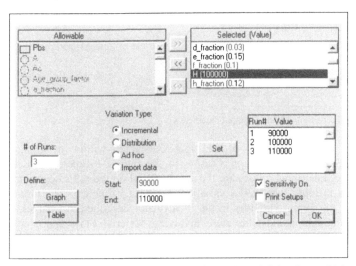

Figure D-1 The Sensi Specs dialogue window in STELLA defining the sensitivity runs for the population of the micro-catchment

Input parameters were initially assessed in STELLA using sensitivity runs and then subsequently classified in terms of their range, reliability and sensitivity as indicated in the Table D.1. The parameters are presented under the four sub-models; the rainfall water balance, the municipal water balance, the household consumption-excretion P-balance and the agricultural soil-plant P-balance. Some degree of judgemental classification depending on the quality of the input data was also applied. For easy reference the input parameters were subdivided into three groups: rough estimate, reliable and accurate on a reliability scale (RI = Reliability Index) of 1 to 3 respectively. Similarly a sensitivity scale of 1 to 3 (SI = Sensitivity Index) was used to group the input parameters into sensitive, less sensitive and non-sensitive in relation to selected P-flux or stock output variable.

This assessment is best described by a 3 × 3 matrix shown in Figure D.2. The value range of the parameter also determines its reliability. Problem input parameters are those which are rough estimates and are also sensitive to selected model output variables (P-flux or stock). These are designated [1, 1] or red whilst those which have been determined accurately and are not sensitive to the selected model output variables are designated [3, 3] or green. Input parameters falling within the main diagonal of the matrix are deemed to be acceptable and can be colour coded orange.

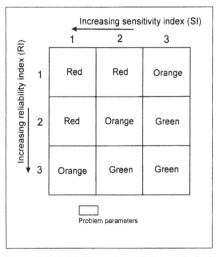

Figure D-2 Reliability and sensitivity matrix for input parameters

This analysis shows that the majority of the input values fall within the orange diagonal (Figure D.2 and Table D.1) in relation to the output P-flux and stock variables. The most critical input values are those, which fall on the red category as their variation results in more than ±20% variation in output variables. The target standard deviation of the output variables in the P-calculator described in Chapter 4 and 5 is ±10%. Although in most literature involving Material Flow and Stock Accounting (MFSA), P-fluxes and stocks have been estimated with 20-50% error margin (Obernosterer *et al.*, 1998; Belevi, 2000).

The input values which are the most sensitive are the population H and area of micro-catchment A, because of their magnitudes and the accuracy involved in their determination. As an example Figure D.3 shows that varying the population H by ±10% results into more than ±20% variation of the calculated P-flux due to organic solid waste (P_{sw}).

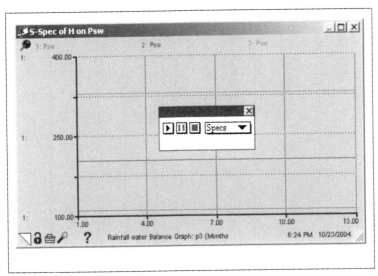

Figure D-4 The Sensi Specs dialogue window in STELLA defining the sensitivity runs for the population of the micro-catchment

Table D-1 Reliability and sensitivity classification of model input values

Input	Description	Unit	Value	Range	RI	SI	Affected output P-flux or stock
Rainfall water balance							
R	monthly average rainfall aggregated from daily totals recorded at three stations, Aspindale, Belvedere and Warren Hills	mm/month	-	0-214	3	1	$P_{ms}, P_{sl}, P_{le}, P_{sr}, P_{ss}$
β	mean rainfall on a rain day and scale parameter of exponential distribution	mm/day	-	0-12	2	2	$P_{ms}, P_{sl}, P_{le}, P_{sr}, P_{ss}$
D	daily interception threshold	mm/day	3	3-5	2	1	$P_{ms}, P_{sl}, P_{le}, P_{sr}, P_{ss}$
S_b/S_{max}	ratio of available soil moisture content at the boundary between moisture constrained transpiration and potential transpiration and maximum available soil moisture for a certain soil type and crop	(-)	0.5	0.5-0.8	2	2	$P_{ms}, P_{sl}, P_{le}, P_{sr}, P_{ss}$
T_{pot}	monthly potential transpiration, which is the sum of the values of daily potential transpiration for all days in a month	mm/month	180	120-240	2	1	$P_{ms}, P_{sl}, P_{le}, P_{sr}, P_{ss}$
B	slope of relation between monthly effective rainfall and monthly transpiration. For calculated $\gamma^o = 0.33$	(-)	0.68	0-1	2	2	$P_{ms}, P_{sl}, P_{le}, P_{sr}, P_{ss}$
C	surface runoff coefficient based on land used and amount of effective rainfall	(-)	0.24	0-1	2	1	$P_{ms}, P_{sl}, P_{le}, P_{sr}, P_{ss}$
h	fraction of storm water entering the foul sewer system	(-)	0.12	0-0.2	1	3	$P_{ms}, P_{sl}, P_{le}, P_{sr}, P_{ss}$
q	constant in $p_{01}= q(R)^r$	(month/mm)^r	0.020	0.020-0.216	2	2	$P_{ms}, P_{sl}, P_{le}, P_{sr}, P_{ss}$
r	power in $p_{01}= q(R)^r$	(-)	0.55	0.22-0.55	2	2	$P_{ms}, P_{sl}, P_{le}, P_{sr}, P_{ss}$
u	constant in $p_{11}= u(R)^v$	(month/mm)^v	0.200	0.017-0.200	2	2	$P_{ms}, P_{sl}, P_{le}, P_{sr}, P_{ss}$
v	power in $p_{11}= u(R)^v$	(-)	0.24	0.24-0.55	2	2	$P_{ms}, P_{sl}, P_{le}, P_{sr}, P_{ss}$
A	area of catchment or system boundary	m^2	6 500 000	±10%	2	1	$P_{ms}, P_{sl}, P_{le}, P_{sr}, P_{ss}$
Municipal water balance							
W	monthly municipal water supply normalised to the catchment area A	mm/month	-	22.9-31.4	3	1	$P_h, P_g, P_{ss}, P_y, P_{ss}$
H	estimated population of Mufakose and Marimba suburbs in year 2000 (1992 census results and using an annual growth rate of 4.5%)	No	99 842	±10%	2	1	$P_h, P_g, P_{ss}, P_y, P_{ss}$

Appendix D Sensitivity and reliability of model parameters

Input	Description	Unit	Value	Range	RI	SI	Affected output P-flux or stock
Age group factor	factor based on age structure in literature of the population: 0-14 years, 43%; 15-64, 54% and 65 years and over, 3 %. Assumed that the age group 0-14 years and >65 years excrete 50% 15-64. Hence (0.43 + 0.3)0.5 + (0.54) = 0.77	-	0.77	±20%	2	1	$P_b, P_g, P_s, P_y, P_{ss}$
N	number of residential stands in Mufakose and Marimba suburbs in year 2000	No	10 100	±10%	3	2	$P_b, P_g, P_s, P_y, P_{ss}$
a	water use fraction which determines the volume of water used for on-plot garden irrigation	-	0.20	±20%	2	2	P_g, P_{gi}, P_{ss}
b	water use fraction which determines the grey water volume generated	-	0.47	±20%	2	2	P_g, P_s, P_{gi}
c	water use fraction which determines the volume of black water arising from toilet flushing	-	0.30	±20%	2	2	P_b, P_s
d	water use fraction which determines the volume of water consumed by an adult person per month either directly or indirectly	-	0.03	±20%	2	2	P_b, P_s, P_y
e	water use fraction which determines the volume of grey water applied on land for garden irrigation purposes	-	0.15	±20%	2	2	P_g, P_{ss}, P_{gi}
f	water use fraction which determines the volume of grey water which evaporates from laundry fabrics, kitchen utensils and human body after washing and bathing	-	0.10	±20%	2	2	P_g, P_s
y	urinary volumetric rate of excretion for an adult person per month in winter or summer season	m³/p. month	0.012	±0.002	3	1	P_y
P balance household subsystem							
λ	quantity of dry organic and biodegradable waste generated per person per month	kg/p. month	1.35	±10%	3	2	P_{sw}, P_{csw}, P_{ss}
M_{fb}	quantity of food and beverage per food group consumed per person per month (total value given)	kg/p. month	26.4	±20%	2	2	$P_b, P_{sw}, P_{csw}, P_y, P_{ss}$
M_{sd}	quantity of soap and detergent used per person per month	kg/p. month	0.81	±20%	2	3	P_g, P_{gi}, P_s, P_{ss}
q'	proportion of organic solid waste which is either deliberately composted or is uncollected and end up being manure on agricultural land	-	0.10	±20%	2	3	P_{sw}, P_{csw}, P_{ss}
σ_{fb}	P-content in food and beverage material as P	kg/kg	-	0.13-4.62	2	2	P_{fb}
σ_{ms}	P-concentration of municipal sewage as P	kg/m³	0.0211	±10%	3	2	P_b, P_s, P_{ms}

Short-cutting the phosphorus cycle in urban ecosystems

Input	Description	Unit	Value	Range	RI	SI	Affected output P-flux or stock
σ_{sd}	P-content in soap and detergent material as P	kg/kg	-	0.45-2.50	3	2	P_g, P_{gb}, P_s, P_{ss}
σ_{sw}	P-content of the organic solid waste fraction as P	kg/kg	0.0016	±10%	2	2	P_{sw}, P_{cyw}, P_{ss}
σ_Υ	P-concentration in urine as P	kg/m³	1.89	±0.3	3	1	P_Υ, P_{mv}, P_y
P balance agricultural subsystem							
θ	proportion of the recommended commercial fertiliser application rate applied by urban farmers as Compound D	-	0.15	0.1-0.2	2	2	P_{mf}, P_{ss}
ρ_b	bulk density of soil	kg/m³	1 400	±300	2	2	P_{ss}
A_c	total area under cultivation	m²	2 900 000	±10%	2	1	P_{bu}, P_{sh}, P_{le}, P_{sr}, P_{ss}, P_{br}, P_{hb}
Effective volume factor	effective volume factor for a specified soil horizon (i.e. less the >2 mm soil fraction expressed as a ratio to the total soil volume)	-	0.88	±20%	3	2	P_{ss}
M_{im}	total mass of imported manure per annum	kg/a	10 100	±20%	1	3	P_{im}, P_{mu}, P_{ss}
M_{mf}	recommended mass of mineral fertiliser applied per annum as '*maizefert*' or Compound D	kg/m². a	0.042	-	-	-	P_{mf}, P_{ss}
Y'_{bu}	combined water-limited dry biomass production potential (economic produce and straw)	kg/m²	1.867	±20%	-	2	P_{bu}, P_{hs}, P_{ss}, P_{hr}
Y'_{hh}	water-limited dry mass economic yield	kg/m²	0.790	±20%	-	2	P_{hu}, P_{hs}, P_{ss}, P_{hr}, P_{hb}
Y_{hb}	economic yield or harvested biomass of maize	kg/m²	0.200	±20%	-	2	P_{hu}, P_{hs}, P_{ss}, P_{hr}, P_{hb}
Y_{sh}	yield of maize straw biomass	kg/m²	0.473	±20%	2	1	P_{hu}
σ_{hh}	P-content of the harvested maize crop biomass as P	kg/kg	0.0011	-	-	1	P_{hu}, P_{hs}, P_{ss}, P_{hr}, P_{hb}
σ_{im}	P-content in imported manure as P	kg/kg	0.0006	±20%	2	3	P_{im}, P_{ss}, P_{mu}
σ_{mf}	phosphorus concentration in fertiliser as P	kg/kg	0.061	-	3	1	P_{mf}, P_{ss}
σ_{sb}	P-content of the maize straw biomass as P	kg/kg	0.0005	-	-	1	P_{hu}, P_{hs}, P_{ss}, P_{hr}, P_{hb}
σ_{sr}	phosphorus concentration in runoff as ortho-P	kg/m³	-	0.315-1.54	2	1	P_{sr}, P_{le}, P_{ss}
σ_{ss}	bio-available phosphorus concentration in soil storage as P (0.2 m deep plough layer)	mg/kg	5.67	1.31-19.63	2	2	P_{ss}

Appendix E

Relevant Websites

The Internet-sites listed below provide information on closed loop (ecological sanitation) hardware component manufactures, grey water handling and reuse systems, rain water harvesting and use and general water conservation systems. The list is not intended to be exhaustive but to indicate a range of manufactures around the world. Technical information, prices and details on how to order are also available on some of these sites. Both dry and wet urine separating WC's and composting WC's can be found on these sites[1].

3P Technik Rainwater systems: http://www.3ptechnik.de/producte/farohr.html
Alascan: http://www.alascanofmn.com/
Aquaplan Germany: http://www.aquaplan-giessen.com
Aquatron: http://www.aquatron.se
ASP UWO-Combimat: http://dbcom.nl/rege-fil.htm
Atlas South Africa: http://www.atlasplastics.co.za
Berger Biotechnik Germany: http://www.berger-biotechnik.de
Biolet: http://www.biolet.com
Clivus Multrum: http://www.clivusmultrum.com
Cotuit Dry Toilets: http://www.cape.com/cdt/
DMA Technology South Africa: http://www.dma-tech.co.za
DryLoo: http://www.dryloo.com/index.htm
Dubletten: http://www.dubbletten.nu
Ecolet: http://www.ekolet.com/ekolet-eng/
Ecotech: http://www.ecological-engineering.com/ecotech.html
Enviro Options South Africa: http://www.eloo.co.za
Harry Rode Tiefbautechniek: http://www.harryrode.de/regenprofi/prod/filt/laubabscheider.htm
Heflex Water Systems: http://www.hewa.nl
http://www.mtlion.com.
Kilian Water rainwater and grey water systems: http://www.kilianwater.nl
Kordes rainwater systems: http:// www.kordes.de
Nature Loo: http://www.nature-loo.com.au/
Nature Loo: http://www.nature-loo.com.au/
Naturum: http://www.naturum.fi/english/index_e.htm
Otterwasser Germany: http://www.otterwasser.de
Phoenix: http://www.compostingtoilet.com/
Rezo rainwater use and infiltration systems: http://www.rezo.nl
Sanivac Vakumtechnik GmbH: http://www.vakutech.de
SBI Swaziland: http://www.sanplat.com
Sun-Mar: http://www.sun-mar.com
Vacuum toilet systems: http://www.wost-man-ecology.se
Water Rhapsody: http://www.water-rhapsody.co.za
Waterless urinals: http://www.ernstsystems.com
Waterless urinals: http://www.uridan.com
Wost Man Ecology: http://www.wost-man-ecology.se

---

[1] The inclusion of suppliers' names, products or concepts in this dissertation does not imply recommendation. Likewise, the omission of names, products or concepts does not imply rejection.

Samenvatting

Conventionele drinkwater- en afvalwatersystemen in stedelijke gebieden danken hun ontwerp aan 19^e–eeuws denken over hygiëne en het bestrijden van ziektes die via het water worden overgebracht. Deze systemen zijn niet verenigbaar met gesloten-kringloopsystemen, waarbij het erom gaat het netto gebruik van hulpbronnen tot een minimum te beperken door het ter plaatse hergebruiken van zowel water als nutriënten.

Het nieuwe denken over menselijk afval richt zich op drie fundamentele aspecten: a) het hygiënisch veilig maken van menselijke uitwerpselen; b) het voorkómen van vervuiling en c) het hergebruik van hygiënisch veilige eindprodukten in de landbouw. Deze "ecologische afvalverwerking" is als benadering in lijn met de zogenaamde Bellagio richtlijnen voor het schoonmaken van het milieu.

In de laatse jaren heeft het concept van "ecologische afvalverwerking" een nieuwe impuls gekregen. Het wordt nu gezien als één van de oplossingen voor een verantwoorde afvalverwerking voor 2,4 miljard mensen in het jaar 2015 (dit is één van de Millennium-ontwikkelingsdoelstellingen). Ecologische afvalverwerking vormt een schakel tussen afvalverwerking, voedsel, landbouw en bodemvruchtbaarheid. Deskundigen op het gebied van afvalverwerking, biologie, landbouw en gezondheid zijn momenteel betrokken bij het (opnieuw) vormgeven van dit concept. Er zijn echter nog een aantal belangrijke onderzoeksvragen, met betrekking tot risico-analyses op het terrein van afvalverwerking, ruimtelijke ordening in steden, volksgezondheid en technologische beperkingen.

Dit proefschrift beschrijft onderzoek naar de haalbaarheid van ecologische afvalverwerking in stedelijke ecosystemen (op de schaal van huishoudens en woonwijken). Het begint met een kritische beoordeling van de kennis en aard van fosfor (P) kringlopen en voorraden (van bron tot verwijdering), en wel op een aflopende geografische schaal van mondiaal, regionaal, nationaal, naar stroomgebied en deelstroomgebieden van rivieren. De hypothese is dat, door het kortsluiten van water en nutriënten-kringlopen op zo klein mogelijk schaal, er een verantwoord gebruik van de beperkte hulpbronnen, zoals P, kan plaatsvinden. Tegelijkertijd worden deze hulpbronnen dan tegen vervuiling beschermd.

Berekeningen geven aan dat de huidige mondiale economische P-reserves nog voldoende zijn voor minstens 100 jaar, bij een verbruik van 14 Mton/jaar. Door mondiale activiteiten zoals het kappen van bossen, intensiveren van landbouw en de afvalverwerking in steden, is het transport van P van terrestrische naar aquatische ecosystemen toegenomen tot, naar schatting, 22 Mton/jaar. Als gevolg hiervan zijn zowel de P toevoeren en fosforgehalten in rivieren, meren en estuaria met een factor 3 toegenomen t.o.v. het pre-industriële (en extensieve landbouw) tijdperk. Door deze activiteiten zijn aanvoer, consumptie en afvalproduktie als het ware losgekoppeld van de natuurlijke fosforkringloop. Hierdoor is een lineaire stroom van fosfor, met een open einde, op gang gekomen, een stroom die wordt bepaald door de economie en door een ongebalanceerde wereldhandel.

Berekeningen in dit proefschrift geven aan dat het totale verbruik van P-meststofen in het jaar 2000 voor Zimbabwe, (zijnde 18.940 Mg/jaar), wordt overtroffen door de totale P-verliezen van 57.160 Mg/jaar, dit ten gevolg van bodemerosie en uitspoeling. In Zimbabwe zijn bovengenoemde verliezen de oorzaak van de belangrijkste milieuproblemen en een verlies van bodemvruchtbaarheid.

Het Chivero-meer, in het verstedelijkte stroomgebied van Harare (de hoofdstad van Zimbabwe), geeft een goed voorbeeld van fosforstromen met een open einde. Uit berekeningen voor het jaar 2000 blijkt dat het meer een fosforinstroom van 676 Mg/jaar ontving, tegen een P-uitstroom van 265 Mg/jaar. Er vindt zo een geleidelijke ophoping van fosfor in het meer plaats, vooral in het sediment. De jaarlijkse P-instroom in het meer komt overeen met zo'n 4% van het P-gebruik in de landbouw van Zimbabwe in het jaar 2000.

Vervolgens is een analytisch model gemaakt om de beheersmogelijkheden voor het kortsluiten van de P-stromen in stedelijke ecosystemen te beschrijven en te analyseren. Dit model is vervolgens toegepast op een klein stroomgebied in een buitenwijk van Harare, met een hoge bevolkingsdichtheid.

In dit onderzoek is uitgegaan van een systeem-benadering waarin Materiaalstroom en Voorraden worden bijgehouden. Hierbij werden twee subsystemen in detail bestudeerd: het "huishoudniveau" (gebruik van grondstoffen en productie van afval) en de "urbane landbouw" (bodem-plant wisselwerking). De berekende stromen en voorraden zijn gebruikt om de fosforbalans op te stellen voor het micro-stroomgebied in de buitenwijk. De verzamelde gegevens en informatie werden vervolgens verwerkt in een mathematisch-conceptuele formulering van de maandelijkse fosforstromen en – voorraden. Hiervoor is de zgn P-calculator ontwikkeld, met behulp van STELLA software. De P-calculator kan worden gebruikt als een instrument voor planning en besluitvorming teneinde de P-kringloop in stedelijke ecosystemen sluitend te maken. Met de P-calculator kunnen ook verschillende opties met elkaar worden vergeleken, met betrekking tot het gebruik van fosfor uit huishoudelijk afval in de landbouw.

Voor het kortsluiten van de fosforkringloop zijn twee opties nader onderzocht, nl. 1) scheiding van urine uit de afvalstroom en vervolgens hergebruik op landbouwgronden en 2) composteren van organisch vast afval uit huishoudens, waarbij het product na compostering kan worden aangewend in de landbouw, om de P-status en de algemene bodemkwaliteit te verbeteren.

De belangrijkste beperking in een 100% hergebruik van urine ligt in de hoeveelheid beschikbare landbouwgrond per hoofd van de bevolking. De berekeningen tonen aan dat, per volwassen persoon in het studiegebied ongeveer 190 m^2 landbouwoppervlakte per jaar nodig is om het fosfor uit de urine te assimileren. Momenteel is de beschikbare landoppervlakte voor stedelijke landbouw slechts 29 m^2 per persoon. Er is daarom slechts een gedeeltelijke P-hergebruik mogelijk binnen de woonwijk. Volgens de "Bellagio richtlijnen" dient de rest van het beschikbare fosfor dan buiten het micro-studiegebied te worden verwerkt.

In de meeste steden is de belangrijkste beperking voor het invoeren van ecologische afvalverwerking een tekort aan geschikte landbouwgrond om de restprodukten van de menselijke stofwisseling te assimileren. Er is daarom behoefte aan een herbezinning op het gebied van stedelijk ontwerp, teneinde de "ecologische voetafdruk" van de steden op het achterland tot een minimum te beperken.

B. Gumbo, 2004

About the author

Bekithemba Gumbo a second son in a family of six was born on 21 April 1968 in Bulawayo, the second largest city in Zimbabwe. He obtained his Bachelor of Science (Civil) Engineering Honours degree at the University of Zimbabwe in 1991. His undergraduate studies were funded through a bursary awarded to him by the Bulawayo City Council.

In 1988 to 1991 he worked for the City of Bulawayo as a student engineer during vacations and after graduating as a water and sewerage engineer (1991 to 1994). During this latter period southern Africa survived the worst drought in living memory. The city of Bulawayo lived through the scare of relocating more than one million people including industry. During this time Mr Gumbo participated in a number of projects on water demand management, wastewater recycling and emergency borehole drilling programme.

In 1994 he was awarded a scholarship by the British Overseas Development Administration (ODA, now Department for International Development, DfID) through the University of Zimbabwe to pursue a Master of Science degree programme in Water and Waste Engineering at Loughborough University in the United Kingdom. He graduated with a distinction in 1995 and returned to Zimbabwe to join the Department of Civil Engineering at the University as a lecturer.

In 1996 Mr Gumbo was nominated to be the Acting Project Manager in the Department of Civil Engineering for the Dutch funded Collaborative Programme for Capacity Building in the Water Sector of Zimbabwe and the Southern Africa Region. One of the major outputs of the programme was the setting up of an MSc programme in Water Resources Engineering and Management (launched in 1998).

It was during this time that he developed an intimate relationship with UNESCO-IHE and many of its staff members. Towards the end of 1997 he was academically admitted at IHE-Delft to pursue a PhD research programme carried out under a sandwich part-time construction. During his research he spent about 16 months in Delft, 2 months in Sweden, 18 months in South Africa and the remainder in Zimbabwe. Mr Gumbo has remained active in other research and educational programmes in water like the WaterNet and PoWER initiatives, and the IUCN Water Demand Management Programme for southern Africa. During this period he published six papers in peer-reviewed journals on the subject of Water Demand Management (Gumbo & Van der Zaag, 2002; Gumbo et al., 2002, 2003b, 2003c, 2004; Gumbo, 2004). These publications are listed in the references even though they have no direct link to this dissertation. He is pleased and satisfied that his PhD research journey opened many doors and that the submission of this somewhat delayed dissertation is a culmination of many other activities not detailed in this report.

Email: bgumbo@eng.uz.ac.zw or gumbo@civil.wits.ac.za
URL: http://www.uz.ac.zw/engineering/civil/gumbo/index.html